Fuel Cell Renewable Hybrid Power Systems

Fuel Cell Renewable Hybrid Power Systems

Editor

Nicu Bizon

MDPI • Basel • Beijing • Wuhan • Barcelona • Belgrade • Manchester • Tokyo • Cluj • Tianjin

Editor
Nicu Bizon
University of Pitesti
Romania

Editorial Office
MDPI
St. Alban-Anlage 66
4052 Basel, Switzerland

This is a reprint of articles from the Special Issue published online in the open access journal *Energies* (ISSN 1996-1073) (available at: https://www.mdpi.com/journal/energies/special_issues/Fuel_Cell_Renewable_Hybrid_Power_Systems).

For citation purposes, cite each article independently as indicated on the article page online and as indicated below:

LastName, A.A.; LastName, B.B.; LastName, C.C. Article Title. *Journal Name* **Year**, *Volume Number*, Page Range.

ISBN 978-3-0365-1307-2 (Hbk)
ISBN 978-3-0365-1308-9 (PDF)

Contents

About the Editor

Nicu Bizon (Senior Member, IEEE) was born in Albesti de Muscel, Arges county, Romania, 1961. He received his 5-year B.S. degree in Electronic Engineering from the University "Polytechnic" of Bucharest, Romania, in 1986, and Ph.D. degree in Automatic Systems and Control from the same university, in 1996. From 1996 to 1989, he was involved in hardware design with Dacia Renault SA, Romania. Since 2000, he has served as Professor with the University of Pitesti, Romania, and received two awards from Romanian Academy, in 2013 and 2016.

He is Editor and author of 5 books published in Springer and the author of 184 scientific papers (including 5 and 74 papers in *IEEE Transactions* and conferences, respectively) published in Scopus which have been cited 1678 times, corresponding to an h-index = 27. His current research interests include power electronic converters, fuel cell and electric vehicles, renewable energy, energy storage system, microgrids, and control and optimization of these systems.

Preface to "Fuel Cell Renewable Hybrid Power Systems"

The incredibly rapid increase in the world's energy demand over the last decade, along with the request for sustainable development, can be addressed using microgrids based on hybrid power systems combining renewable energy sources and fuel cell systems. This book includes innovative solutions and experimental research as well as state-of-the-art studies in the following challenging fields: fuel cell (FC) systems—modeling, control, optimization, and innovative technologies to improve the fuel economy, lifetime, reliability, and safety in operation; hybrid power systems (HPSs) based on renewable energy sources (RESs) (RES HPS)—optimized RES HPS architectures; global maximum power point tracking (GMPPT) control algorithms to improve energy harvesting from RESs; advanced energy management strategies (EMSs) to optimally ensure the power flow balance on DC (and/or AC bus) for standalone RES HPSs or grid-connected RES HPSs (microgrids); RES HPS with an FC system as a backup energy source (FC RES HPS)—innovative solutions to mitigate RES power variability and load dynamics to energy storage systems (ESSs) by controlling the generated FC power, DC voltage regulation, and/or load pulse mitigation by active control of the power converters from hybrid ESS; FC vehicles (FCVs)—FCV powertrain, ESSs topologies and hybridization technologies, and EMSs to improve the fuel economy; optimal sizing of FC RES HPSs and FCVs; the changes in climate that are visible today and are a challenge for the global research community.

The stationary applications sector is one of the most important energy consumers. Harnessing the potential of renewable energy worldwide is currently being considered to find alternatives for obtaining energy by using technologies that offer maximum efficiency and minimum pollution. In this context, new energy generation technologies are needed to both generate low carbon emissions as well as identifying, planning, and implementing the directions in which the potential of renewable energy sources can be harnessed. Hydrogen fuel cell technology represents one of the alternative solutions for future clean energy systems. Hence, the first chapter presents the potential applications of hydrogen energy in hybrid power system using SWOT (strengths, weaknesses, opportunities, threats) analysis. The main strategies to be used for integrating the hydrogen-based and classical energy sources in the hybrid power system were identified and detailed. In addition to research, technical, and implementation factors, hydrogen integration also depends on legislative and energy decision-makers, potential investors, and final beneficiaries.

In Chapter 2 is an analysis of fuel cell (FC) system integration in a hybrid power system using three control variants of the FC power based on the required-power-following (RPF) control mode, which ensures the load demand under variable renewable energy. One variant is to control FC power via the FC boost converter, and the other two variants are via air regulator and the fuel regulator. The FC system will compensate the power flow balance on the DC bus and operate the battery stack in charge-sustained mode. Thus, the FC power will be mainly given by the positive difference between the load demand and renewable power. If renewable power is higher than the load demand, then this excess will power an electrolyzer to maintain the charge-sustained mode of the battery. Seven control architectures have been investigated using a fuel economy optimization function, resulting in 15% fuel savings for the best RPF-based strategy compared to the commercial strategy based on feed-forward control.

Chapter 3 compares power losses in the case of current operating conditions of electricity distribution networks (EDNs) and modern microgrids based on renewable energy sources. Optimal allocation of capacitor banks is performed using five metaheuristic algorithms to minimize the power losses on the IEEE 33-bus system and a real 215-bus EDN from Romania.

Sensitivity analysis to evaluate the critical parameters in the design of a new fuel economy strategy for a proton-exchange membrane fuel cell (PEMFC)-based hybrid power system is presented in Chapter 4. The fuel economy strategy uses load-following control and the global extremum seeking (GES) algorithm to minimize fuel consumption. The multimodal behavior in dither frequency and parameter keff is highlighted for the optimization function defined as a mix of the FC net power and fuel efficiency, with the latter being weighted with parameter keff. The results show that the best fuel economy can be obtained for 100 Hz dither frequency and keff = 20.

Chapter 5 addresses the optimal sizing of a PEMFC-powered electric truck to minimize vehicle production and use costs. Property costs were estimated for various design parameters such as the cost of hydrogen and powertrain components, and mileage.

In Chapter 6, a low-power station based on direct methanol fuel cell (DMFC) stacks is designed and tested in a real environment. The generated power can be increased by using multiple DMFC stacks in parallel and an appropriate energy management strategy for the power flows.

Chapter 7 presents the advantages of using a high-temperature (HT) proton-exchange membrane fuel cell (PEMFC) and a carbon capture/liquefaction system in a hydrogen-fueled ship application. The steam methane reforming and steam methanol reforming technologies are evaluated from the point of view of the energetic and exergetic performances, respectively of the occupied space. Compared to fuel cell vehicles (FCVs), the use of electric vehicles (EVs) in the transport of goods and passengers has rapidly increased in number and manufacturer diversity. Hence, it is necessary to optimally plan the location of charging stations near work offices in mall parking areas and in certain locations on frequently used routes.

Chapter 9 presents and tests a charging station according to the standard of the International Electrotechnical Commission (IEC) 61851-1. In this chapter, it is demonstrated that a Simulink dynamic model based on an open source for the proton-exchange membrane fuel cell (PEMFC) system can be easily used in real-time design solutions by exporting the generated code as C/C++. The calculation time is reduced taking into account only important PEMFC parameters in the modeling without significantly depreciating the modeling performance.

As reported in recent studies on microgrids based on renewable energy, wind and photovoltaic energy are the most widely used renewable energy sources used in hybrid energy systems grid-connected. Chapter 10 presents an advanced fuzzy logic control for wind turbines to improve their behavior during transient regimes after grid failures. To improve the efficiency of solar energy conversion, the last chapter studies the design and safe use of a Stirling engine combined with a solar concentrator. As this book presents the latest solutions in the implementation of fuel cells and renewable energy in mobile and stationary applications such as hybrid and microgrid power systems, we hope the chapters within will be of interest to readers working in related fields.

Nicu Bizon

Editor

Review

Hydrogen Fuel Cell Technology for the Sustainable Future of Stationary Applications

Raluca-Andreea Felseghi [1,*], Elena Carcadea [2], Maria Simona Raboaca [1,2,*], Cătălin Nicolae TRUFIN [3] and Constantin Filote [1]

[1] Faculty of Electrical Engineering and Computer Science, Stefan cel Mare University of Suceava, Universității Street, No.13, 720229 Suceava, Romania; filote.constantin@gmail.com

[2] National Research and Development Institute for Cryogenic and Isotopic Technologies—ICSI Rm. Valcea, Uzinei Street, No. 4, P.O. Box 7 Raureni, 240050 Valcea, Romania; elena.carcadea@icsi.ro

[3] Research and Development Department, ASSIST Software, Romania, 720043 Suceava, Romania; catalin.trufin@assist.ro

* Correspondence: Raluca.FELSEGHI@insta.utcluj.ro (R.-A.F.); simona.raboaca@icsi.ro (M.S.R.)

Received: 29 September 2019; Accepted: 29 November 2019; Published: 3 December 2019

Abstract: The climate changes that are becoming visible today are a challenge for the global research community. The stationary applications sector is one of the most important energy consumers. Harnessing the potential of renewable energy worldwide is currently being considered to find alternatives for obtaining energy by using technologies that offer maximum efficiency and minimum pollution. In this context, new energy generation technologies are needed to both generate low carbon emissions, as well as identifying, planning and implementing the directions for harnessing the potential of renewable energy sources. Hydrogen fuel cell technology represents one of the alternative solutions for future clean energy systems. This article reviews the specific characteristics of hydrogen energy, which recommends it as a clean energy to power stationary applications. The aim of review was to provide an overview of the sustainability elements and the potential of using hydrogen as an alternative energy source for stationary applications, and for identifying the possibilities of increasing the share of hydrogen energy in stationary applications, respectively. As a study method was applied a SWOT analysis, following which a series of strategies that could be adopted in order to increase the degree of use of hydrogen energy as an alternative to the classical energy for stationary applications were recommended. The SWOT analysis conducted in the present study highlights that the implementation of the hydrogen economy depends decisively on the following main factors: legislative framework, energy decision makers, information and interest from the end beneficiaries, potential investors, and existence of specialists in this field.

Keywords: alternative energy; energy efficiency; fuel cell; hydrogen energy; stationary application

---

1. Introduction

Unconventional energy sources have gained and will continue to gain an increasing share in energy systems around the world [1], due to both the research and political efforts [2–8] involved in their development, as well as due to the price increases of energy obtained by traditional methods [9]. The primary energy sources, generally called renewable, are those sources found in the natural environment, available in virtually unlimited quantities or regenerated through natural processes, at a faster rate than they are consumed. Officially recognized renewable energies originate from the Sun's rays, the internal temperature of the Earth or the gravitational interactions of the Sun and the Moon with the oceans. The processes and methods of producing or capturing these types of alternative energy are in the process of being improved, the lower costs of infrastructure investments and the improved efficiency of conversion processes have made renewable energy sources provide a small part of the energy needs on a planetary scale [10]. The more optimistic forecasts estimate that renewable energy production will enjoy a 30–50% share of the total energy market by around 2050, but this depends on reducing production costs and finding massive energy storage possibilities [11]. In addition, none of these forms of energy can also provide fuels in satisfactory quantities for use in various stationary, mobile or industrial applications [5].

In this context, we are currently looking for alternatives for obtaining energy by using technologies that offer maximum efficiency, high reliability and minimum pollution. Such a technology, considered at the moment the cleanest, through which sustainable energy can be obtained, is based on fuel cells [12]. As fuel cells develop, hydrogen-based energy production has become a reality [13]. The future hydrogen-based economy presents hydrogen as an energy carrier within a secure and sustainable energy system [14]. Humanity is on the verge of a new era characterized by advanced technologies and new fuels. We will witness new and completely different ways of producing and using energy. The energy could be generated by sources with virtually zero pollution. Hydrogen can be considered as a synthetic fuel, carrying secondary energy in a future era after the fossil fuel economy [15–18].

In order to outline an overview of the sustainability elements, the potential of using hydrogen as an alternative energy source for stationary applications and for identifying the possibilities of increasing the share of hydrogen energy in stationary applications, in this paper, a SWOT analysis was performed.

The work was structured as follows: the first part introduces the topic, presents and briefly describes the issues addressed. In Section 2 the Materials and Methods used in the present study are described. The data of the theme regarding the technical aspects and the sustainability elements are presented in Section 3—Hydrogen energy and Fuel cells—Hydrogen conversion technology. For the identification of the own potential of harnessing the hydrogen energy in stationary applications, but also of the opportunities and possible threats from the external environment, a SWOT analysis was developed in Section 4 and all aspects involved are discussed and critically analyzed, with the aim to evaluate the options and establish strategies to address the issues that best align the resources and capacities of hydrogen energy to the requirements of the stationary applications domain. The conclusions are presented is Section 5 as short bullet points that convey the essential conclusions of this paper.

2. Materials and Methods

The documentation for carrying out the study is based on the specialized scientific literature, articles in journals, papers presented at conferences on the hydrogen applcation topic and on-line scientific databases and web pages, including Google Academic, Google Scholar, MDPI, Science Direct, Scopus and research platforms or topic-specific web pages. In addition, this paper utilizes and analyzes a large number of reports, informations regarding the hydrogen & fuel cell strategic research agenda and documents published by the European Union (EU), the United Nations Organization (ONU), the International Energy Agency (IEA) [19–21] and other important dates from research and development institutions that are relevant to hydrogen economy, including E4Tech [22,23], International Association for Hydrogen Energy (IAHE) [24], National Research and Development Institute for Cryogenic and Isotopic Technologies (ICSI) Râmnicu Vâlcea, Romania.

The instrument used in this paper in order to verify and analyze the overall position with respect to general acceptance status regarding the harnessing energy potential of hydrogen technology and its use as an alternative energy source for stationary applications is the SWOT analysis.

SWOT analysis provides an overview of the characteristics specific to the objective/domain of analysis and the environment in which it will be implemented. The SWOT analysis functions as an x-ray of the concept of hydrogen energy implementation in stationary applications and at the same time evaluates the internal and external influence factors of the concept, as well as its position in the applicability environment in order to highlight the strengths and weaknesses of the concept, in relation to the opportunities and threats existing at the moment [25].

The steps to perform the SWOT analysis are shown schematically in the diagram in Figure 1 in order to identify the strengths, weaknesses, opportunities and threats characteristics of the concept of hydrogen energy for stationary applications.

Figure 1. The SWOT process.

As a rule, SWOT analysis allows investigators to improve the performance of current strategies by using new opportunities or by neutralizing potential threats [25]. Therefore, this analysis could be useful in helping decision makers and stakeholders to have a better overview of the concept of hydrogen energy used in stationary applications, facilitating the improvement of the current situation. As a result, SWOT analysis can be considered as an appropriate instrument for this research with scope to identify significant elements and advantages regarding the use of hydrogen energy in stationary applications, research/implementation/solutions/market status and possible changes, challenges, perspectives and improvements.

In order to perform the proposed SWOT analysis, the development in the form of the schematic matrix illustrated in Figure 2, the following stages were required:

Stage 1: Documentation, collection, interpretation of materials and data, and critical analysis.

Stage 2: Discussions with several experts, researchers and PhD students on the topic of hydrogen economy, but also of civil engineers given the scope of the concept, namely stationary applications.

Stage 3: Based on the data and results obtained from the previous stages, all the significant elements regarding the strengths, weaknesses, opportunities and threats are critically discussed, analyzed, classified and the SWOT matrix was drawn.

Stage 4: The SWOT matrix was used to determine strategies for strengths-opportunities (S&O), strategies for weaknesse-opportunities (W&O), strategies for strengths-threats (S&T) and strategies for weaknesses-threats (W&T). To develop S&O strategies, internal strengths are correlated with external opportunities, and strengths are key elements that will take advantage of opportunities. W&O strategies were developed by matching them internal weaknesses with external opportunities and overcoming weaknesses by taking advantage of opportunities. S&T strategies correlate internal strengths with external threats, and strengths are used to avoid threats. W&T strategies correlate internal weaknesses with external threats, and weaknesses are minimized to avoid threats.

	Strengths *internal*	Weaknesses *internal*
Opportunities *external*	S&O	W&O
Threats *external*	S&T	W&T

Figure 2. SWOT analysis matrix [25].

3. Considerations Regarding Hydrogen Fuel Cell Technology

3.1. Hydrogen—Energy Vector within a Sustainable Energy System

Today hydrogen is recognized as a non-polluting energy carrier because it does not contribute to global warming if it is produced from renewable sources [26]. In addition, hydrogen is the only secondary energy carrier that is suitable for a wide range of applications in the market [27]. At the center of attention is the fact that hydrogen can be obtained from a wide range of primary energies [28]. It can be used advantageously for a wide range of applications, ranging from transport and portable to stationary use [29]. In addition, hydrogen can also be used in decentralized systems without emitting carbon dioxide [30]. Hydrogen is already a part of today's chemical industry, but as a source of energy its rare benefits can only be achieved with technologies such as fuel cells [31].

Since hydrogen can be produced from a wide range of primary energies and can be consumed in a larger number of applications, it will become an energy center, just as electricity today is [32]. The advantage of hydrogen over electricity is that it can be stored in the medium and long term [33]. As a consequence, an energy carrier helps to increase the stabilization of energy security and price, giving rise to competition between different energy sources [34].

Veziroglu [35,36], editor of the journal specialized in hydrogen technology and energy, the *International Journal of Hydrogen Energy* summarizes several features that recommend the use of hydrogen as a secondary energy vector produced using unconventional technologies:

- Hydrogen is a concentrated primary energy sources, which can be made available to the consumer in a convenient way.
- It offers the possibility of conversion into different forms of energy through high efficiency conversion processes.
- It is an inexhaustible source, if it is obtained electrolytically from water; hydrogen production and consumption represents a closed cycle, the source of production—water—is kept constant and represents a classic cycle of recirculation of this type of raw material.
- Is the easiest and cleanest fuel; the burning of hydrogen is almost entirely devoid of pollutant emissions.
- It has a much higher gravimetric energy density compared to other fuels.
- Hydrogen can be stored in various ways, such as gas at normal pressure or at high pressure, in the form of liquid hydrogen or as solid hydride.
- It can be transported over long distances, stored in the native form or in one of the modalities presented above.
- Hydrogen-based fuel cells have efficiencies of up to 60% [35,36].

Hydrogen is considered by more and more specialists to be a true fuel of the future. Hydrogen is condensed to −252.77 °C, and the specific weight of liquefied hydrogen is 71 g/L, which gives it the highest energy density per unit of mass between all fuels and energy carriers: 1 kg of hydrogen contains as much energy as 2.1 kg of natural gas or 2.8 kg of oil. This characteristic of it, made of hydrogen the fuel used in the propulsion and energy supply of the spaceships. Unlike other fuels, such as oil, natural gas and coal, hydrogen is renewable and non-toxic when used in fuel cells. Hydrogen has a very high potential as environmentally friendly fuel and in reducing the import of energy resources [37–42].

Even if the use of hydrogen at present seems unprofitable, due to the improvement of the technologies, we could assist in the not too distant future in the development of a hydrogen-based economy [43].

Etymologically, the word hydrogen is a combination of two Greek words, meaning "to make water" [44]. Produced from non-fossil sources and raw materials, using different forms of alternative energy (solar, wind, hydroelectric, geothermal, biomass, etc.), hydrogen is considered to be a prime fuel in the supply of so-called "green energy" [45]. Thus, hydrogen-fueled systems can be considered as the best solution for accelerating and ensuring global energy stability [43]. Hydrogen is expected to play an important role in the future energy scenarios of the world, the most important factor that will determine the specific role of hydrogen will probably be the demand for clean growing energy [35]. At the same time, hydrogen can, to some extent, replace fossil fuels and become the preferred clean, non-toxic energy carrier in the near future [36]. The main characteristics of hydrogen [35,37], presented in Table 1, recommend it as an alternative fuel to the classic ones.

Table 1. Hydrogen characteristics.

Characteristics	Unit	Value
Density	kg/m^3	0.0838
Higher Heating Value (HHV)/liquid hydrogen (LH$_2$)	MJ/kg	141.90–119.90
HHV/cryogenic hydrogen gas (CGH$_2$)	MJ/m^3	11.89–10.05
Boiling point	K	20.41
Freezing point	K	13.97
Density (liquid)	kg/m^3	70.8
Air diffusion coefficient	cm^2/s	0.61
Specific heat	kJ/kg K	14.89
Ignition limits in air	% (volum)	4–75
Ignition energy in the air	Millijoule	0.02
Ignition temperature	K	585.00
Flame temperature in air	K	2318.00
Energy in explosion	kJ/g TNT	58.823
Flame emissivity	%	17–25
Stoichiometric mixture in air	%	29.53
Air/fuel stoichiometry	kg/kg	34.30/1
Burning speed	cm/s	2.75
Power reserve factor	-	1.00

In order to highlight the advantages that hydrogen has, compared to other fuels, the main properties of the various fuels currently used are presented in Table 2.

Table 2. Comparison between the main properties of hydrogen and other fuels [46,47].

Fuel Type	Energy/Mass Unit (J/kg)	Energy/Volume Unit (J/m^3)	Energy Reserve Factor	Carbon Emission Specific (kgC/kg Fuel)
Liquid hydrogen	141.90	10.10	1.00	0.00
Hydrogen gas	141.90	0.013	1.00	0.00
Fuel oil	45.50	38.65	0.78	0.84
Gasoline	47.40	34.85	0.76	0.86
Jet fuel	46.50	35.30	0.75	-
GPL	48.80	24.40	0.62	-
GNL	50.00	23.00	0.61	-
Methanol	22.30	18.10	0.23	0.50
Ethanol	29.90	23.60	0.37	0.50
Biodiesel	37.00	33.00	-	0.50
Natural gases	50.00	0.04	0.75	0.46
Coal	30.00	-	-	0.50

Analyzing the table information it can be concluded that the main arguments in favor of the use of hydrogen as synthetic fuel, obtained from renewable sources, are the following: it has the highest energy/mass unit of all the fuel types; it is ecologically friendly, as from its combustion produces only water vapor, indicating that for hydrogen the amount of carbon emissions is zero; it has the highest energy reserve factor, and the largest conversion factor into electricity, respectively, and for this reason it is considered the best of the fuels presented, and the energy efficiency is very high. Hydrogen is expected to play an important role in future energy scenarios globally [46,47]. The advantages that promote it as an energy vector in relation to other forms of energy are:

- Hydrogen can be transported remotely through pipes, in safe conditions.
- Hydrogen is a non-toxic energy carrier, with a high specific energy per mass unit (for example, the energy obtained from 9.5 kg of hydrogen is equivalent to that of 25 kg of gasoline).
- Hydrogen can be generated from various energy sources, including renewable ones.
- Compared to electricity or heat, hydrogen can be stored for relatively long periods of time.
- Hydrogen can be used advantageously in all sectors of the economy (as a raw material in the industry, as a fuel for cars and as an energy carrier in sustainable energy systems to generate electricity and heat through fuel cells).

Barriers to be overcome refer to issues regarding:

- Hydrogen burns in the presence of air, which can cause operational safety problems.
- Storing hydrogen in liquid form is difficult because very low temperatures are required to liquefy hydrogen.
- High costs of hydrogen technologies and processes.
- High costs of hydrogen energy conversion technologies through fuel cells.
- The viability/cost ratio of hydrogen and fuel cell technologies is relatively low.
- The current lack of logistics, transport infrastructure and distribution of hydrogen to final consumers requires costly investments.

Schematically the pro/against hydrogen arguments are shown in Figure 3.

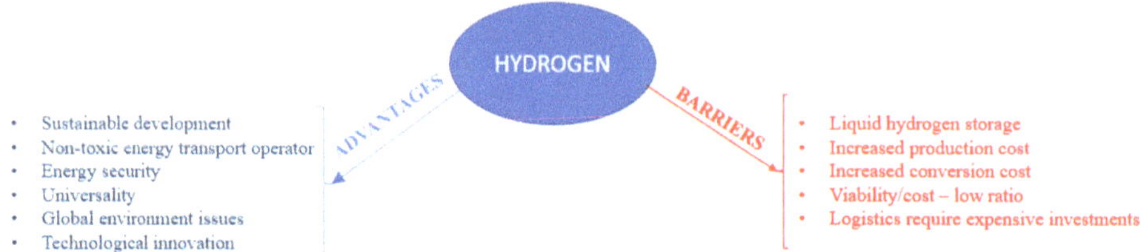

Figure 3. Advantages and barriers regarding the use of hydrogen as energy vector [10].

In view of the research and development programs supported in this field, the technical problems regarding the production, storage and distribution of hydrogen, together with the reduction of costs and the increase of the life of the equipment used in the generation of energy based on hydrogen, will be solved shortly time, and hydrogen will become a possible solution for providing fuels, at the same time being an alternative energy resource to the traditional ones.

3.2. Fuel Cells—Hydrogen Conversion Technology

Science has shown that there are two alternatives for sustainable energy supply: renewable sources and fuel cells—hydrogen-based energy—which will play a complementary role in ensuring global energy resource security [36].

By promoting the use of hydrogen-based energy technologies as clean energy technologies for stationary applications [48], at the level of local communities, industrial and commercial communities, the research topic in this field will help the practical development of sustainable and clean energy systems.

3.2.1. General Aspects

Short History

The beginning of the 19th century represents the starting point of research in the field of fuel cells. In 1801 Humphry Davy demonstrated the principle that underlies the functioning of the fuel cell, and later in 1839 the lawyer and amateur scientist William Grove accidentally discovered the principle of the fuel cell during an electrolysis expereriment, when he disconnected the battery from the electrolysis device and touched the two electrodes [49]. He called this cell "gas battery", which consisted of platinum electrodes placed in tubes containing hydrogen gas and oxygen, respectively, tubes submerged in dilute sulfuric acid. The generated voltage was around 1 V. Sometime later, Grove connected several such "gas batteries" in series and used this thus obtained voltage source to supply the electrolyzer that separates the hydrogen from the oxygen. Due to electrode corrosion problems and the instability of the materials used, Grove's fuel cell did not produce practical results [50]. Langer and Mond developed Grove's invention by observing that over time the reactivity phenomenon of platinum black in contact with electrolytes is diminished, and the life of the fuel cell is prolonged by keeping the electrolyte in a non-conducting porous material [42,51], and using in 1889 for the first time the term fuel cell.

Subsequently, the studies were deepened with significant progress, and the 1960s–1970s brought the first practical applications of fuel cells, these being used by NASA for spacecraft. At the same time, the General Electric company was developing fuel cells with proton exchanger membranes, which were the basis of the fuel cells used for the generation of electricity in the Gemini program missions and the Apollo space program [51,52]. Starting in 1990, the research aimed to implement fuel cell technology in stationary applications, and fuel cells of different capacities for use in this field have been developed. Progress on membrane durability and improved energy performance of fuel cell assemblies has prompted the use of phosphoric acid fuel cells (PAFCs) in cogeneration applications and their widespread use [19,31,51].

Also, proton exchange membrane fuel cell (PEMFC) and solid oxide fuel cell (SOFC) technology has been developed for this field, especially for small stationary applications. The year 2000 also marks significant contributions in the field of portable devices, when fuel cell technology with direct methanol (DMFC) is widely adopted and used in this type of applications [19,31,51].

In the last two decades there has been a rapid acceleration in the increase of the use of fuel cells covering a wide diversity of applications in the portable, mobile and stationary fields. These increases are due, on the one hand, to technical progress in the field of fuel cells, and on the other hand, they are driven by global concerns about energy security, efficiency, energy sustainability, reducing greenhouse gas emissions and not least, decreasing dependence on the use of fossil fuels.

Fuel Cell Concept

Fuel cells are now increasingly being researched, considering that they are revolutionizing the ways in which energy is produced. They use hydrogen as a fuel, while also ensuring the possibility of generating clean energy, with the protection and even improvement of the environmental parameters [17,18].

By definition, the fuel cell is an electric cell, which, unlike battery cells, can be continuously fed with fuel, so that the electrical power from the output of this electric cell can be maintained indefinitely [38,39,42]. Therefore, the fuel cell converts hydrogen or hydrogen-based fuels directly into electricity and heat through the electrochemical reaction of hydrogen with oxygen. The process carried out at the fuel cell level is the inverse of electrolysis:

$$2H_2 + O_2 \rightarrow 2H_2O + \text{energy} \tag{1}$$

where, the conversion of the chemical energy of the fuel (hydrogen) and the oxidant (oxygen) into continuous current, heat and water as reaction products [42] takes place. Due to the fact that in the fuel cell the hydrogen and oxygen gases are transformed by an electrochemical reaction into water, this has considerable advantages over the thermal engines: higher efficiency, practically silent operation, lack of pollutant emissions where the fuel is even hydrogen, and if hydrogen is produced from renewable energy sources, the electrical power thus obtained is indeed sustainable [38,39,42].

Component Elements

In order to analyze the importance of all the phenomena that take place in a fuel cell, it is necessary to know the component elements (Figure 4).

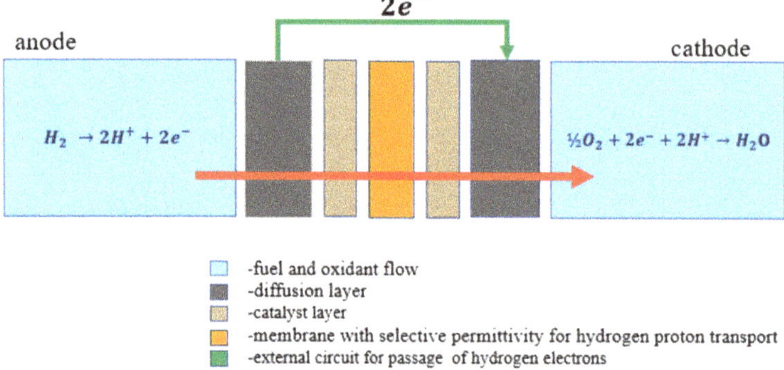

Figure 4. Components of the fuel cell [24,42].

There are some distinct elements in a fuel cell [42], namely:

Electrolyte. In order for a substance to fulfill the role of electrolyte in a fuel cell it must fulfill several conditions, as follows: high chemical stability with respect to the two electrodes, in order not to react with them; high melting and boiling points for cells operating at high temperatures; predominantly ionic conductivity and absence of electronic conductivity. The main types of electrolyte are liquid electrolytes (the most common are basic or acidic solutions, the ion transport phenomenon being similar to the one from the electrolysis of aqueous solutions), solid electrolytes (ion exchange membranes and crystalline solid electrolytes are used), melted electrolytes and liquid electrolytes with dissolved fuel [42].

Catalyst layer (at the anode and cathode). All electrochemical reactions in the fuel cells take place on the surface of catalyst layers. To increase the reaction rate of the cell, the electrodes (catalyst layers) contain catalyst particles. Thus, the electrode of a fuel cell is made of a porous carbon support on which the catalyst is deposited. The thickness of an electrode is usually between 5–15 μm, and the charge of the catalyst is between 0.1 and 0.3 mg/cm^2 [42].

Bipolar plate (at the anode and cathode). Bipolar plates play a dual role, guiding the reactant gases to the electrolytic exchange surface of the fuel cell and driving the obtained electric current. Materials used for bipolar plates must have a high conductivity and be gas-tight. They must also be corrosion resistant and chemically inert. Taking into account these considerations, graphite or steel can be used, but also composite materials. The gas flow channels are "engraved" in the bipolar plates, which otherwise should be as thin as possible to reduce the weight and volume of the battery.

The geometry of the flow channels has an influence on the flow velocity of the reactants and the mass transfer, implicitly they have a determining influence on the performance of the fuel cells, being necessary to optimize the flow surface so that the reaction surface is as large as possible [42].

Operating Principle

Although there are different types of fuel cells, they all work on the same principle:

- Hydrogen or a hydrogen rich fuel is introduced to the anode, where the anode-coated catalyst separates electrons from positive ions (protons).
- At the cathode, oxygen is combined with electrons and, in some cases, with protons or water, resulting in hydrated water or ions.
- The electrons that form at the fuel cell anode cannot pass directly through the electrolyte to the cathode, but only through an electrical circuit. This movement of electrons determines the electric current [38,39,42,52].

The electrochemical conversion consists of the direct conversion into electrical energy of the chemical energy stored in various active materials. This type of conversion is called direct because no other intermediate form is interposed between the initial and final energy forms. Indirect energy conversion systems contain several transformation stages, between which the form of thermal or mechanical energy is obligatory. Direct energy conversion eliminates the "link" thermal or mechanical energy by achieving higher efficiency, which does not depend on the limited efficiency of the thermal machines. The idea of obtaining electricity by direct conversion of chemical energy arose when the problem of unfolding and reverse of the phenomenon of water electrolysis (which results in its components), that is to say, to obtain electric current from the reaction between hydrogen and oxygen [53]. The schematic in Figure 5 illustrates a comparison regarding the operating principle of fuel cells—direct systems of conversion of energy forms and the classical technologies devoted to conversion — indirect systems [54,55].

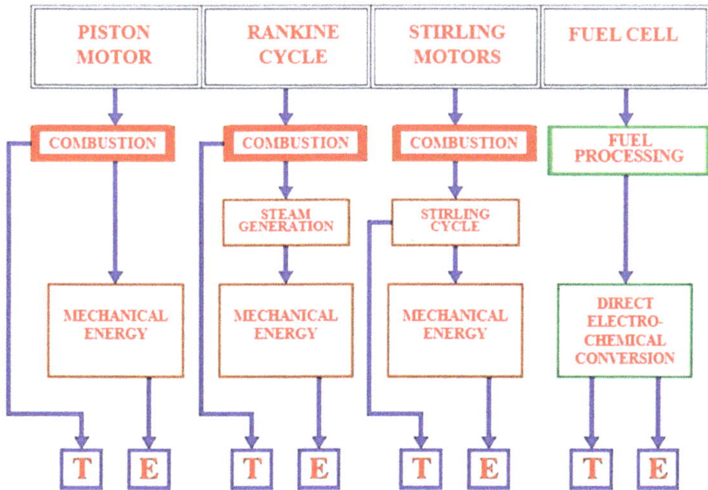

Figure 5. Fuel conversion process. Comparison of the operation principles of various technologies.

Fuel cells are electrochemical electricity generators characterized by a continuous supply of reactants to the two electrodes. The fuel rating comes from the fact that they use as sources of chemical energy, natural or synthetic fuel substances, which are subjected to oxidation and reduction reactions.

The anode, or fuel electrode, is the place where the oxidation of the fuel (H_2, CH_3OH, N_2H_4, hydrocarbons, etc.) takes place. The cathode, or oxygen (air) electrode, is the place where molecular oxygen reduction occurs [31,42,48].

Electrochemical oxidation of hydrogen is carried out at an anode of a conducting material (eg platinum dispersed on activated carbon) constituting the negative pole of the cell [42,44]:

$$\text{acid electrolyte:} \qquad H_2 \rightarrow 2H^+ + 2e^- \tag{2}$$

$$\text{alkaline electrolyte:} \qquad H_2 + 2OH^- \rightarrow 2H_2 + 2e^- \tag{3}$$

The electrochemical reduction of oxygen occurs at a catalytic cathode constituting the positive pole of the cell:

$$\text{acid electrolyte:} \qquad \frac{1}{2}O_2 + 2H^+ + 2e^- \rightarrow H_2O \tag{4}$$

$$\text{alkaline electrolyte:} \qquad \frac{1}{2}O_2 + H_2O + 2e^- \rightarrow 2OH^- \tag{5}$$

The catalytic functions of the electrodes are very important, namely: the hydrogen electrode (the anode) must ensure the adsorption of the hydrogen molecule, its activation, promoting the reaction with the hydroxyl ion; the oxygen electrode (cathode) must allow molecular oxygen adsorption, promoting reaction with water.

The anode and cathode are separated by an ionic conductor, electrolyte, and/or a membrane that prevents the reactants from mixing and the electrons pierce the heart of the cell. Initially, the energy released from the oxidation of conventional fuels, generally used in the form of heat, can be converted directly into electricity with excellent efficiency, in a fuel cell. Because in almost all oxidation reactions an electron transfer between fuel and oxidant occurs, it is obvious that the chemical oxidation energy can be converted directly into electricity. There is an oxidation-reduction reaction in which the oxidation of the fuel and the oxidant reduction occur with a loss on the one side and an electron gain on the other. Any galvanic element involves oxidation to the negative pole (loss of electrons) and reduction to the positive one (gain of electrons) and, as in all galvanic elements, the fuel cells tend to separate the two partial reactions in the sense that the changed electrons pass through an external use circuit [39,42,44].

The Main Types of Fuel Cell

The classification of fuel cells can be done according to several criteria that take into account certain common features. Thus, depending on the type of fuel (gases, solids, liquids), depending on the electrolyte used (liquid or solid), depending on how the fuel is consumed (directly and indirectly), but the most widely used classification method is the one takes into account the operating temperature:

- Low temperature (cold) fuel cells, operating between 20–100 °C;
- Medium temperature (hot) fuel cells, operating between 200–300 °C;
- High temperature fuel cells, operating between 600–1500 °C [41,42].

Currently in (or close to) current usage, there are six types of fuel cells, with various operating temperatures, as follows [38,42]:

- Alkaline fuel cells (AFCs), with operating temperatures around 70 °C.
- Direct methanol fuel cells (DMFCs), operating at temperatures between 60–130 °C.
- Molten carbon fuel cells (MCFCs) at temperatures up to 650 °C.
- Phosphoric acid fuel cells (PAFCs) at temperatures of 180–200 °C.
- Proton exchange membrane fuel cells (PEMFCs). These work at low temperatures of 100 °C, but also at temperatures of 150 °C to 200 °C.
- Solid oxide fuel cells (SOFCs). They operate at high temperatures from 800–1000 °C.

Each of these fuel cell types has specific characteristics. They have obvious advantages, but also disadvantages (in fact limited possibilities of use, now or in the near future) [24,41,42,56]. Apart from these six types there are others, more or less different from the first ones, which will probably be used in the near future: the zinc-air fuel cell (ZAFC) which uses a zinc anode similar to a battery; the ceramic proton exchanger fuel cell (PCFC), a relatively new type of fuel cell; the regenerative fuel cell (RFC) which is based on the most attractive way of generating hydrogen and oxygen by electrolysis, having the Sun as an energy source; in the fuel cell hydrogen and oxygen produce electricity, heat and water, which is then recycled and used for electrolysis.

With regard to the applicability of these typologies of energy conversion technologies for the stationary applications, five of them have been identified in the specialized literature which are used to serve stationary consumers, as follows: DMFC is particularly suited to the supply of electricity in small domestic applications, while MCFC, PAFC, PEMFC, SOFC are used to serve the large consumer market in the stationary field [45,56–58].

3.2.2. Practical Applications of the Fuel Cell

Based on the literature, especially the most current reports prepared in 2014–2018 by *Fuel Cell Today*, which is the main source of information, studies and analysis covering the global fuel cell market and *Fuel Cells and Hydrogen Joint Undertaking - New Energy World*, which represents the organization whose main objective is hydrogen and its technology at the European Union level, aspects of recent information from the last five years, regional and global developments regarding the implementation of these equipments has been synthesized [19–24].

Various pilot projects aiming at hydrogen technology are being validated, and the performances of fuel cell systems operating under real conditions are analyzed and reports on technology performance, progress and new challenges are being prepared. This analysis includes fuel cell assemblies of various types, namely proton exchange membrane fuel cells, solid oxide fuel cells, phosphoric acid fuel cells and molten carbon fuel cells, having dimensions of power generation systems with generated power ranging from 5 kW up to 2.8 MW, and the equipment has nominal powers ranging from 0.5 kW up to 400 kW [20,21]. Fuel cell systems are used in stationary applications [59–63] where they can be used for various purposes, namely as back-up power supplies, power generation for remote locations, stand-alone power stations for one or more consumers, distributed generation for buildings and cogeneration (in which excess thermal energy resulting from the production of electricity is used to provide the thermal agent) [24,40].

In commercial markets, fuel cells destined to the stationary sector show an increasing trend of technology transfer from the producer to the final consumer (Figure 6), being currently recognized as a feasible option compared to the conventional technologies of generator type, internal combustion engines or batteries. At the level of 2014, this technology transfer amounted to a value of 395,000 units regarding the delivery to the stationary field of fuel cell equipment, and for 2018 it increased to 575,000 units [22–24], this being possible due to the increased use of fuel cells as a practical application in which they play the role of energy backup (back-up system), but also due to the success of the residential fuel cell program "ENE-FARM" developed in Japan.

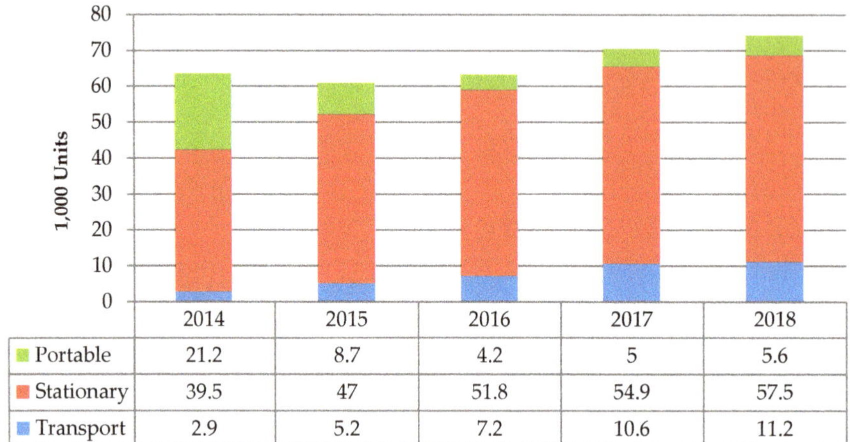

Figure 6. Shipments by application.

The distribution by region of this number of units transferred to practical applications is graphically illustrated in Figure 7 [22–24].

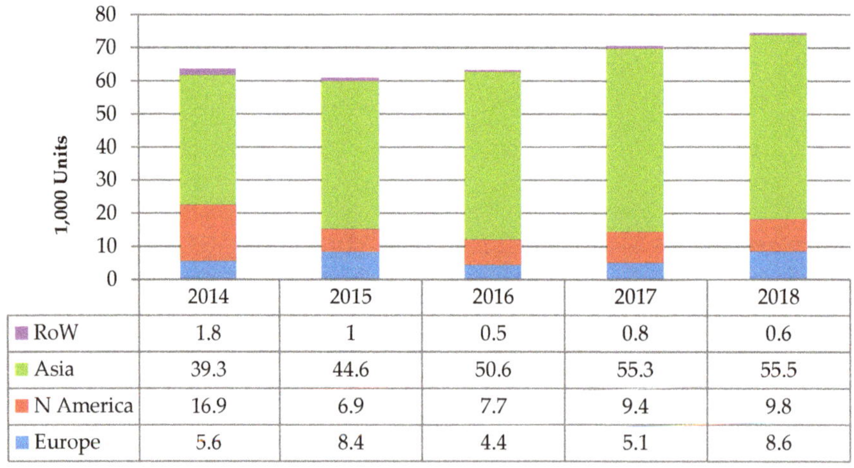

Figure 7. Shipments by region of implementation.

It is noted that Asia is the region with the largest number of fuel cell units in practical applications over the last five years, with an increase due to the commercial development of micro-cogeneration fuel cells produced in Japan, so for 2018 a number of 55,500 units was reported, being 29.20% higher than in 2014. The North American region reported in 2014 a number of 16,900 units transferred to practical applications, and in 2015 there was a 59.20% decrease compared to the previous year, followed by an increasing trend, in 2018, 9800 units were reported to practical applications. With regard to Europe, they remained relatively constant during the period analyzed, except for a decline in 2016, but also in 2017 they maintain the same stagnation trend.

When analyzing the value data reported over the last five years regarding the distribution of the technological transfer according to the fuel cell typology (Figure 8), it is observed that the fuel cell with proton exchange membrane is dominant. This is due to the possibility of using this type of fuel cell for a wide range of applications for all three segments (portable, stationary and transport), from small applications, micro-cogeneration systems to centralized power generation through high power applications.

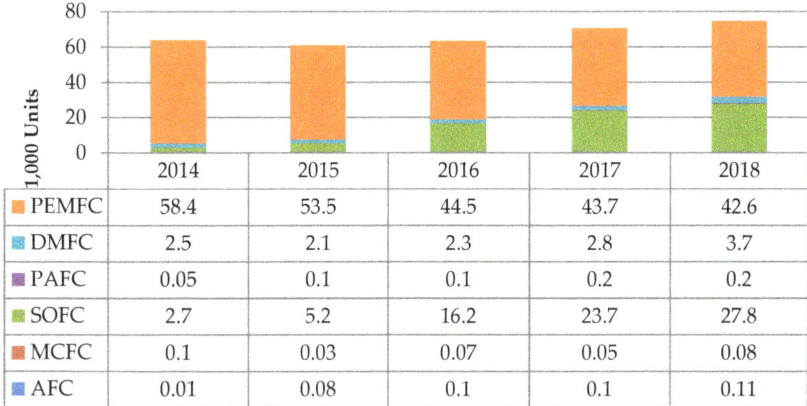

1,000 Units	2014	2015	2016	2017	2018
PEMFC	58.4	53.5	44.5	43.7	42.6
DMFC	2.5	2.1	2.3	2.8	3.7
PAFC	0.05	0.1	0.1	0.2	0.2
SOFC	2.7	5.2	16.2	23.7	27.8
MCFC	0.1	0.03	0.07	0.05	0.08
AFC	0.01	0.08	0.1	0.1	0.11

Figure 8. Shipments by fuel cell type.

It is noted that with the development and extensive use of PEMFC the other types of fuel cells, like DMFC, MCFC, and PAFC, recorded during the five years small ascending or decreasing variations, but this is also due to the fact that most types of cells are integrated into projects and programs that are in pilot phase, where the results are being validated.

With regard to SOFC technology, there was also a significant increase in the number of units transferred to practical applications since 2016, this is due to the transfers to the stationary area that are the subject of the Japanese Ene-Farm project and scheme. If in 2014, approx. 2700 units were reported, in 2018, a significant increase was achieved, reaching 27,800 SOFC units transferred to their practical applications [19,22–24].

Based on the number of units transferred to practical applications, *Fuel Cell Today* made a calculation regarding the sum of the fuel cell capacities that were installed to support these applications. The total capacity obtained is schematically illustrated in Figure 9.

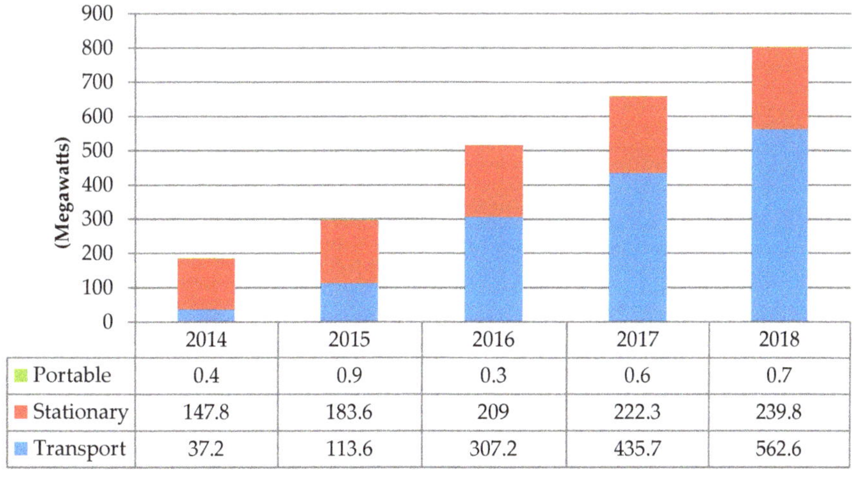

(Megawatts)	2014	2015	2016	2017	2018
Portable	0.4	0.9	0.3	0.6	0.7
Stationary	147.8	183.6	209	222.3	239.8
Transport	37.2	113.6	307.2	435.7	562.6

Figure 9. Megawatts by application.

The analysis of the data regarding the stationary sector shows a tendency of continuous growth starting with 2014. At the level of 2018 a total capacity transferred from the producer to the practical applications of 239.8 MW worldwide has been reported [19,22–24]. This was due in particular to the large number of micro-cogeneration units implemented in Asia, but the development of large capacity applications of central type of distributed hydrogen energy generation within which they are integrated and fuel cells with high power, totaling a significant number of megawatts is also worth noting. It should be also noted that the applications of hydrogen energy in the field of electromobility and transport registered a spectacular growth in 2018 compared to 2014.

Stationary applications for hydrogen fuel cells refer to fuel cell units designed to provide power at a 'fixed' location. They include small, medium and large stationary prime power, backup and uninterruptable power supplies (UPS), combined heat and power (CHP) and combined cooling and power. On-board APUs 'fixed' to larger vehicles such as trucks and ships are also included [22–24].

The distribution of total capacities installed by regions is graphically illustrated in Figure 10. Over the last five years, America and Asia have competed for the leading region in terms of implementation and adoption of fuel cells in stationary applications.

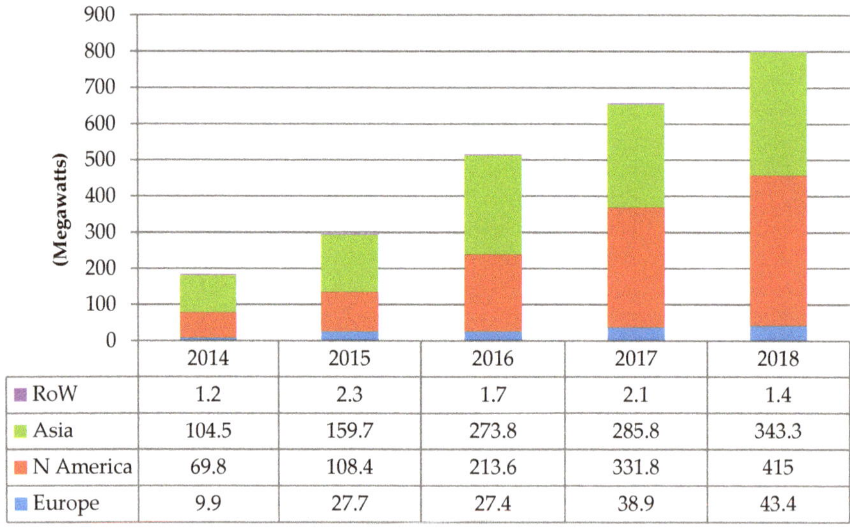

(Megawatts)	2014	2015	2016	2017	2018
RoW	1.2	2.3	1.7	2.1	1.4
Asia	104.5	159.7	273.8	285.8	343.3
N America	69.8	108.4	213.6	331.8	415
Europe	9.9	27.7	27.4	38.9	43.4

Figure 10. Megawatts by region of implementation.

If in 2014 Asia surpassed America, with 34.7 MW total installed capacity, in 2018 America became the leader because it already has fuel cell systems in place, reaching a total capacity of 415 MW, whereas in Asia, compared to 2014, a growing trend was reported with only 343.3 MW installed. Europe is experiencing moderate growth in this sector, but compared to America which currently holds the leading position, values are reported as 90% lower in Europe, respectively 43.8 MW.

If one looks at the aspect of the total installed capacity in relation to the fuel cell typology (Figure 11) it is found that the PEMFC technology is used in a wide range of segments of the practical applications, therefore it contributes with the greatest number of megabytes of total installed capacity from 2014 to 2018. Considering the validation and demonstration of the large size capabilities of many PAFC and SOFC units implemented in the stationary field starting with 2014, there is an increasing trend in the use and implementation in stationary applications of these types of fuel cells [19,22–24].

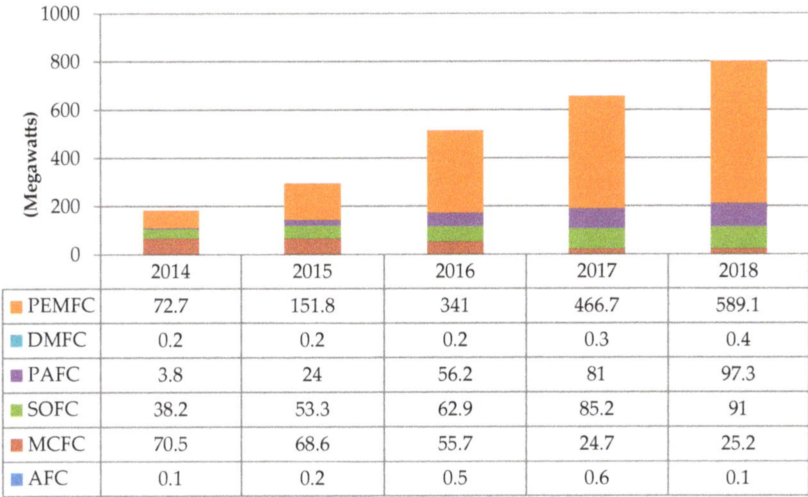

Figure 11. Megawatts by fuel cell type.

	2014	2015	2016	2017	2018
PEMFC	72.7	151.8	341	466.7	589.1
DMFC	0.2	0.2	0.2	0.3	0.4
PAFC	3.8	24	56.2	81	97.3
SOFC	38.2	53.3	62.9	85.2	91
MCFC	70.5	68.6	55.7	24.7	25.2
AFC	0.1	0.2	0.5	0.6	0.1

The prospect of significant growth in the number of fuel cell applications for the stationary field has real potential in the coming years. On a large scale, fuel cells for the power generation sector based on hydrogen as a raw material for the production of electricity in the centralized system (power plants) have shown real success in Japan, Korea and North America, and Europe offers, also, a number of opportunities in this sector [19–24].

From a sustainability point of view, the reduction of carbon dioxide emissions in all fields of activity by using fuel cells and the production of hydrogen from renewable electricity is advantageous for reducing the level of carbon dioxide emissions in the electricity sector. For the purpose of stability of the electricity distribution network, hydrogen-based power plants will contribute to their balance and will locally provide the necessary energy supply near the renewable energy production sites [34].

A number of major plans targeting this segment are announced at global and regional levels, and governments and partner organizations specialized in hydrogen and fuel cells make a concerted joint effort to ensure centralized production, storage and distribution infrastructure and supply end-users with hydrogen, as well as a wide variety of technical, financial and management issues are being reviewed to develop a future hydrogen-based economy [24,43]. The main arguments worth highlighting regarding fuel cell power generation systems in stationary applications are as follows:

- By using hydrogen technology in the generation of electricity, a degree of autonomy of 100% can be obtained compared to the national centralized network for the supply of electricity.
- Hydrogen and fuel cells meet 100% of the consumer's energy needs, with no unmeted energy demand.
- Renewable sources can be better harnessed by completely eliminating the deficiencies related to their meteorological intermittency, but also to the issues related to the storage in batteries, eliminating totally the losses associated to these disadvantages by the technology of hydrogen, particularly electrolytic production of hydrogen based on renewable energies and storage via hydrogen, a secondary energy carrier, which can release this energy stored by the electrochemical conversion carried out by the fuel cell.
- The excess energy resulting from the operation of the systems can be harnessed by hydrogen either as green energy exported to the centralized electricity network or as a useful fuel for other types of applications.

- Electrolytic hydrogen production is directly influenced by the availability of renewable energy resources, having a variable character over time, which implicitly influences the electricity production of the fuel cell, which is also directly proportional to the availability of hydrogen.
- The carbon dioxide emissions in the case of energy systems with fuel cells are much lower, registering an average of over 80% decrease compared to the conventional energy systems that support standard applications.
- The costs of the equipment components of the energy systems regarding the hydrogen technology and the costs with the purchase of the hydrogen fuel have a high influence in the diagram of the total costs of these systems, but the technology of electricity generation based on hydrogen and the methods of production, storage and distribution of hydrogen are the object of continuous research and development, and over time a number of pilot projects currently in progress in this field will be validated, which will influence and determine cost reductions in the near future, and this equipment, and also hydrogen fuel, will be competitive with the classic technologies in the field of energy production and storage [12–24,35–43].
- The optimization of the fuel cell systems in order to increase the overall efficiency of the hybrid power generation systems can be performed by nonlinear control [64], extremum seeking [65], and global optimization [66,67] techniques.

3.2.3. The Main Modalities to Energy Supply through Hydrogen Fuel Cell Technologies

The main types of fuel cell technologies presented above, having different operating characteristics and principles, can serve different segments of the energy generation market, either in CHP or power generation. Each type of technology has advantages and disadvantages that motivate its end use in specific applications and domains. In brief, the applications for which different types of fuel cells are suitable, but also their advantages and disadvantages, are presented in Table 3.

Table 3. Suitability in practical applications of the fuel cells types adapted from [68].

Fuel Cell Type	Typical Electrical Efficiency (LHV)	Power (kW)	Applications	Advantages	Disadvantages
AFC	60%	1–100	Back-up power; Electromobility; Military; Space.	Stable materials allow lower cost components; Low temperature; Quickly start-up.	Sensitive to CO_2 in fuel and air; Electrolyte management (aqueous); Electrolyte conductivity (polymer).
MCFC	50%	300–3000	Electric utility; Distributed generation.	Fuel flexibility; High efficiency; Suitable for hybrid/gas turbine cycle; Suitable for carbon capture; Suitable for CHP.	High temperature corrosion and breakdown of cell components; Long start-up time; Low power density.
PAFC	40%	5–400	Distributed generation.	Suitable for CHP; Increased tolerance to fuel impurities	Expensive catalysts; Long start-up time; Sulfur sensitivity.
PEMFC	60% direct H_2 40% reformed fuel	1–100	Back-up power; Distributed generation; Electromobility; Grid support; Portable power; Power to power (P2P).	Solid electrolyte reduces corrosion & electrolyte management problems; Low temperature; Quickly start-up.	Expensive catalysts; Sensitive to fuel impurities.
SOFC	60%	1–2000	Auxiliary power; Distributed generation; Electric utility.	Fuel flexibility; High efficiency; Potential for reversible operation; Solid electrolyte; Suitable for CHP; Suitable for hybrid/gas turbine cycle.	High temperature corrosion and breakdown of cell components; Long start-up time; Limited number of shutdowns.

The main modalities of energy support of the stationary applications by hydrogen fuel cell technology that have been identified in the specialized literature refer to the following aspects:

CHP With Fuel Cells in the Buildings Domain

Fuel cells are suitable for micro-cogeneration and CHP because the technology inherently produces electricity and heat from a single source of fuel such as hydrogen, and systems can also run on traditional fuels, such as natural gas. Currently, the CHP fuel cell units are installed in buildings, being functional in individual regimes, but such systems with low power capacities are under development, the projects having the objectives oriented towards the energetic support of the collective houses with several apartments. In this type of application, the fuel cell with proton exchanger membrane is commonly used, which works and ensures the energy demand both during the day when peaks are recorded and at night. Solid oxide fuel cells can also be used in residential micro-cogeneration systems, having a relatively efficiency equal to that of PEMFCs. Because SOFCs use higher operating temperatures than PEMFCs, they are more tolerant to carbon monoxide in the fuel, and this allows for some simplification in terms of system configuration.

All CHP technologies offer increased combined efficiency compared to traditional solutions for separate generation of electricity and thermal energy. Cogeneration with fuel cells can exceed the value of the "traditional frontier" in terms of energy efficiency due to the special performances that this type of technology achieves (Table 4) [69,70]. PEMFC and SOFC are usually used for energy supply systems for small residential applications, and SOFC, PAFC and MCFC for systems that energetically support large commercial and industrial applications.

Table 4. A brief summary of the CHP performance of fuel cells adapted from [69,70].

	MCFC	PAFC	PEMFC	SOFC
Electrical capacity (kW)	300+	100–400	0.75–2	0.75–250
Electrical efficiency (LHV)	47%	42%	35–39%	45–60%
Thermal capacity (kW)	450+	110–450	0.75–2	0.75–250
Thermal efficiency (LHV)	43%	48%	55%	30–45%
Application	Residential & Commercial	Commercial	Residential	Residential & Commercial
Degradation rate (per year)	1.5%	0.5%	1%	1–2.5%
Expected lifetime (hours)	20,000	80,000–130,000	60,000–80,000	20,000–90,000

The starting point for this sector is the project initiated in the 1990s by the Japanese government which supported the research activity in order to develop a city-gas-derived hydrogen system that would generate both electric and thermal energy for individual residential buildings, following which was developed the system of residential micro-cogeneration recognized worldwide under the name of Ene-Farm [71]. At the end of 2018, 200,000 PEMFC units were reported as being implemented as part of the Ene-Farm project [23]. In the future, a system similar to the one developed by the Ene-Farm project is planned by Japan to be installed in collective apartment buildings. The success of Ene-Farm has inspired various demonstration projects in other parts of the world, including Korea, Denmark, Germany, the USA and the UK [23,71].

Backup Power Systems wich Using Renewable Energy Sources or Converting Waste into Energy

This type of function involves storing and increasing the degree of use by avoiding the losses associated with the excess energy produced in the power plants that operate by exploiting the renewable energy sources. Various concerns in this sector have laid the foundations for the research and development projects of these systems, being currently in progress or in validation of the obtained results. As an example: Solar to Hydrogen—MYRTE combines solar energy with electrolyzers, hydrogen storage and fuel cell usage, and the project was a partnership between French Nuclear and Alternative Energy Commission, the energy company AREVA and the University of Corsica [72].

It is worth mentioning the recent initiative undertaken by the European Commission that funded through the public-private partnership Fuel Cells and Hydrogen Joint Undertaking (FCH JU) within the Assessment of the Potential, the Actors and Relevant Business Cases for Large Scale and Seasonal Storage of Renewable Electricity by Hydrogen Underground Storage in Europe project (HyUnder) [73]. The idea behind the project was to establish a European initiative for the implementation of energy technologies based on hydrogen generated by increasing the percentage of the use of renewable resources. This project aimed at researching large-scale hydrogen storage in underground caverns, an aspect related to the energy market and existing storage technologies, and aims to identify and analyze the areas of applicability, potential stakeholders, safety rules, the regulatory framework and the societal impact on public acceptance. Within this project, case studies were provided in several areas of Europe. Each case study analyzed the competitiveness of hydrogen storage compared to other types of energy storage, the geological potential of hydrogen storage and the way in which this hypothesis hydrogen storage can be implemented on the energy market. National Research and Development Institute for Cryogenic and Isotopic Technologies—ICSI Rm. Valcea, Romania, participated in the mentioned project through the National Center for Hydrogen and Fuel Cells.

Hydrogen production by water electrolysis leads to the consumption of water resources. In some areas, this is not a problem, but elsewhere it is a huge barrier to the implementation of hydrogen fuel cell technology. For these reasons, a series of studies have been directed towards methods of obtaining hydrogen from various wastes of which can be exemplified: preparation and catalytic steam reforming of crude bio-ethanol obtained from fir wood [74], pyrolysis-catalytic steam reforming of agricultural biomass wastes and biomass components for production of hydrogen/syngas [75], methodology for treating biomass, coal, msw/any kind of wastes and sludges from sewage treatment plants to produce clean/upgraded materials for the production of hydrogen, energy and liquid fuels-chemicals [76], biohydrogen production from solid wastes [77], not least, carbohydrate-to-hydrogen production technologies [78]. Table 5 presents the main hydrogen production methods in terms of efficiency and energy consumption.

Table 5. Hydrogen production methods-efficiency and energy consumption adapted from [70,79,80].

	Energy Consumption (kWh/kgH$_2$)	Efficiency (LHV)
Biomass gasification	69–76	44–48
Coal gasification	51–74	45–65%
Electrolysis	50–65	51–67%
Methane reforming	44–51	65–75%

The production of hydrogen from fossil fuels and biomass, including the catalytic reforming of natural gas, appear to be environmentally unresponsible methods [81], especially due to carbon emissions that qualify the processes as a negative emission technology. In this regard, concentrated research efforts are being made to develop cleaner hydrogen production systems. Thus, a series of high purity hydrogen production installations, which work with carbon capture and storage (CCS) or post-combustion carbon capture (PCC), are demonstrated. Noteworthy in this direction is the research activity supported by Graz University of Technology, Institute of Chemical Engineering and Environmental Technology, Austria by Bock, Zacharias and Hacker. They studied and demonstrated the production of high purity hydrogen (99.997%) with the co-production of pure nitrogen (98.5%) and carbon dioxide (99%) with a raw material use of up to 60% in the largest loops with fixed beds worldwide [82].

Prime Power Generation Large Capacity Electric Power Stations

Several types of fuel cells find applicability in power generation for large stationary applications. AFC, PAFC, PEMFC, SOFC and MCFC systems are used worldwide for the generation of distributed electricity for local use [83]. Figure 12 illustrates the relative weight of the various high-capacity fuel cell technologies installed by the end of 2018 [22–24]. It is noted that the sector is dominated by three types of technologies, MCFC having the highest weight, followed by SOFC and PAFC. To date, only a small number of high-capacity installations based on PEMFC and AFC technologies have been implemented.

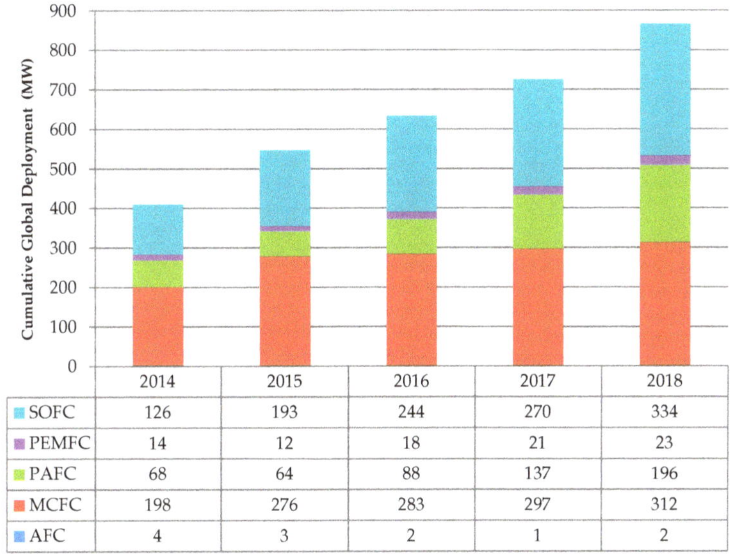

Figure 12. Large scale stationary fuel cells.

4. Results and Discussion

The important elements to be discussed regarding hydrogen fuel cell technology were schematically presented in Figure 13, being widely developed within the SWOT analysis [84] and refer to technological, environmental, social and economic factors.

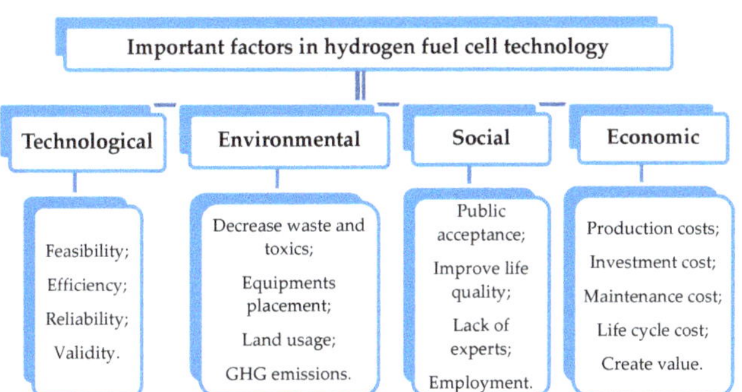

Figure 13. Hierarchy of important factors in hydrogen fuel cell technology.

4.1. Strengths–Weaknesses–Opportunities–Threats (SWOT) Analysis

Based on the data from the specialized literature collected, studied, analyzed critically, the SWOT matrix was developed (Table 6), the specific main characteristic elements of hydrogen fuel cell technology being presented, its worthy to be considered if we talk about energy for stationary applications towards a green to green paradigm.

Table 6. Strengths–Weaknesses–Opportunities–Threats (SWOT) analysis.

Strengths	Weaknesses
S1. Technical strengths: - hydrogen has the highest energy/mass unit of all fuel types; - 1 kg of hydrogen contains as much energy as 2.1 kg of natural gas or 2.8 kg of oil; - fuel cell technolohy has an overall efficiency of up to 60%; - the fuel cell converts hydrogen directly into electricity + heat through the electrochemical reaction of hydrogen with oxygen; - hydrogen concentrates primary renewable energy sources, which it makes available to the consumer in a convenient way; - can be produced on-site at the consumer's place - reduce the dependence on long distance pipelines; - hydrogen can be stored for the medium and long term; - integration with smart grid - hydrogen can stabilize power fluctuations in the grid; - can be transported over long distances stored in various forms or has potential of utilizing current fuel transportation infrastructure. **S2. Environmental strengths:** - it is an inexhaustible source, if it is obtained by electrolysis of water, hydrogen production and consumption is a closed cycle; - the burning of hydrogen is almost entirely devoid of pollutant emissions; - hydrogen is a non-toxic energy transport operator; - environmental friendly and reduces the amount of GHG emissions from the energy system; - low noise pollution compared to other energy production methods. **S3. Sustainable development strengths:** - consequent energy supply and energy security (hydrogen could be considered as a never ending source of energy); - hydrogen as a non-polluting energy carrier; - huge development potential; - possibility of production on site for stand alone applications and remote areas; - hydrogen can be obtained from a wide range of primary renewable energies; - hydrogen is the only secondary energy carrier that is suitable for wide applications in stationary, transport and portable domains; - sustainable transportable energy source; - favorable research and development theme; - stimulates and creates new jobs; - hydrogen will become an energy center, just as electricity is now. **S4. Diversity in resources harnessing:** - hydrogen can be obtained from a wide range of primary energies; - various methods of obtaining hydrogen; - harnessing waste as it is possible to produce hydrogen from waste as a by-product; - can be used as a feedstock in other industries; - hydrogen has potential to integrate in the energy system of the intermittent renewable energies; - allows remote communities to manage their own energy supply; - decrease dependence to fossil classic fuels and increase alternative energy diversity.	**W1. Unavailability of an efficient hydrogen infrastructure:** - lack of points of hydrogen production; - lack of an efficient transport, distribution and storage systems; - incomplete hydrogen infrastructure; - limited access and availability (unavailability) of enough hydrogen refill stations; - necessity to change current distribution system in residential buildings; - lack of plans for the development and implementation of the hydrogen economy. **W2. Introduction risks:** - the complexity of the hydrogen economy; - the integration of hydrogen as energy carrier into the energy system is not tested on an industrial scale; - lack of effective instruments for introduction of hydrogen in the existing transmission and distribution natural gas networks; - development of support services assistance in the hydrogen energy domain is still very immature; - hydrogen burns in the presence of air, which can cause operational safety problems; - hydrogen energy as a new product presents uncertainties related to its public acceptance. **W3. Lack of support from the government:** - insufficient cooperation between political authorities and professional associations in the field of hydrogen energy and economic operators, producers of hydrogen fuel cell technologies. **W4. System integration:** - lack of codes, technical design regulations, implementation procedures, technical standards for hydrogen economy in general, stationary applications of hydrogen energy in particular; - lack of widespread awareness of capabilities and potential benefits of hydrogen fuel cell technologies used to supply clean energy for stationary applications; - weak development of hydrogen supply network; - uncertainties and lack of information related to problems of exploitation under conditions of stability and safety of hydrogen; - unavailability of clear marketing policies and strategies to promote hydrogen energy as clean energy for stationary applications; - unclear plans for a future economy based on hydrogen energy. **W5. High costs:** - high initial investment installation costs; - high production costs of hydrogen; - high production costs of fuel cell; - high production costs of systems based on hydrogen fuel cell technologies; - high price of energy generated by hydrogen-based energy systems; - high costs for hydrogen storage; - high costs for adaptation of the hydrogen economy; - lack of focused research and development works from major companies to develop the equipment and reduce costs.

Table 6. *Cont.*

Opportunities	Threats
O1. Development potential - hydrogen is a key element for a future green sustainable development of energy systems; - hydrogen energy can be considered as an object of innovations and technological development along the lines of energy efficiency; - hydrogen fuel cells technology enables investment in sustainable energy infrastructure; - encouraging the generation of green energy from indigenous unconventional sources; - hydrogen fuel cell technology can be considered next energy efficiency solution for supply energy in stationary applications; - developing social policies that respond to the challenges generated by the implementation of clean energy policies; - development of human capital in order to ensure the implementation of the energy strategy in a future hydrogen-based economy; - hydrogen and fuel cell technology stimulates research, innovation and development in the energy systems domain. **O2. Improve energy security** - hydrogen is expected to play an important role in global future energy scenarios; - an energy carrier helps to increase the stabilization of energy security and price, giving rise to competition between different energy sources; - increasing the energy efficiency through the efficient use of energy resources throughout the energy cycle - production, transport, storage, distribution and final consumption; - decarbonisation of the energy sector at minimum costs; - energy diversification; - integrating hydrogen into the energy mix that responds to the sustainable development desire and which ensures the reduction of energy import dependency. **O3. Increase cooperation** - opportunities for collaboration between academic institutions, research institutes on line of knowledge transfer to economic operators; - educational opportunity for universities with energetic and environmental learning profile to development of a new teaching discipline; - improve cooperation with governments, local authorities and make alliance with local political administrations or economic operators as investors in the implementation of green energy systems; - collaboration opportunities among line ministries, departments and other energy system actors; - international interconnection. **O4. New business opportunity** - emergence of hydrogen market; - emergence of a new commercialization plans; - emergence of potential suppliers, demanders and end-users; - involving several companies in the energy sector and setting up new ones; - emergence and development of new business models; - emergence of new jobs; - development of the blockchain model for hydrogen economy.	**T1. Technical** - low stimulation of hydrogen competitiveness in field of stationary applications regarding the generation of energy from alternative sources to the classical ones; - immaturity of some technologies for the conversion of hydrogen into electric and thermal energy despite the efforts stimulated on the one hand by the technical progress in the field of fuel cells, and on the other hand driven by the global concerns regarding energy security, efficiency, energy sustainability, reduction of greenhouse gases emissions and last but not least, the reduction of dependence on the use of fossil fuels; - lack of specialists and experts in the field regarding the implementation of hydrogen energy projects for stationary applications; - insufficient storage capacity in large quantities of hydrogen: Increasing production of fluctuating renewable energy intensifies the need for electricity storage to ensure network reliability and flexibility. Using hydrogen as a mean to store energy in the long run may in the future help address the challenge of grid balancing when large quantities of fluctuating renewable electricity are introduced in the energy mix; - immature solutions for massive hydrogen storage, which are not widely tested (e.g. underground hydrogen storage, potentially attractive solution, but still needs to be evaluated thoroughly from a technical, economic and societal standpoint); - limited practical experience in both producers and consumers; - lack technical information of potential investors regarding hydrogen new technologies for power generation and energy efficiency of fuel cell technology, which it generates a low degree of interest from them. **T2. Social** - negative influence from other energy actors; - public acceptance of the widespread use of hydrogen in stationary applications is unclear; - technical regulations, standards and procedures deficiencies for the applicability of hydrogen energy in stationary applications; - immaturity of the legislative framework; - weak support from authorities and government to shift to a hydrogen-based economy; - non-recognition of hydrogen-based power generation systems and hydrogen economy as strategic infrastructure; - the results of the research projects cannot be adequately replicated due to the various difficulties at the legislative level. **T3. Economic** - its were not developed sufficient fiscal instruments to support the investment programs in the energy efficiency sector and the use of hydrogen energy in stationary applications; - lack of potential suppliers, potential investors and demanders; - competitions with other renewable resources; - the difficulty to compete with the current fossil fuel market; - strong position of fossil fuel producers; - deficient organization and financing of the hydrogen economy.

This type of instrument/analysis method is validated by numerous studies carried out in the energy and the environment domain [85–87]. The objectives of the analysis were to highlight the strengths and weaknesses, in relation to the opportunities and threats existing or potential regarding the conditions of the implementation of hydrogen as the source of energy in stationary applications.

4.2. Strategies Proposed for the Use in Stationary Applications of Hydrogen Energy

In order to outline an overview of the possibilities of implementing hydrogen-based energy systems to power stationary applications, respectively to identify the possibilities for increasing the share of the use of hydrogen as alternative resource, a series of strategies have been proposed with a character of recommendation (Figure 14).

	Strengths	**Weaknesses**
Opportunities	**S&O1:** promote the utilization of hydrogen energy in stationary applications to make it more popular; **S&O2:** develop hydrogen economy with comprehensive legislation; **S&O3:** stimulate public acceptance.	**W&O1:** stimulate developments, innovations and research; **W&O2:** stimulate development in hydrogen infrastructure; **W&O3:** governmental and funding programmes.
Threats	**S&T1:** continued R&D funding to explore the potential applications; **S&T2:** strategies between hydrogen energy and other renewables to decrease competition between them; **S&T3:** cooperation between energy actors, R&D centers and politicians.	**W&T1:** promote regulated hydrogen economy; **W&T2:** absorb private and foreign investments to financially support hydrogen economy projects; **W&T3:** implementation of specific laws for safety and stability in use.

Figure 14. Established and recommended strategies.

At EU level, there are a large number of projects [88] that are already facing some of the strategies highlighted in Figure 14. To assist and promote the EU's commitment to the "hydrogen challenge", it is worth highlighting some of these significant projects in the area of Hydrogen Fuel Cell Technology for sustainable future of Stationary Applications: TriSOFC—Durable Solid Oxide Fuel Cell Tri-generation system for low carbon Buildings [89]; C3SOFC—Cost Competitive Component Integration for Stationary Fuel Cell Power, application area: stationary power production and CHP [90]; STAGE-SOFC—Innovative SOFC system layout for stationary power and CHP applications [91]; Remote area Energy supply with Multiple Options for integrated hydrogen-based Technologies—demonstration of fuel cell-based energy storage solutions for isolated micro grid or off-grid remote areas [92]; Demo4Grid—Demonstration of 4MW Pressurized Alkaline Electrolyser for Grid Balancing Services [93]; ELECTROU—MW Fuel Cell micro grid and district heating at King's Cross [94], ene.field—European-wide field trials for residential fuel cell micro-CHP [95], H2 Future-Hydrogen meeting future needs of low carbon manufacturing value chains [96].

Energy strategies in the context of sustainable development refer both to the present and to the future, as they define the vital interests and establish the lines of action to meet the present and future needs while managing the evolutions in the field. When discussing energy security, it must be viewed as a vital component and includes: security of energy sources, securing the existing energy routes, identifying alternative energy routes, identifying alternative energy sources, environment protection. As a result, the topics discussed and analyzed in the present article fall into the current national, european and international context, the importance of the problem being topical both from a scientific, technological, but also from a socio-economic or cultural point of view.

In this context, hydrogen, as an energy vector or environmentally friendly synthetic fuel, together with the fuel cell, its conversion technology, can play an important role in energy strategies regarding the efficiency and decarbonisation of energy generation systems in stationary applications. Technologies using low-carbon footprint hydrogen can be valuable in various end-use stationary applications.

5. Conclusions

Hydrogen and fuel cell technology have advanced considerably over the last fifteen years. At the global level, this area continues to face significant challenges—technical, commercial and infrastructure-related—that need to be overcome before fuel cells can realize the full potential of which they are capable. Policy makers have included hydrogen and fuel cell on the map of future energy strategies and have already taken into account the fact that fuel cells have great real potential and can successfully meet the technical, social, economic and environmental objectives in the context of the multidisciplinary concept of sustainable development.

In this paper, the review of literature and agencies' reports on specialized metrics in the domain of the hydrogen fuel cell technologies, highlights the essential considerations regarding stationary applications, as follows:

- More than 850 MW of large stationary fuel cell systems with a (> 200 kw) nominal power have been installed worldwide for power generation and CHP applications up until 2018.
- Worldwide, the use of three types of fuel cell technologies is prevalent: MCFC, SOFC and PAFC.
- AFC and PEMFC are relatively new technologies under development and implementation within stationary applications.
- The main modalities of integrating hydrogen fuel cell technology into stationary applications are in the form of CHP units with fuel cells for small individual residential buildings, back-up power systems and large capacity electric power stations or distributed generation systems.
- The key factors that influencing development include: energy and climate policies, fuel cell funding programmes, concurrent technologies, the attendance of fuel cell system producers and energy costs.

The SWOT analysis conducted in the present study highlights that the implementation of the hydrogen economy depends decisively on the following main factors: legislative framework, energy decision makers, information and interest from the end beneficiaries, potential investors, and existence of specialists in this field.

Author Contributions: conceptualization, R.-A.F.; methodology, R.-A.F. and E.C.; validation, E.C., M.S.R. and C.F.; formal analysis, C.N.T.; investigation, M.S.R. and E.C.; data curation, C.T.; writing—Original draft preparation, R.-A.F., M.S.R. and C.N.T.; writing—Review and editing, E.C. and C.F.; visualization, R.-A.F.; supervision, C.F.; project administration, R.-A.F. and E.C.

Acknowledgments: This work was supported by a grant from the Romanian Ministry of Research and Innovation, CCCDI-UEFISCDI, project number PN-III-P1-1.2-PCCDI-2017-0776/No. 36 PCCDI/15.03.2018, within PNCDI III.

Conflicts of Interest: The authors declare no conflict of interest.

References

1. Burke, M.J.; Stephens, J.C. Political power and renewable energy futures: A critical review. *Energy Res. Soc. Sci.* **2018**, *35*, 78–93. [CrossRef]
2. Paris Agreement on Climate Change. Available online: http://www.investenergy.ro/acordul-de-la-paris-privind-schimbarile-climatice-intrat-vigoare/ (accessed on 26 December 2018).

3. Ministry of Energy. The National Market Policy Framework for Alternative Fuels in the Transport Sector and the Installation of Relevant Infrastructure in Romania. Available online: http://www.mmediu.ro/app/webroot/uploads/files/Cadrul-National-de-Politica_Combustibilii-Alternativi-in-Sectorul%20Transporturilor%281%29.pdf (accessed on 26 December 2018).

4. Energy Policy of the European Union, Romanian Ministry of Foreign Affairs. Available online: http://www.europarl.europa.eu/ftu/pdf/ro/FTU_2.4.7.pdf (accessed on 18 July 2019).

5. Specialized Consulting Services Needed for the Elaboration of the National Energy Strategy for the Period 2014–2035 and Prospects for 2050. Available online: File:///C:/Documents%20and%20Settings/Home/Desktop/Anexa0511.pdf (accessed on 20 July 2019).

6. Energy Security Strategy and Energy Policy. Current Approaches in the European Union and Internationally. Available online: http://www.fnme.ro/ (accessed on 20 July 2019).

7. Energy Strategy of Romania, Government of Romania. Department for Energy. Available online: http://media.hotnews.ro/media_server1/document-2014-12-5-18755546-0-strategia-energetica-analiza-stadiului-actual.pdf (accessed on 20 July 2019).

8. European Parliament, Directive 2009/28/EC on the Promotion of the Use of Energy from Renewable Sources. Available online: https://eur-lex.europa.eu/LexUriServ/LexUriServ.do?uri=OJ:L:2009:140:0016:0062:EN:PDF (accessed on 22 July 2019).

9. Şoimoşan, T.M.; Felseghi, R.A.; Firescu, V. Aspecte de management privind valorificarea surselor regenerabile de energie la producerea căldurii în zonele urbane din România (Management aspects regarding the use of renewable energy sources for heat generation in urban areas of Romania). *Rev. Manag. Econ. Eng.* **2016**, *15*, 512–525.

10. Owusu, P.A.; Asumadu-Sarkodie, S. A review of renewable energy sources, sustainability issues and climate change mitigation. *Cogent Eng.* **2016**, *3*, 1167990. [CrossRef]

11. Srinivasan, S.S.; Stefanakos, E.K. Clean Energy and Fuel Storage. *Appl. Sci.* **2019**, *9*, 3270. [CrossRef]

12. Midilli, A.; Dincer, I. Key strategies of hydrogen energy systems for sustainability. *Int. J. Hydrogen Energy* **2007**, *32*, 511–524. [CrossRef]

13. Midilli, A.; Ay, M.; Dincer, I.; Rosen, M.A. On hydrogen and hydrogen energy strategies. I: Current status and needs. *J. Renew. Sustain. Energy Rev.* **2005**, *9*, 255–271. [CrossRef]

14. Dincer, I.; Acar, C. Smart Energy Solutions with Hydrogen Options. *Int. J. Hydrogen Energy* **2018**, *43*, 8579–8599. [CrossRef]

15. Viktorsson, L.; Heinonen, J.T.; Skulason, J.B.; Unnthorsson, R.A. Step towards the Hydrogen Economy—A Life Cycle Cost Analysis of A Hydrogen Refueling Station. *Energies* **2017**, *10*, 763. [CrossRef]

16. Badea, G.; Felseghi, R.A.; Aşchilean, I.; Răboacă, S.M.; Şoimoşan, T.M. The role of hydrogen as a future solution to energetic and environmental problems for residential buildings. In Proceedings of the 11th Int. Conference on Processes in Isotopes and Molecules—PIM, Cluj-Napoca, Romania, 27–29 September 2017. [CrossRef]

17. Andrews, J.; Shabani, B. Where Does Hydrogen Fit in a Sustainable Energy Economy? *Procedia Eng.* **2012**, *49*, 15–25. [CrossRef]

18. Ball, M.; Weeda, M. The Hydrogen Economy—Vision or reality? *Int. J. Hydrogen Energy* **2015**, *40*, 7903–7919. [CrossRef]

19. International Energy Agency (IEA) Hydrogen, Global Trends and Outlook for Hydrogen, December 2017 Available online: https://ieahydrogen.org/pdfs/Global-Outlook-and-Trends-for-Hydrogen_Dec2017_WEB.aspx (accessed on 27 July 2019).

20. International Energy Agency, Technology Roadmap Hydrogen and Fuel Cells. Available online: https://www.iea.org/publications/freepublications/publication/TechnologyRoadmapHydrogenandFuelCells.pdf (accessed on 18 July 2019).

21. International Energy Agency, The Future of Hydrogen Report Prepared by the IEA for the G20, Japan. Available online: https://www.g20karuizawa.go.jp/assets/pdf/The%20future%20of%20Hydrogen.pdf (accessed on 12 August 2019).

22. E4tech, The Fuel Cell Industry Review 2017. Available online: http://www.fuelcellindustryreview.com/archive/TheFuelCellIndustryReview2017.pdf (accessed on 24 August 2019).

23. E4tech, The Fuel Cell Industry Review 2018. Available online: https://www.californiahydrogen.org/wp-content/uploads/2019/01/TheFuelCellIndustryReview2018.pdf (accessed on 24 August 2019).

24. European Hydrogen and Fuel Cell Technology Platform, Strategic Research Agenda. Available online: https://www.fch.europa.eu/sites/default/files/documents/hfp-sra004_v9-2004_sra-report-final_22jul2005.pdf (accessed on 23 July 2019).

25. SWOT Manual. Available online: www.austincc.edu/oiepub/services/SWOT%20Manual%20AY13_v1.pdf (accessed on 12 August 2019).

26. Afgan, N.; Veziroğlu, A. Sustainable resilience of hydrogen energy system. *Int. J. Hydrogen Energy* **2012**, *372013*, 5461–5467. [CrossRef]

27. Fraile, D.; Lanoix, J.C.; Maio, P.; Rangel, A.; Torres, A. Overview of the Market Segmentation for Hydrogen Across Potential Customer Groups, Based on Key Application Areas. Available online: https://www.fch.europa.eu/sites/default/files/project_results_and_deliverables/D%201.2.%20Overview%20of%20the%20market%20segmenatation%20for%20hydrogen%20across%20potential%20customer%20groups%20based%20on%20key%20application%20areas.pdf (accessed on 18 August 2019).

28. Dincer, I. Green Paths for Hydrogen Production. *Int. J. Hydrogen Energy* **2012**, *37*, 1954–1971. [CrossRef]

29. Jeon, H.; Kim, S.; Yoon, K. Fuel Cell Application for Investigating the Quality of Electricity from Ship Hybrid Power Sources. *J. Mar. Sci. Eng.* **2019**, *7*, 241. [CrossRef]

30. İnci, M.; Türksoy, Ö. Review of fuel cells to grid interface: Configurations, technical challenges and trends. *J. Clean. Prod.* **2019**, *213*, 1353–1370. [CrossRef]

31. Sharaf, O.Z.; Orhan, M.F. An overview of fuel cell technology: Fundamentals and applications. *J. Renew. Sustain. Energy Rev.* **2014**, *32*, 810–853. [CrossRef]

32. Hamacher, T. Hydrogen as a Strategic Secondary Energy Carrier. In *Hydrogen and Fuel Cell*; Töpler, J., Lehmann, J., Eds.; Springer: Berlin/Heidelberg, Germany, 2016.

33. Pötzinger, C.; Preißinger, M.; Brüggemann, D. Influence of Hydrogen-Based Storage Systems on Self-Consumption and Self-Sufficiency of Residential Photovoltaic Systems. *Energies* **2015**, *8*, 8887–8907. [CrossRef]

34. Midilli, A.; Ay, M.; Dincer, I.; Rosen, M.A. On hydrogen and hydrogen energy strategies. II: Future projections affecting global stability and unrest. *J. Renew. Sustain. Energy Rev.* **2005**, *9*, 273–287. [CrossRef]

35. Momirlan, M.; Veziroğlu, T.N. The properties of hydrogen as fuel tomorrow in sustainable energy system for a cleaner planet. *Int. J. Hydrogen Energy* **2005**, *30*, 795–802. [CrossRef]

36. Veziroğlu, T.N.; Şahin, S. 21st Century's energy: Hydrogen energy system. *J. Energy Convers. Manag.* **2008**, *49*, 1820–1831. [CrossRef]

37. Stolten, D. *Hydrogen and Fuel Cells, Fundamentals, Technologies and Applications (Hydrogen Energy)*; WILEY-VCH Verlag GmbH&Co. KgaA: Weinheim, Germany, 2010.

38. Appleby, A.J.; Foulkes, F.R. *Fuel Cell Handbook*; Van Nostrand: New York, NY, USA, 1989.

39. Larminie, J.; Dicks, A. *Fuel Cell Systems Explained*, 2nd ed.; Wiley: New York, NY, USA, 2003.

40. Felseghi, R.A. Contributions Regarding the Application of Fuel Cell in the Passive Houses Domain. Ph.D. Thesis, Building Services Engineering Department, Technical University of Cluj-Napoca, Cluj-Napoca, Romania, March 2015.

41. Hoogers, G. *Fuel Cell Technology Handbook*; CRC Press: Boca Raton, FL, USA, 2003.

42. Ştefănescu, I. *Pile de combustibil—Între teorie şi practică (Fuel cells–between theory and practice-book)*; Editura CONPHYS: Râmnicu Vâlcea, Romania, 2010.

43. Dincer, I.; Zamfirescu, C. Sustainable hydrogen production options and the role of IAHE. *Int. J. Hydrogen Energy* **2012**, *37*, 16266–16286. [CrossRef]

44. Iordache, I.; Ştefănescu, I. *Tehnici Avansate De Tratare a Apei Şi Separare a Hidrogenului (Advanced Techniques for Water Treatment and Hydrogen Separation)*; Editura AGIR: Bucureşti, Romania, 2010.

45. Felseghi, R.A.; Şoimoşan, T.M.; Megyesi, E. *Overview of Hydrogen Production Technologies from Renewable Resources, Renewable Energy Sources & Clean Technologies*; STEF92 Technology Ltd: Sofia, Bulgaria, 2014; pp. 377–384. [CrossRef]

46. Mekhilef, S.; Saidur, R.; Safari, A. Comparative study of different fuel cell technologies. *J. Renew. Sustain. Energy Rev.* **2012**, *16*, 981–989. [CrossRef]

47. Tabak, J. *Natural Gas and Hydrogen*; Facts On File, Series Energy and the Environment: New York, NY, USA, 2009.

48. Torrero, E.; McClelland, R. *National Rural Electric Cooperative Association Cooperative Research, Residential Fuel Cell Demonstration Handbook*; National Renewable Energy Laboratory: Golden, CO, USA, 2002.

49. Grove, W.R. On Voltaic Series and the Combination of Gases by Platinum. *Lond. Edinb. Philos. Mag. J. Sci.* **1839**, *14*, 127–130. [CrossRef]

50. Grove, W.R. On a Gaseius Voltaic Battery. *Lond. Edinb. Philos. Mag. J. Sci.* **1842**, *21*, 417–420.

51. Bockris, J.O.M. Review: The hydrogen economy: Its history. *Int. J. Hydrogen Energy* **2013**, *38*, 2579–2588. [CrossRef]

52. Barbir, F. *PEM Fuel Cells: Theory and Practice*; Elsevier: London, UK, 2005.

53. Felseghi, R.A.; Şoimoşan, T.M.; Safirescu, O.C.; Aşchilean, I.; Roman, M.D.; Iacob, G.D. *Estimation of Hydrogen and Electrical Energy Production by Using Solar and Wind Resources for a Residential Building from Romania, Monitoring, Controlling and Architecture of Cyber Physical Systems*; Trans Tech Publications Ltd: Stafa-Zurich, Switzerland, 2014; pp. 543–551. [CrossRef]

54. Şoimoşan, T.M.; Moga, L.M.; Danku, G.; Căzilă, A.; Manea, D.L. Assessing the Energy Performance of Solar Thermal Energy for Heat Production in Urban Areas: A Case Study. *Energies* **2019**, *12*, 1088. [CrossRef]

55. Seyed Mahmoudi, S.M.; Sarabchi, N.; Yari, M.; Rosen, M.A. Exergy and Exergoeconomic Analyses of a Combined Power Producing System including a Proton Exchange Membrane Fuel Cell and an Organic Rankine Cycle. *Sustainability* **2019**, *11*, 3264. [CrossRef]

56. Badea, G.; Felseghi, R.A.; Giurca, I.; Safirescu, O.C.; Aşchilean, I.; Şoimoşan, T.M. Choosing the optimal fuel cell technology for application in the field of passive houses, Recent Advances in Urban Planning and Construction. In *Energy, Environmental and Structural Engineering Series|20, Proceedings of the 4th WSEAS International Conference USCUDAR, Budapesta, Hungary, 10–12 December 2013*; Ungaria: Budapesta, Hungary, 2013; pp. 125–130.

57. Fereshteh, S.; Mohammad, R.R. Chapter 14–Direct Methanol Fuel Cell. In *Methanol*; Science and Engineering: Auckland, New Zealand, 2018; pp. 381–397. [CrossRef]

58. Damo, U.M.; Ferrari, M.L.; Turan, A.; Massardo, A.F. Solid oxide fuel cell hybrid system: A detailed review of an environmentally clean and efficient source of energy. *Energy* **2019**, *168*, 235–246. [CrossRef]

59. Sima, C.A.; Lazaroiu, G.C.; Dumbrava, V.; Tirsu, M. A Hybrid System Implementation for Residential Cluster. In Proceedings of the 11th International Conference on Electromechanical and Power Systems SIELMEN 2017, Chisinau, Moldova, 12–13 October 2017; pp. 1–6.

60. Aschilean, I.; Rasoi, G.; Raboaca, M.S.; Filote, C.; Culcer, M. Design and Concept of an Energy System Based on Renewable Sources for Greenhouse Sustainable Agriculture. *Energies* **2018**, *11*, 1201. [CrossRef]

61. Felseghi, R.A. Fuel cell as solution for power supply of passive house. Case study. In *Progress of Cryogenics and Isotopes Separation*; National Research-Development Institute for Cryogenic and Isotopic Technologies: Râmnicu Vâlcea, Romania, 2015; Volume 18; p. 53.

62. Naghiu, G.S.; Giurca, I.; Aschilean, I.; Badea, G. Comparative analysis on the solutions of hydrogen production using solar energy with and without connection to the power network. *Procedia Technol.* **2016**, *22*, 781–788. [CrossRef]

63. Sdanghi, G.; Maranzana, G.; Celzard, A.; Fierro, A. Review of the current technologies and performances of hydrogen compression for stationary and automotive applications. *Renew. Sustain. Energy Rev.* **2019**, *102*, 150–170. [CrossRef]

64. Bizon, N. Nonlinear control of fuel cell hybrid power sources: Part II—Current control. *Appl. Energy* **2011**, *88*, 2574–2591. [CrossRef]

65. Bizon, N.; Thounthong, P.; Raducu, M.; Constantinescu, L.M. Designing and modelling of the asymptotic perturbed extremum seeking control scheme for tracking the global extreme. *Int. J. Hydrogen Energy* **2017**, *42*, 17632–17644. [CrossRef]

66. Bizon, N.; Thounthong, P. Fuel Economy using the Global Optimization of the Fuel Cell Hybrid Power Systems. *Energy Convers Manag.* **2018**, *173*, 665–678. [CrossRef]

67. Bizon, N.; Mazare, A.G.; Ionescu, L.M.; Enescu, F.M. Optimization of the Proton Exchange Membrane Fuel Cell Hybrid Power System for Residential Buildings. *Energy Convers Manag.* **2018**, *163*, 22–37. [CrossRef]

68. U.S. Department of Energy, Fuel Cell Technology Office, Comparison of Fuel Cell Technologies. Available online: https://www.energy.gov/sites/prod/files/2016/06/f32/fcto_fuel_cells_comparison_chart_apr2016.pdf (accessed on 30 October 2019).

69. Staffell, I. Zero carbon infinite COP heat from fuel cell CHP. *Appl. Energy* **2015**, *147*, 373–385. [CrossRef]

70. Stolten, D.; Samsun, R.C.; Garland, N. *Fuel Cells—Data, Facts and Figures*; Wiley-VCH: Weinheim, Germany, 2015.

71. Ministry of Economy, Trade and Industry JAPAN, New Era of a Hydrogen Energy Society. Available online: http://injapan.no/wp-content/uploads/2018/03/180227_METI-New-Era-of-a-Hydrogen-Energy-Society.pdf (accessed on 30 October 2019).

72. International Energy Agency, R&D platform MYRTE: Coupling Photovoltaic Plant—Hydrogen and Fuel Cells. Available online: http://ieahydrogen.org/Activities/National-Projects-(1)/France/08-10-Hydrogen-on-Islands_MYRTE-project_Poggi.aspx (accessed on 31 October 2019).

73. HyUnder Project. Available online: www.hyunder.eu (accessed on 30 October 2019).

74. Dan, M.; Senila, L.; Roman, M.; Mihet, M.; Lazar, M.D. From wood wastes to hydrogen—Preparation and catalytic steam reforming of crude bio-ethanol obtained from fir wood. *Renew. Energy* **2015**, *74*, 27–36. [CrossRef]

75. Akubo, K.; Mohamad, A.N.; Williams, P.T. Pyrolysis-catalytic steam reforming of agricultural biomass wastes and biomass components for production of hydrogen/syngas. *J. Energy Inst.* **2019**, *92*, 1987–1996. [CrossRef]

76. Arvelakis, S. Methodology for Treating Biomass, Coal, Mswiany Kind of Wastes and Sludges from Sewage Treatment Plants to Produce Clean/Upgraded Materials for the Production of Hydrogen, Energy and Liquid Fuels-Chemicals. U.S. Patent 2019/0010414 A1, 10 January 2019. Available online: https://patentimages.storage.googleapis.com/2a/2a/33/ce0704f3503d75/US20190010414A1.pdf (accessed on 17 November 2019).

77. Keskin, T.; Abubackar, H.N.; Arslan, K.; Azbar, N. Chapter 12—Biohydrogen Production from Solid Wastes, Biohydrogen. In *Biomass, Biofuels, Biochemicals: Biohydrogen*, 2nd ed.; Elsevier: Amsterdam, The Netherlands, 2019; pp. 321–346.

78. Sharma, K. Carbohydrate-to-hydrogen production technologies: A mini-review. *Renew. Sustain. Energy Rev.* **2019**, *105*, 138–143. [CrossRef]

79. Schmidt, O.; Gambhir, A.; Staffell, I.; Hawkes, A.; Nelson, J.; Few, S. Future cost and performance of water electrolysis: An expert elicitation study. *Int. J. Hydrogen Energy* **2017**, *42*, 30470–30492. [CrossRef]

80. Lehner, M.; Tichler, R.; Steinmu'ller, H.; Koppe, M. *Power-to-Gas: Technology and Business Models*; Springer International Publishing: Cham, Switzerland, 2014.

81. Maier, D.; Sven-Joachim, I.; Fortmüller, A.; Maier, A. Development and operationalization of a model of innovation management system as part of an integrated quality-environment-safety system. *Amfiteatru Econ.* **2017**, *19*, 302–314.

82. Zacharias, R.; Visentin, S.; Bock, S.; Hacker, V. High-pressure hydrogen production with inherent sequestration of a pure carbon dioxide stream via fixed bed chemical looping. *Int. J. Hydrog Energy* **2019**, *44*, 7943–7957. [CrossRef]

83. European Comission, JRC Technical Reports, Global Deployment of Large Capacity Stationary Fuel Cells. Available online: https://publications.jrc.ec.europa.eu/repository/bitstream/JRC115923/jrc115923_stationary_fuel_cells_16042019_final_pubsy_online.pdf (accessed on 2 November 2019).

84. Maier, D.; Maftei, M.; MaieR, A.; Bitan, G.E. A review of product innovation management literature in the context of organization sustainable development. *Amfiteatru Econ.* **2019**, *21*, 816–829.

85. Igliński, B.; Piechota, G.; Iglińska Anna Cichosz, M.; Buczkowski, R. *The Study on the SWOT Analysis of Renewable Energy Sector on the Example of the Pomorskie Voivodeship (Poland), Clean Techn Environ Polici*; Springer: Berlin/Heidelberg, Germany, 2015.

86. Baycheva-Merger, T.; Wolfslehner, B. Evaluating the implementation of the Pan-European Criteria and indicators for sustainable forest management—A SWOT analysis. In *Ecological Indicators*; Elsevier Ltd.: London, UK, 2015; Volume 60, pp. 1192–1199.

87. Bull, J.W.; Jobstvogt, N.; Böhnke-Henrichs, A.; Mascarenhas, A.; Sitas, N.; Baulcomb, C.; Lambini, C.K.; Rawlins Baral, H.; Zähringer, J.; Carter-Silk, E.; et al. Strenghts, Weaknesses, Opportunities and Threats: A SWOT analysis of the ecosystem services framework. *Elsevier Ecosyst. Serv.* **2016**, *17*, 99–111. [CrossRef]

88. Fch Ju Projects. Available online: https://www.fch.europa.eu/fchju-projects (accessed on 18 November 2019).

89. TriSOFC Project—Durable Solid Oxide Fuel Cell Tri-Generation System for Low Carbon Buildings. Available online: http://www.trisofc.com (accessed on 18 November 2019).

90. C3SOFC Project—Cost Competitive Component Integration for Stationary Fuel Cell Power. Available online: https://www.fch.europa.eu/project/cost-competitive-component-integration-stationary-fuel-cell-power (accessed on 18 November 2019).

91. STAGE-SOFC Project—Innovative SOFC System Layout for Stationary Power and CHP Applications. Available online: http://www.stage-sofc-project.eu/ (accessed on 18 November 2019).

92. REMOTE Project—Remote Area Energy Supply with Multiple Options for Integrated Hydrogen-Based Technologies. Available online: https://www.fch.europa.eu/project/remote-area-energy-supply-multiple-options-integrated-hydrogen-based-technologies (accessed on 18 November 2019).
93. Demo4Grid Project—Demonstration of 4MW Pressurized Alkaline Electrolyser for Grid Balancing Services. Available online: https://www.demo4grid.eu/ (accessed on 18 November 2019).
94. ELECTROU Project—MW Fuel Cell Micro Grid and District Heating at King's Cross. Available online: https://www.fch.europa.eu/project/mw-fuel-cell-micro-grid-and-district-heating-king%E2%80%99s-cross (accessed on 18 November 2019).
95. Ene. Field Project—European-Wide Field Trials for Residential Fuel Cell Micro-CHP. Available online: www.enefield.eu (accessed on 18 November 2019).
96. H2Future Project—Hydrogen Meeting Future Needs of Low Carbon Manufacturing Value Chains. Available online: https://cordis.europa.eu/project/rcn/207465/en (accessed on 18 November 2019).

Article

Optimization of the Fuel Cell Renewable Hybrid Power System Using the Control Mode of the Required Load Power on the DC Bus

Nicu Bizon [1,2,*], Valentin Alexandru Stan [2] and Angel Ciprian Cormos [2]

[1] Faculty of Electronics, Communication and Computers, University of Pitesti, 1 Targu din Vale, 110040 Pitesti, Romania
[2] Polytechnic University of Bucharest, 313 Splaiul Independentei, 060042 Bucharest, Romania; valentin.stan@upb.ro (V.A.S.); angel.cormos@upb.ro (A.C.C.)
* Correspondence: nicu.bizon@upit.ro

Received: 9 March 2019; Accepted: 14 May 2019; Published: 17 May 2019

Abstract: In this paper, a systematic analysis of seven control topologies is performed, based on three possible control variables of the power generated by the Fuel Cell (FC) system: the reference input of the controller for the FC boost converter, and the two reference inputs used by the air regulator and the fuel regulator. The FC system will generate power based on the Required-Power-Following (RPF) control mode in order to ensure the load demand, operating as the main energy source in an FC hybrid power system. The FC system will operate as a backup energy source in an FC renewable Hybrid Power System (by ensuring the lack of power on the DC bus, which is given by the load power minus the renewable power). Thus, power requested from the batteries' stack will be almost zero during operation of the FC hybrid power system based on RPF-control mode. If the FC hybrid power system operates with a variable load demand, then the lack or excess of power on the DC bus will be dynamically ensured by the hybrid battery/ultracapacitor energy storage system for a safe transition of the FC system under the RPF-control mode. The RPF-control mode will ensure a fair comparison of the seven control topologies based on the same optimization function to improve the fuel savings. The main objective of this paper is to compare the fuel economy obtained by using each strategy under different load cycles in order to identify which is the best strategy operating across entire loading or the best switching strategy using two strategies: one strategy for high load and the other on the rest of the load range. Based on the preliminary results, the fuel consumption using these best strategies can be reduced by more than 15%, compared to commercial strategies.

Keywords: hybrid power systems; renewable energy sources; fuel cell systems; required-power-following control; optimization; fuel economy

---

1. Introduction

The very fast increase of global energy demand over recent decades calls for a new approach to energy sustainable development based on Hybrid Power Systems (HPS) combining Renewable Energy Sources (RESs) and Fuel Cell (FC) systems [1–3]. Therefore, innovative solutions based on experimental research have been proposed for the implementation of the Fuel Cell Hybrid Power Systems (FCHPS) with or without support from the RESs [4–6].

The state-of-the-art studies in this field have identified the following challenging topics for the next stage of research [7–9]:

- Modeling, control, and optimization of the FC system to improve the fuel economy [10–12];
- Developing innovative solutions and advanced technologies to improve the lifetime, reliability and safety in operation of the FC system [13,14];
- Numerical models for the control of hydrogen and thermal/electric energies productions through Solid Oxide Electrolyzer/Fuel Cells [15]:
- Proposal of innovative stand-alone or grid-connected RES HPS architectures, which can be optimized based on advanced Energy Management Strategies (EMSs) [16–18] and Global Maximum Power Point Tracking (GMPPT) control algorithms [19–21] applied to available RESs (photovoltaic systems, wind turbines etc.) in order to optimally ensure the power flow balance on the DC bus (and/or the AC bus) [22–24] and improve the harvested energy from the RESs [25–27];
- Hybridization of the RES HPS with an FC system as backup energy source (FC/RES HPS) to mitigate the RES power variability and load dynamics by controlling the generated FC power at the level of the required power on the DC bus [28–30].
- Use of hybrid Energy Storage Systems (ESSs) with advanced control of the ESS bidirectional power converters to ensure that the DC voltage regulation and mitigation of the power pulses on the DC bus as well [31–33];
- Improving the fuel economy of the FC vehicles (FCVs) by using innovative FCV powertrain [34–36], advanced EMSs [37–39], and hybrid technologies [40–42];
- Optimal sizing of the FC RES HPSs [43–45], FCVs [46–48], and hybrid ESSs [49,50].
- Electrical and thermal analysis of the HPS [51].

In this study, the optimization of the FCHPS is approached using the control mode of the required load power on the DC bus, named the Required-Power-Following (RPF) control-mode of the FC system. The FC system using RPF-control mode will generate, on the DC bus, the needed power to compensate the DC power flow balance for a Hybrid Power System operating with a variable load demand. The RPF-control mode will use one from the three inputs variables of the FC system that can control the FC power: the reference input for the FC boost controller, the air regulator, or the fuel regulator. So, the other two inputs or only one input can be used to optimize the operation of the FC system in order to improve the fuel consumption based on the optimization function chosen. Thus, the RPF-control loop and two optimization loops controlling all three reference inputs of the FC system means three optimization strategies. In addition, beside one needed loop for the RPF-control, only one optimization loop controlling two from the three reference inputs of the FC system means the other four optimization strategies. Consequently, seven optimization strategies can be set for an FC Hybrid Power System to be analyzed.

The real-time searching and tracking of the optimum is mandatory for the optimization algorithm used in this study [52–54]. Furthermore, the optimization function is time-dependent and could become a multimodal type by controlling the FCHPS in different operating modes [55]. So, a Real-Time Optimization (RTO) algorithm must be selected to search the global optimum. The Global Extremum Seeking (GES) scheme proposed in [20] will be considered here, with minor changes of the parameters' values to improve the searching performance.

Summarizing, the novelty of this study is that seven RTO strategies will be analyzed for a FCHPS under the same load profile in order to estimate the fuel consumption compared to that obtained using the Static Feed-Forward (sFF) strategy, which is commercially implemented [56].

The goal of this study is to identify which is the best RTO strategy in full range of load demand or the best two RTO strategies that can be used for high and low values of the load demand, respectively.

The obtained results reveal that the fuel economy is better (in comparison with the sFF strategy) for only two RTO strategies, if the FCHPS is operated using the same strategy in the full range of load demand. If the FCHPS is operated using the best RTO strategy for high values of the load demand and another one which is best in the rest of the load range, then more combinations are possible (which we refer to as the switching strategies). This is because two RTO strategies are identified as best for high levels of load and two others as best for low load.

The rest of this study is organized as follows. The modeling of the FCHPS, the Energy Management Unit implementing the RPF-control mode and the optimization loops, and setting of the RTO strategies are presented in Section 2. The fuel economy obtained using a RTO strategy for the FCHPS is discussed in Section 3. Th final section concludes the performed study by highlighting the main findings and next work.

2. The Energy Management Unit of the Fuel Cell Hybrid Power System

The diagram of the FCHPS Energy Management Unit (EMU) is presented in Figure 1. The setting block will select one of the RTO strategies proposed in the literature [17,29,38,44,56–59] (see the right side of the Figure 1 and Table 1, where the references are mentioned in the last column) in order to evaluate the fuel economy under the same load profile.

Table 1. Real-Time Optimization (RTO) strategies.

Strategy	$I_{ref(Boost)}$	$I_{ref(Air)}$	$I_{ref(Fuel)}$	Ref.
sFF	$I_{ref(RPF)}$	I_{FC}	I_{FC}	[56]
RTO1	$I_{ref(RPF)}$	I_{FC}	$I_{ref(GES2)} + I_{FC}$	[57]
RTO2	$I_{ref(RPF)}$	$I_{ref(GES1)} + I_{FC}$	I_{FC}	[58]
RTO3	$I_{ref(GES1)}$	$I_{ref(RPF)}$	I_{FC}	[17]
RTO4	$I_{ref(GES2)}$	I_{FC}	$I_{ref(RPF)}$	[38]
RTO5	$I_{ref(GES1)}$	$I_{ref(RPF)}$	$I_{ref(GES2)} + I_{FC}$	[29]
RTO6	$I_{ref(GES2)}$	$I_{ref(GES1)} + I_{FC}$	$I_{ref(RPF)}$	[59]
RTO7	$I_{ref(RPF)}$	$I_{ref(GES1)} + I_{FC}$	$I_{ref(GES2)} + I_{FC}$	[44]

The performance under constant load could be measured by different indicators, such as the generated FC net power (P_{FCnet}) or the fuel consumption efficiency ($Fuel_{eff} \cong P_{FCnet} / FuelFr$), which are both important to be optimized. Thus, the optimization function used in this study will linearly mix these performance indicators as below:

$$f(x, AirFr, FuelFr, p_{load}) = P_{FCnet} + k_{fuel} \cdot Fuel_{eff} \tag{1}$$

where x is the state vector, p_{load} the disturbance, and the fuel flow rate ($FuelFr$) and the air flow rate ($AirFr$) are the control variables of the FC power. Also, note that another way to control the generated FC power on the DC bus is via the boost DC–DC converter, which is usually used as an interface for the FC system to the DC bus. So, three input control variables of the FC system will be considered in this study to control the FC net power generated on the DC bus.

The objective is to maximize the optimization function (1). The weighting coefficient k_{fuel} [liters per minute (lpm)/W] will be adjusted to explore the best fuel savings obtained for all RTO strategies using constant load and variable profile for the load.

The dynamics of the FCHPS is set by the smooth function g:

$$\dot{x} = g(x, AirFr, FuelFr, p_{Load}), \ x \in X \tag{2}$$

where the time constants of the 6 kW/45 V FC system and the 100 Ah/200 V batteries' stack were set to 0.2 s and 10 s, and the equivalent series resistor of the 100 F ultracapacitors' stack to 0.01 Ω. The DC voltage reference is set to 200 V, the ultracapacitors' stack has 100 V initial voltage and 10 kΩ parallel resistor (being connected via a bidirectional DC–DC to the DC bus and controlled to regulate the DC voltage), and the batteries' stack has an 80% initial State-of-Charge (SOC) (being connected directly to the 200 V DC bus).

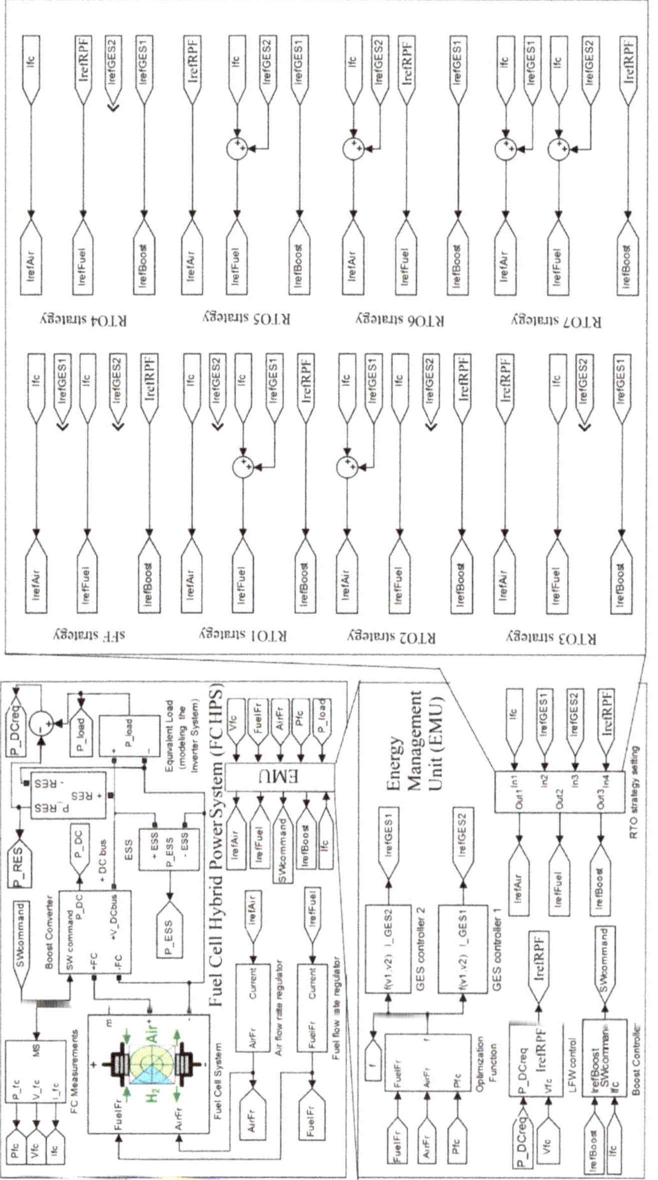

Figure 1. The Fuel Cell Hybrid Power Systems (FCHPS), Energy Management Unit (EMU) and RTO strategy setting block.

The losses in the FC system are mainly given by the air compressor, so the FC net power is:

$$P_{FCnet} \cong P_{FC} - P_{cm} \tag{3}$$

where P_{RES} is the FC power, the P_{cm} is the air compressor power [60]:

$$P_{cm} = I_{cm} \cdot V_{cm} = \left(a_2 \cdot AirFr^2 + a_1 \cdot AirFr + a_0\right) \cdot \left(b_1 \cdot I_{FC} + b_0\right)$$

and the coefficients are [60,61]: $a_0 = 0.6$ $a_1 = 0.04$ $a_2 = -0.00003231$ $b_0 = 0.9987$, and $b_1 = 46.02$. The dynamics of the air compressor is modeled through a second order system with 100 Hz natural frequency and 0.7 damping ratio [60].

Note that in a Renewable FCHPS the variable profile from the available Renewable Energy Sources (P_{RES}) can also be considered as disturbance as well. But both variable power profiles can be considered together by using a single disturbance, referred to as the power requested on the DC bus ($P_{DCreq} = P_{load} - P_{RES} > 0$). When $P_{DCreq} = P_{load} - P_{RES} < 0$, an excess of RES power is available on the DC bus, which, for example, can be used to supply an electrolyzer in order to generate hydrogen and then store it in tanks. So, during the excess of power on the DC bus ($P_{RES} > P_{load}$), the FC system will operate in standby mode, but during a lack of power on the DC bus ($P_{RES} < P_{load}$) then the FC system must compensate the power flow balance on the DC bus by generating on average (AV) the FC net power, $P_{FCnet(AV)}$, instead of taking this power ($P_{DCreq(AV)}$) from the batteries' stack, where:

$$P_{FCnet(AV)} = I_{FC(AV)}V_{FCnet(AV)} \cong \frac{P_{DCreq(AV)}}{\eta_{boost}} \tag{4}$$

Consequently, based on the power flow balance on the DC bus (5):

$$C_{DC}u_{DC}\frac{du_{DC}}{dt} = \eta_{boost}p_{FCnet} + p_{ESS} - p_{DCreq} \tag{5}$$

the average of the p_{ESS} power is zero ($P_{ESS(AV)} \cong 0$), where p_{ESS} is the output power of the hybrid battery/ultracapacitors ESS.

So, the size of the batteries stack can be further reduced or even be eliminated, with only the ultracapacitors' stack remaining to dynamically compensate the lack or excess of power on the DC bus, due to load pulses or sharp variation in the load demand or the RES power. Note that the capacitor C_{DC} connected on DC bus is used only to filter the ripple of the DC voltage (u_{dc}). The regulation of the DC voltage to 200 V is easily done by controlling the bidirectional DC–DC converter of the ultracapacitors' stack.

So, summarizing the above presented RTO strategy, $P_{ESS(AV)} \cong 0$ if the FC system is to be operated in the RPF-control mode using the reference current $I_{ref(RPF)}$ to control the FC power (as much as necessary to comply the power flow balance on the DC bus), then:

$$I_{ref(RPF)} \cong I_{FC(AV)} = \frac{P_{DCreq(AV)}}{\eta_{boost}V_{FCnet(AV)}} \tag{6}$$

Note that promising results have been achieved for the Renewable FCHPSs using the RPF- control mode on the FC system [55,61]. For a clear comparison of the RTO strategies but without losing the generality of this study, the proposed RPF-control mode of the FC system will be analyzed for a FCHPS without variable RES power ($P_{RES} = 0$). So, as explained before, the disturbance in the FCHPS will be represented by the load power ($P_{DCreq} = P_{load}$).

The RPF-control mode for the FC system can be obtained by controlling the boost DC–DC converter ($I_{ref(RPF)} = I_{ref(boost)}$), the *AirFr* regulator ($I_{ref(RPF)} = I_{ref(Air)}$), or the *FuelFr* regulator ($I_{ref(RPF)} = I_{ref(Fuel)}$).

For example, controlling the boost DC–DC converter, the 0.1 A hysteretic controller of the boost converter will ensure $I_{ref(RPF)} = I_{ref(boost)} \cong I_{FC}$, so the FC system will be operated in the RPF-control mode based on (6).

The other two control variables, *FuelFr* and *AirFr*, will be set based on (7) [56]:

$$FuelFr = \frac{60000 \cdot R \cdot (273 + \theta) \cdot N_C \cdot I_{ref(Fuel)}}{2F \cdot (101325 \cdot P_{f(H2)}) \cdot (U_{f(H2)}/100) \cdot (x_{H2}/100)} \tag{7a}$$

$$AirFr = \frac{60000 \cdot R \cdot (273 + \theta) \cdot N_C \cdot I_{ref(Air)}}{4F \cdot (101325 \cdot P_{f(O2)}) \cdot (U_{f(O2)}/100) \cdot (y_{O2}/100)} \tag{7b}$$

where the signals $I_{ref(Fuel)}$ and $I_{ref(Air)}$ are the references used to optimize the FCHPS.

The FC parameters ($N_C, \theta, U_{f(O2)}, P_{f(H2)}, P_{f(O2)}, x_{H2}, y_{O2}$) are set to default values of the 6 kW/45 V FC system, and $R = 8.3145$ J/(mol K) and $F = 96{,}485$ As/mol are two well-known constants [62].

For safety reasons in the FC system operation, the slope of the signals $I_{ref(Fuel)}$ and $I_{ref(Air)}$ have been limited to 100 A/s. This will limit the response time of the FC system to generate the requested DC power set by (4) in order to compensate the power flow balance on the DC. For example, if the FC system operates in a stationary regime generating 5 kW at about 50 V, and it must pass in 8 kW generating regime due to step-up in load demand, then the transitory regime will be of about 0.6 s (3000 W/50 V = 60 A and 60 A/100 A/s = 0.6 s). In this case, the response time of the FC system is limited by the 100 A/s slope, not by the 0.2 s FC constant time. So, the ESS must be appropriately designed to ensure such transitory regimes, which could arise in driving the FC vehicles. But if this design is performed considering all potential cases, then the FCHPS will operate properly under an unknown profile of load demand and then the optimization problem can be approached in real-time based on Global Extremum Seeking (GES) control.

For example, in the RTO7 strategy, both control variables, *FuelFr* and *AirFr*, are used to optimize the FCHPS, considering the GES references $I_{ref(GES1)}$ and $I_{ref(GES2)}$ generated by the two GES controllers. The searching signals $I_{ref(Fuel)}$ and $I_{ref(Air)}$ will be given by $I_{ref(Fuel)} = I_{ref(GES2)} + I_{FC}$ and $I_{ref(Air)} = I_{ref(GES1)} + I_{FC}$. The optimum on the optimization surface is close to the operating point of the FCHPS using the reference strategy, which is usually implemented in FC systems and is referred to in the literature as the Static Feed-Forward (sFF) strategy [56]. Thus, the searching area is limited around the FC current set by the RPF-control mode ($I_{FC} \cong I_{ref(RPF)}$) in order to speed up the searching process. So, in less than 10 dither's periods, the optimum will be found, which means up to 100 milliseconds for 100 Hz sinusoidal dither. The searching time is, therefore, in the order of the FC time constant and the FC system reacting to changes in both inputs control variables, *FuelFr* and *AirFr*.

Instead, only one control variable is used by the strategies RTO1 and RTO2 in search of the optimum, as follows: $I_{ref(Fuel)} = I_{ref(GES2)} + I_{FC}$ and $I_{ref(Air)} = I_{ref(GES1)} + I_{FC}$. The other control variable is set to $I_{FC} \cong I_{ref(RPF)}$ in order to speed up the searching process.

The RPF-control mode of the FC system is obtained by setting $I_{ref(Air)} = I_{ref(RPF)} \cong I_{FC}$ for the strategies RTO3 and RTO5. In the same manner for the strategies RTO4 and RTO6, the RPF-control mode of the FC system is obtained by setting $I_{ref(Fuel)} = I_{ref(RPF)} \cong I_{FC}$. Only the boost controller is used in search of the optimum by the RTO3 strategy and the RTO4 strategy, so $I_{ref(boost)} = I_{ref(GES1)} \cong I_{FC}$ and $I_{ref(boost)} = I_{ref(GES2)} \cong I_{FC}$ due to low value of the hysteresis band of 0.1 A. The other control variable in the RTO3 strategy and the RTO4 strategy, which is $I_{ref(Fuel)}$ and $I_{ref(Air)}$, respectively, is set to $I_{FC} \cong I_{ref(RPF)}$ in order to speed up the searching process. So, the searching for the optimum is performed on the optimization curve, which is a multimodal function to the FC current.

On the other hand, the strategies RTO5 and RTO6 use a second control variable in searching for the optimum (which is $I_{ref(Fuel)} = I_{ref(GES2)} + I_{FC}$ and $I_{ref(Air)} = I_{ref(GES1)} + I_{FC}$), besides the control variable $I_{ref(boost)}$. So, in this case, the searching for the optimum is performed on the multimodal optimization surface.

Thus, seven possible RTO strategies will be analyzed in this study, compared to sFF strategy as a reference. All RTO strategies use the GES control scheme proposed in [20] to find the optimum of the optimization function in case of k_{fuel} set to 0, 25, and 50 [lpm/W].

The interested reader can search the following references [17,29,38,44,56–59] for the hybrid system configuration, the control block diagrams, and the setting used for each strategy analyzed in this study.

The GES control has been implemented based on [20]. The interested reader, in the design of the GES control, can find design examples in [63,64]. In this study, the following values are used for the GES parameters: $k_1 = 1$, $k_2 = 2$, $k_{Np} = 20$, $k_{Ny} = 1/1000$, $f_{d1} = 100$ Hz and $f_{d2} = 200$ Hz, and $b_h = 0.1$ and $b_l = 1.5$ for the cut-off frequencies ($b_h f_d$ and $b_l f_d$) of the band-pass filter.

3. Results

Each RTO strategy has been analyzed using the FCHPS diagram shown in Figure 1 and control loops setting listed in Table 1.

The hydrogen consumption during a load cycle is evaluated for each RTOk strategy (k = 1 ÷ 7) using the performance indicator $Fuel_T = \int FuelFr(t)dt$ (measured in liter [L]) and the fuel economy compared to sFF strategy, $\Delta Fuel_{Tk} = Fuel_{Tk} - Fuel_{T0}$, is listed in Tables 2–4.

Table 2. Fuel economy for $k_{fuel} = 0$ and different P_{load}.

P_{load} [kW]	$Fuel_{T0}$ [L]	$\Delta Fuel_{T1}$ [L]	$\Delta Fuel_{T2}$ [L]	$\Delta Fuel_{T3}$ [L]	$\Delta Fuel_{T4}$ [L]	$\Delta Fuel_{T5}$ [L]	$\Delta Fuel_{T6}$ [L]	$\Delta Fuel_{T7}$ [L]
2	34.02	1.22	1.2	11.26	−0.46	8	−0.42	−0.22
3	56.3	0.13	0.79	4.14	−1.22	6.16	−1.7	−0.38
4	74.88	−0.13	0.77	2.08	−2.28	1.94	−3.1	−0.54
5	98.6	−0.38	0.55	−0.08	−5.6	−5.18	−5.24	−0.72
6	125.58	−1.38	0.42	−2.28	−7.66	−11.56	−8.48	−0.9
7	158.34	−4.34	−0.14	−12.16	−13.56	−24.48	−14.04	−1.08
8	176	−11.8	−4	−28.48	−22.92	−43.34	−27.36	−1.38

Table 3. Fuel economy for $k_{fuel} = 25$ and different P_{load}.

P_{load} [kW]	$Fuel_{T0}$ [L]	$\Delta Fuel_{T1}$ [L]	$\Delta Fuel_{T2}$ [L]	$\Delta Fuel_{T3}$ [L]	$\Delta Fuel_{T4}$ [L]	$\Delta Fuel_{T5}$ [L]	$\Delta Fuel_{T6}$ [L]	$\Delta Fuel_{T7}$ [L]
2	34.02	1.22	−0.09	12.14	−0.644	6.78	−0.56	−0.352
3	56.3	−0.25	−0.24	5.548	−3.876	1.76	−2	−0.6
4	74.88	−0.71	−0.25	1.2	−5.176	−3.72	−3.76	−0.88
5	98.6	−1.03	−0.46	−6.44	−8.76	−11.42	−6.52	−1.2
6	125.58	−2.08	−1.58	−14.14	−12.54	−17.82	−11.28	−1.6
7	158.34	−10.56	−4.24	−28.42	−24.26	−30.24	−20.76	−2.08
8	176	−22.92	−18.48	−31.08	−26	−47.72	−37.98	−2.92

Table 4. Fuel economy for $k_{fuel} = 50$ and different P_{load}.

P_{load} [kW]	$Fuel_{T0}$ [L]	$\Delta Fuel_{T1}$ [L]	$\Delta Fuel_{T2}$ [L]	$\Delta Fuel_{T3}$ [L]	$\Delta Fuel_{T4}$ [L]	$\Delta Fuel_{T5}$ [L]	$\Delta Fuel_{T6}$ [L]	$\Delta Fuel_{T7}$ [L]
2	34.02	1.28	1.22	7.628	−0.1	8.56	−0.42	−0.28
3	56.3	0.1	0.56	2.764	−3.7	4	−2	−0.48
4	74.88	−0.23	0.42	0.288	−5.264	1.1	−3.66	−0.64
5	98.6	−0.48	0.28	−5.8	−8.76	−6.34	−6.28	−0.84
6	125.58	−1.08	0.22	−13.02	−13.98	−13	−9.42	−1.04
7	158.34	−3.56	−1.14	−24.82	−20.74	−23.9	−14.48	−1.32
8	176	−6.8	−8.48	−29.8	−25	−45.52	−23.44	−1.8

3.1. Constant Load

The fuel savings for each RTOk strategy compared to sFF strategy ($\Delta Fuel_{Tk} = Fuel_{Tk} - Fuel_{T0}$, k = 1 ÷ 7) is listed in Tables 2–4 for $k_{fuel} = 0$, $k_{fuel} = 25$, and $k_{fuel} = 50$ and different values of the constant load (P_{load} mentioned in first column of the Tables 2–4). The second column of the Tables 2–4 mentions for the sFF strategy the hydrogen consumption ($Fuel_{T0}$) during a load cycle.

The fuel economy ($\Delta Fuel$) for all RTOk strategy (k = 1 ÷ 7) are presented in Figures 2–4 for $k_{fuel} = 0$, $k_{fuel} = 25$, and $k_{fuel} = 50$.

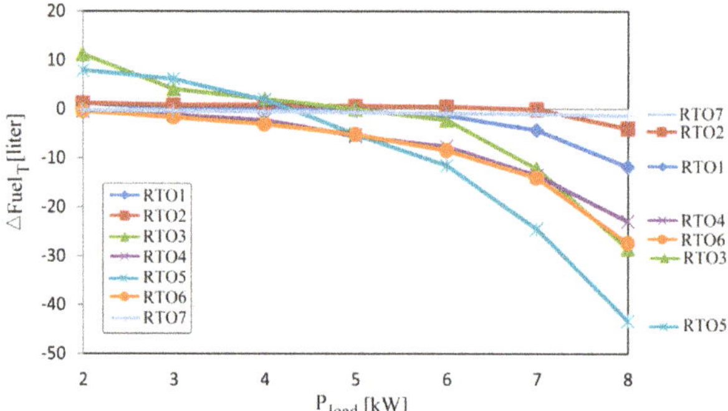

Figure 2. Fuel economy for $k_{fuel} = 0$ and different P_{load}.

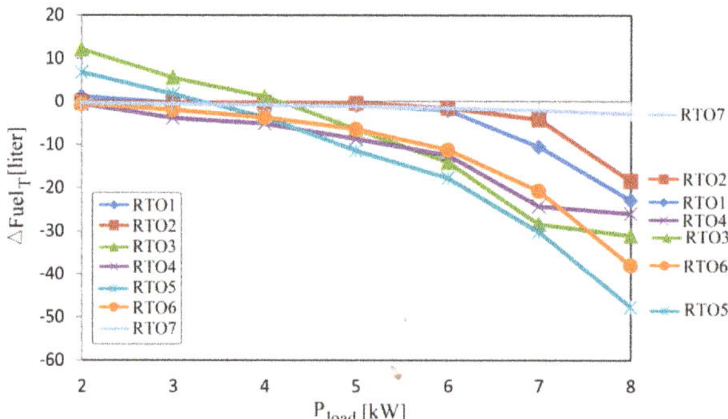

Figure 3. Fuel economy for $k_{fuel} = 25$ and different P_{load}.

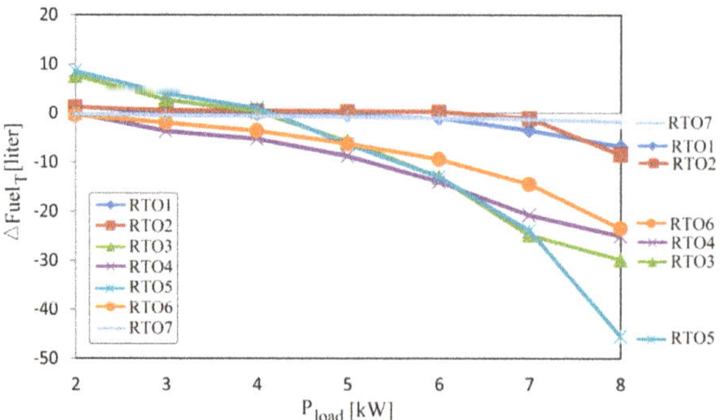

Figure 4. Fuel economy for $k_{fuel} = 50$ and different P_{load}.

In case of $k_{fuel} = 0$ (see Figure 2), the optimization function will be $f(x, AirFr, FuelFr, P_{Load}) = P_{FCnet}$, so the optimization objective is to increase the FC net power. Thus, the fuel savings will be higher in the case of $k_{fuel} \neq 0$ (see Tables 3 and 4 compared to Table 2) because the optimization function $f(x, AirFr, FuelFr, P_{Load}) = P_{FCnet} + k_{fuel} \cdot Fuel_{eff}$ takes in consideration the efficiency of the fuel consumption ($Fuel_{eff}$). It is worth mentioning that the fuel economy does not increase by increasing the value of the k_{fuel} parameter (see Table 4 compared to Table 3). In fact, a sensitivity analysis of the fuel economy to values of the k_{fuel} parameter in range 0 to 50 has been revealed the multimodal variation of the fuel economy for all RTO strategy. Based on this analysis performed in other work, the best economy can be obtained for a value of the k_{fuel} parameter in range 20 to 30, but this is not the goal of the work shown here. As mentioned before, the goal is to identify which is the best RTO strategy in the full range of load demand or the best two RTO strategies that can be used: one in the range of high load demand and the other in the rest of range (of low load demand), respectively.

The analysis of Figures 2–4 highlights the best strategies for high load demand ($P_{load} = 8$ kW), rated load demand ($P_{load} = 6$ kW), and low load demand ($P_{load} \leq 5$ kW).

In the case of $k_{fuel} = 25$ (see Figure 3) the order of the RTO strategies (starting with that which gives the best fuel economy) is as follows:

- RTO5, RTO6, RTO3, and RTO4 for $P_{load} = 8$ kW;
- RTO5, RTO3, RTO4, and RTO6 for $P_{load} = 6$ kW;
- RTO4 and RTO6 for $P_{load} \leq 5$ kW;

It is worth mentioning this order in the case of $k_{fuel} = 0$ and $k_{fuel} = 50$.

In the case of $k_{fuel} = 0$ (see Figure 2) the order of the RTO strategies (starting with that which gives the best fuel economy) is as follows:

- RTO5, RTO3, RTO6, and RTO4 for $P_{load} = 8$ kW;
- RTO5, RTO6, RTO4, and RTO3 for $P_{load} = 6$ kW;
- RTO6 and RTO4 for $P_{load} \leq 5$ kW;

In the case of $k_{fuel} = 50$ (see Figure 4) the order of the RTO strategies (starting with that which gives the best fuel economy) is as follows:

- RTO5, RTO3, RTO4, and RTO6 for $P_{load} = 8$ kW;
- RTO4, RTO5, RTO3, and RTO6 for $P_{load} = 6$ kW;
- RTO4 and RTO6 for $P_{load} \leq 5$ kW;

Note that the same RTO strategies are in the first four positions for $P_{load} = 8$ kW and $P_{load} = 6$ kW when $k_{fuel} = 0$, $k_{fuel} = 25$, and $k_{fuel} = 50$. The strategies RTO4 and RTO6 are between these four positions, but also are in the first two positions for $P_{load} \leq 5$ kW. So, the strategies RTO4 and RTO6 can be used to optimize the FCHPS operating in the entire loading range. It is difficult to say which is the recommended strategy for variable load demand. So, more tests will be performed in the next section.

Also, it is worth mentioning that the strategies RTO5 and RTO3 cannot be used for $P_{load} \leq 5$ kW due to low performance compared to other RTO strategies (such as the strategies RTO4 and RTO6, for example). So, it is clear that a switching (SW) strategy (which will use the strategies RTO4 and RTO6 for $P_{load} \leq 5$ kW and the strategies RTO5 and RTO3 for $P_{load} > 5$ kW) can improve the obtained fuel economy, as compared to the case of one RTO strategy being used in the full range of load demand. The following SW strategies are identified based on the above results:

- SW1 strategy, which will use the strategies RTO6 and RTO5 for $P_{load} \leq 5$ kW and $P_{load} > 5$ kW, respectively;
- SW2 strategy, which will use the strategies RTO4 and RTO5 for $P_{load} \leq 5$ kW and $P_{load} > 5$ kW, respectively;
- SW3 strategy, which will use the strategies RTO6 and RTO3 for $P_{load} \leq 5$ kW and $P_{load} > 5$ kW, respectively;

- SW4 strategy, which will use the strategies RTO4 and RTO3 for $P_{load} \leq 5\,kW$ and $P_{load} > 5\,kW$, respectively;

The strategies SW1, SW2, and SW3 request two GES controllers, due to the use of strategies RTO6 and RTO5, instead of one GES controller as for SW4 strategy.

3.2. Variable Load

The profile chosen for the variable load is of the scale up and down type with the following constant levels within 4 s: 0.75 $P_{load(AV)}$, 1.25 $P_{load(AV)}$, and $P_{load(AV)}$. Thus, the average value (AV) of the variable load profile is $P_{load(AV)} = (0.75\,P_{load(AV)} \cdot 4\,s + 1.25\,P_{load(AV)} \cdot 4\,s + P_{load(AV)} \cdot 4\,s)/12\,s$, where $P_{load(AV)}$ is mentioned in the first column of Tables 5–7. The second column of the Tables 5–7 mentions the hydrogen consumption during the variable load cycle for the sFF strategy ($Fuel_{T0}$).

Table 5. Fuel economy for $k_{fuel} = 0$ and different $P_{load(AV)}$.

$P_{load(AV)}$ [kW]	$Fuel_{T0}$ [L]	$\Delta Fuel_{T1}$ [L]	$\Delta Fuel_{T2}$ [L]	$\Delta Fuel_{T3}$ [L]	$\Delta Fuel_{T4}$ [L]	$\Delta Fuel_{T5}$ [L]	$\Delta Fuel_{T6}$ [L]	$\Delta Fuel_{T7}$ [L]
2	34.14	1.3	1.35	5.26	0.7	8.26	−0.18	0
3	53.92	0.71	0.6	4.28	−0.56	6.7	−1.32	−0.62
4	75.8	0.07	0.52	2.4	−0.82	2.84	−2.84	−1.36
5	100.62	−1.6	0.4	−4.38	−4.72	−2.84	−4.52	−2.26
6	130.2	−3.8	−0.2	−15.08	−11.46	−29.08	−8.22	−3.7

Table 6. Fuel economy for $k_{fuel} = 25$ and different $P_{load(AV)}$.

$P_{load(AV)}$ [kW]	$Fuel_{T0}$ [L]	$\Delta Fuel_{T1}$ [L]	$\Delta Fuel_{T2}$ [L]	$\Delta Fuel_{T3}$ [L]	$\Delta Fuel_{T4}$ [L]	$\Delta Fuel_{T5}$ [L]	$\Delta Fuel_{T6}$ [L]	$\Delta Fuel_{T7}$ [L]
2	34.14	0.5	−0.51	7.18	1.92	10.8	−0.1	−0.44
3	53.92	−0.48	−0.75	7.24	0.82	8.74	−1.04	−1.44
4	75.8	−1.8	−1	3.32	−0.64	−0.26	−3.84	−2.58
5	100.62	−3	−1.2	−3.16	−4.16	−12.96	−9.3	−3.98
6	130.2	−5.3	−1.8	−13.28	−10.08	−42.54	−18.56	−6.18

Table 7. Fuel economy for $k_{fuel} = 50$ and different $P_{load(AV)}$.

$P_{load(AV)}$ [kW]	$Fuel_{T0}$ [L]	$\Delta Fuel_{T1}$ [L]	$\Delta Fuel_{T2}$ [L]	$\Delta Fuel_{T3}$ [L]	$\Delta Fuel_{T4}$ [L]	$\Delta Fuel_{T5}$ [L]	$\Delta Fuel_{T6}$ [L]	$\Delta Fuel_{T7}$ [L]
2	34.14	0.51	−0.5	14.5	3.4	10.14	−0.38	−0.34
3	53.92	−0.47	−0.74	12.7	2.9	8.82	−1.18	−1.22
4	75.8	−1.58	−0.97	3.5	−0.5	4.92	−2.88	−2.24
5	100.62	−2.99	−1.25	−2.34	−3.98	−9.66	−6.2	−3.38
6	130.2	−5.23	−1.72	−12.08	−9.46	−37.26	−16.38	−5.3

The fuel economy $\Delta Fuel_{Tk} = Fuel_{Tk} - Fuel_{T0}$ for each RTOk strategy (k = 1 ÷ 7) compared to commercial strategy, such as the sFF strategy is listed in Tables 4–6 for $k_{fuel} = 0$, $k_{fuel} = 25$, and $k_{fuel} = 50$.

The fuel economy ($\Delta Fuel$) mentioned in Tables 4–6 for all RTOk strategy (k = 1 ÷ 7) are presented in Figures 5–7 for $k_{fuel} = 0$, $k_{fuel} = 25$, and $k_{fuel} = 50$, using $P_{load(AV)}$ as variable.

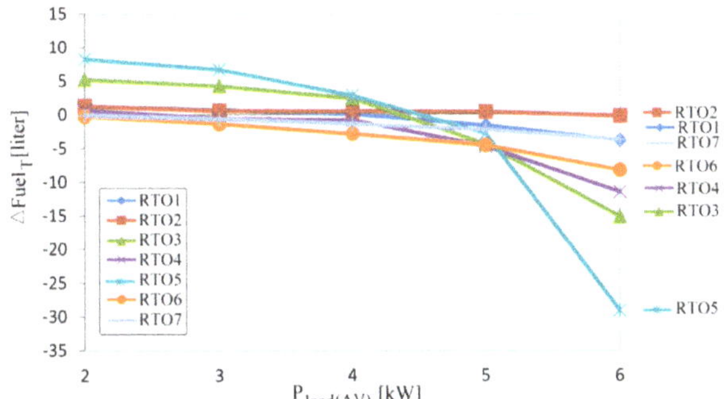

Figure 5. Fuel saving for $k_{fuel} = 0$ and different $P_{load(AV)}$.

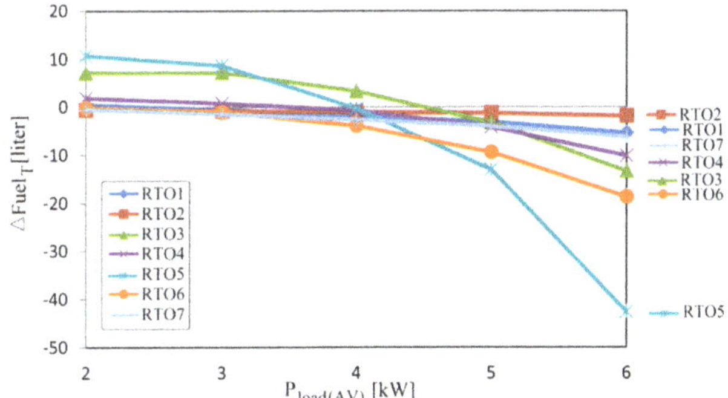

Figure 6. Fuel saving for $k_{fuel} = 25$ and different $P_{load(AV)}$.

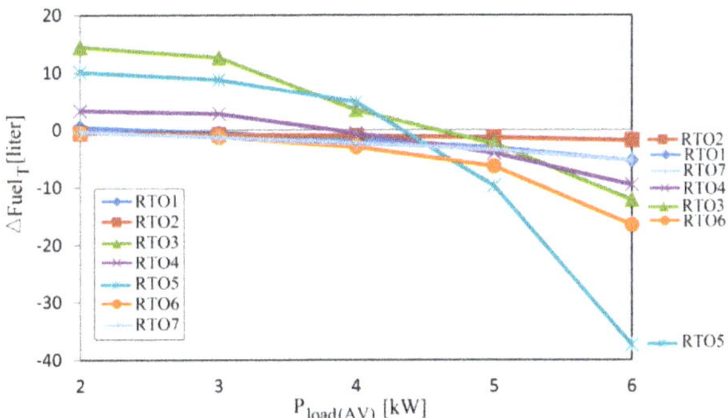

Figure 7. Fuel saving for $k_{fuel} = 50$ and different $P_{load(AV)}$.

Analysis of the Figures 5–7 highlights the same best four strategies for the load profile with $P_{load(AV)} > 5$ kW (as in case of $P_{load} > 5$ kW). This validates the results obtained for constant load.

Besides the strategies RTO5 and RTO3, it also appears that the RTO4 strategy cannot be used for $P_{load(AV)} \leq 5$ kW due to low performance compared to strategy RTO6 for variable load. So, the SW strategies that remain to be tested, compared to strategy RTO6 (which still ensure fuel saving for different $P_{load(AV)}$) are as follows:

- SW1 strategy, which will use the strategies RTO6 and RTO5 for $P_{load} \leq 5$ kW and $P_{load} > 5$ kW, respectively;

- SW3 strategy, which will use the strategies RTO6 and RTO3 for $P_{load} \leq 5\,\text{kW}$ and $P_{load} > 5\,\text{kW}$, respectively;

These results validate the aforementioned conclusions for constant load.

4. Conclusions

Seven control topologies are identified based on the three ways (inputs) to regulate the FC power: via the FC boost converter, and via the fueling regulators. One of these inputs will control the generated FC net power, ensuring the DC power flow balance under variable load. The other two or only one are used for optimization the FC system in order to minimize the fuel consumption. Thus, seven strategies are analyzed for an FC system under constant and variable load demand.

The conclusions of the analysis performed for all strategies are as follows:

- The strategies RTO3 and RTO5 have been identified as best strategies to operate the FC hybrid power system in the high range of the load demand; both strategies control the air flow regulator to obtain the RPF-control mode for the FC system;
- The strategies RTO4 and RTO6 have been identified as best strategies to operate the FC hybrid power system in the low range of the load demand; both strategies control the fuel flow regulator to obtain the RPF-control mode for the FC system;
- The RTO6 strategy has been identified as best strategy to operate the FC hybrid power system under variable load in the entire loading range;
- The fuel economy will be improved by using the switching strategies under variable load in the entire loading range.

The next work will be focused on comparative analysis of fuel economy for the four switching strategies identified, using the RTO6 strategy as reference. Besides the RTO6 strategy (which uses two GES controllers), the RTO4 strategy (which uses only one GES controller) can become a possible strategy in the low range of load demand, due to simple implementation and minor differences in fuel economy compared to RTO6 strategy under constant load. The cases of strategies SW2 and SW4 remain to be further analyzed in order to decide which strategy is best from different points of view: performance, cost, complexity of implementation. For example, the cost and complexity of implementation is lower for the SW4 strategy compared to strategies SW1, SW2, and SW3 due to the use of one GES controller instead of two GES controllers. The threshold of the load demand (which will be used in switching strategies to split the loading range) will be set after a sensitivity analysis for all switching strategies using a threshold in the range of 4 kW to 6 kW.

Author Contributions: Conceptualization, Methodology, and Validation: N.B., V.A.S., A.C.C.; Writing—Review & Editing: V.A.S.; A.C.C., N.B.

Acknowledgments: This work was supported by a grant of the CNCS/CCCDI-UEFISCDI within "Brancusi" European and international cooperation, PN-III P3-3.1, project "Global Maximum Power Point Tracking algorithms for Photovoltaic Systems under rapid changes in climatic conditions".

Conflicts of Interest: The authors declare no conflict of interest

References

1. Gielen, D.; Boshell, F.; Saygin, D.; Bazilian, M.D.; Wagner, N.; Gorini, R. The role of renewable energy in the global energy transformation. *Energy Strateg. Rev.* **2019**, *24*, 38–50. [CrossRef]
2. Liu, Z.; Kendall, K.; Yan, X. China Progress on Renewable Energy Vehicles: Fuel Cells, Hydrogen and Battery Hybrid Vehicles. *Energies* **2019**, *12*, 54. [CrossRef]
3. Ou, K.; Yuan, W.-W.; Choi, M.; Yang, S.; Jung, S.; Kim, Y.-B. Optimized power management based on adaptive-PMP algorithm for a stationary PEM fuel cell/battery hybrid system. *Int. J. Hydrogen Energy* **2018**, *43*, 15433–15444. [CrossRef]
4. Rafique, M.K.; Haider, Z.M.; Mehmood, K.K.; Zaman, M.S.U.; Irfan, M.; Khan, S.U.; Kim, C.-H. Optimal Scheduling of Hybrid Energy Resources for a Smart Home. *Energies* **2018**, *11*, 3201. [CrossRef]

5. Liu, T.; Zhang, D.; Wang, S.; Wu, T. Standardized modelling and economic optimization of multi-carrier energy systems considering energy storage and demand response. *Energy Convers. Manag.* **2019**, *182*, 126–142. [CrossRef]

6. Chen, H.; Yang, C.; Deng, K.; Zhou, N.; Wu, H. Multi-objective optimization of the hybrid wind/solar/fuel cell distributed generation system using Hammersley Sequence Sampling. *Int. J. Hydrogen Energy* **2017**, *42*, 7836–7846. [CrossRef]

7. Bizon, N. Hybrid power sources (HPSs) for space applications: Analysis of PEMFC/Battery/SMES HPS under unknown load containing pulses. *Renew. Sustain. Energy Rev.* **2019**, *105*, 14–37. [CrossRef]

8. Wang, F.-C.; Hsiao, Y.-S.; Yang, Y.-Z. The Optimization of Hybrid Power Systems with Renewable Energy and Hydrogen Generation. *Energies* **2018**, *11*, 1948. [CrossRef]

9. Huang, Y.; Wang, H.; Khajepour, A.; Li, B.; Ji, J.; Zhao, K.; Hu, C. A review of power management strategies and component sizing methods for hybrid vehicles. *Renew. Sustain. Energy Rev.* **2018**, *96*, 132–144. [CrossRef]

10. Wu, W.; Partridge, J.S.; Bucknall, R.W.G. Simulation of a stabilised control strategy for PEM fuel cell and supercapacitor hybrid propulsion system for a city bus. *Int. J. Hydrogen Energy* **2018**, *43*, 19763–19777. [CrossRef]

11. Wang, F.-C.; Lin, K.-M. Impacts of Load Profiles on the Optimization of Power Management of a Green Building Employing Fuel Cells. *Energies* **2019**, *12*, 57. [CrossRef]

12. Guo, C.; Lu, J.; Tian, Z.; Guo, W.; Darvishan, A. Optimization of critical parameters of PEM fuel cell using TLBO-DE based on Elman neural network. *Energy Convers. Manag.* **2019**, *183*, 149–158. [CrossRef]

13. Araya, S.S.; Zhou, F.; Sahlin, S.L.; Thomas, S.; Jeppesen, C.; Kær, S.K. Fault Characterization of a Proton Exchange Membrane Fuel Cell Stack. *Energies* **2019**, *12*, 152. [CrossRef]

14. Ahmadi, P.; Torabi, S.H.; Afsaneh, H.; Sadegheih, Y.; Ganjehsarabi, H.; Ashjaee, M. The effects of driving patterns and PEM fuel cell degradation on the lifecycle assessment of hydrogen fuel cell vehicles. *Int. J. Hydrogen Energy* **2019**. [CrossRef]

15. Fragiacomo, P.; De Lorenzo, G.; Corigliano, O. Performance Analysis of an Intermediate Temperature Solid Oxide Electrolyzer Test Bench under a CO_2-H_2O Feed Stream. *Energies* **2018**, *11*, 2276. [CrossRef]

16. Bukar, A.L.; Tan, C.W. A Review on Stand-alone Photovoltaic-Wind Energy System with Fuel Cell: System Optimization and Energy Management Strategy. *J. Clean. Prod.* **2019**. [CrossRef]

17. Bizon, N. Energy optimization of Fuel Cell System by using Global Extremum Seeking algorithm. *Appl. Energy* **2017**, *206*, 458–474. [CrossRef]

18. Sulaiman, N.; Hannan, M.A.; Mohamed, A.; Ker, P.J.; Majlan, E.H.; Wan Daud, W.R. Optimization of energy management system for fuel-cell hybrid electric vehicles: Issues and recommendations. *Appl. Energy* **2018**, *228*, 2061–2079. [CrossRef]

19. Gherairi, S. Hybrid Electric Vehicle: Design and Control of a Hybrid System (Fuel Cell/Battery/Ultra-Capacitor) Supplied by Hydrogen. *Energies* **2019**, *12*, 1272. [CrossRef]

20. Bizon, N. Global Extremum Seeking Control of the Power Generated by a Photovoltaic Array under Partially Shaded Conditions. *Energy Convers. Manag.* **2016**, *109*, 71–85. [CrossRef]

21. Bizon, N.; Thounthong, P. Fuel Economy using the Global Optimization of the Fuel Cell Hybrid Power Systems. *Energy Convers. Manag.* **2018**, *173*, 665–678. [CrossRef]

22. You, Z.; Wang, L.; Han, Y.; Zare, F. System Design and Energy Management for a Fuel Cell/Battery Hybrid Forklift. *Energies* **2018**, *11*, 3440. [CrossRef]

23. De Lorenzo, G.; Andaloro, L.; Sergi, F.; Napoli, G.; Ferraro, M.; Antonucci, V. Numerical simulation model for the preliminary design of hybrid electric city bus power train with polymer electrolyte fuel cell. *Int. J. Hydrogen Energy* **2014**, *39*, 12934–12947. [CrossRef]

24. Ahmadi, S.; Bathaee, S.M.T.; Hosseinpour, A.H. Improving fuel economy and performance of a fuel-cell hybrid electric vehicle (fuel-cell, battery, and ultra-capacitor) using optimized energy management strategy. *Energy Convers. Manag.* **2018**, *160*, 74–84. [CrossRef]

25. Bizon, N. Real-time optimization strategies of FC Hybrid Power Systems based on Load-following control: A new strategy, and a comparative study of topologies and fuel economy obtained. *Appl. Energy* **2019**, *241C*, 444–460. [CrossRef]

26. Arsalis, A.; Georghiou, G.E. A Decentralized, Hybrid Photovoltaic-Solid Oxide Fuel Cell System for Application to a Commercial Building. *Energies* **2018**, *11*, 3512. [CrossRef]

27. Tiar, M.; Betka, A.; Drid, S.; Abdeddaim, S.; Tabandjat, A. Optimal energy control of a PV-fuel cell hybrid system. *Int. J. Hydrogen Energy* **2017**, *42*, 1456–1465. [CrossRef]

28. Wu, W.; Partridge, J.; Bucknall, R. Development and Evaluation of a Degree of Hybridisation Identification Strategy for a Fuel Cell Supercapacitor Hybrid Bus. *Energies* **2019**, *12*, 142. [CrossRef]

29. Bizon, N.; Mazare, A.G.; Ionescu, L.M.; Enescu, F.M. Optimization of the Proton Exchange Membrane Fuel Cell Hybrid Power System for Residential Buildings. *Energy Convers. Manag.* **2018**, *163*, 22–37. [CrossRef]

30. Vivas, F.J.; De las Heras, A.; Segura, F.; Andújar, J.M. A review of energy management strategies for renewable hybrid energy systems with hydrogen backup. *Renew. Sustain. Energy Rev.* **2017**, *82*, 126–155. [CrossRef]

31. Oberoi, A.S.; Nijhawan, P.; Singh, P. A Novel Electrochemical Hydrogen Storage-Based Proton Battery for Renewable Energy Storage. *Energies* **2019**, *12*, 82. [CrossRef]

32. Bizon, N. Effective Mitigation of the Load Pulses by Controlling the Battery/SMES Hybrid Energy Storage System. *Appl. Energy* **2018**, *229*, 459–473. [CrossRef]

33. Olatomiwa, L.; Mekhilef, S.; Ismail, M.S.; Moghavvemi, M. Energy management strategies in hybrid renewable energy systems: A review. *Renew. Sustain. Energy Rev.* **2016**, *62*, 821–835. [CrossRef]

34. Spanoudakis, P.; Tsourveloudis, N.C.; Doitsidis, L.; Karapidakis, E.S. Experimental Research of Transmissions on Electric Vehicles' Energy Consumption. *Energies* **2019**, *12*, 388. [CrossRef]

35. Won, J.S.; Langari, R. Intelligent energy management agent for a parallel hybrid vehicle-part II: Torque distribution, charge sustenance strategies, and performance results. *IEEE Trans. Veh. Technol.* **2005**, *54*, 935–953. [CrossRef]

36. Das, V.; Padmanaban, S.; Venkitusamy, K.; Selvamuthukumaran, R.; Siano, P. Recent advances and challenges of fuel cell based power system architectures and control—A review. *Renew. Sustain. Energy Rev.* **2017**, *73*, 10–18. [CrossRef]

37. Mumtaz, S.; Ali, S.; Ahmad, S.; Khan, L.; Hassan, S.Z.; Kamal, T. Energy Management and Control of Plug-In Hybrid Electric Vehicle Charging Stations in a Grid-Connected Hybrid Power System. *Energies* **2017**, *10*, 1923. [CrossRef]

38. Bizon, N. Real-time optimization strategy for fuel cell hybrid power sources with load-following control of the fuel or air flow. *Energy Convers. Manag.* **2018**, *157*, 13–27. [CrossRef]

39. Ettihir, K.; Boulon, L.; Agbossou, K. Optimization-based energy management strategy for a fuel cell/battery hybrid power system. *Appl. Energy* **2016**, *163*, 142–153. [CrossRef]

40. Ijaodola, O.; Ogungbemi, E.; Khatib, F.N.; Wilberforce, T.; Ramadan, M.; Hassan, Z.E.; Thompson, J.; Olabi, A.G. Evaluating the Effect of Metal Bipolar Plate Coating on the Performance of Proton Exchange Membrane Fuel Cells. *Energies* **2018**, *11*, 3203. [CrossRef]

41. Bizon, N. Energy Efficiency of Multiport Power Converters used in Plug-In/V2G Fuel Cell Vehicles. *Appl. Energy* **2012**, *96*, 431–443. [CrossRef]

42. Liso, V.; Zhao, Y.; Yang, W.; Nielsen, M.P. Modelling of a Solid Oxide Fuel Cell CHP System Coupled with a Hot Water Storage Tank for a Single Household. *Energies* **2015**, *8*, 2211–2229. [CrossRef]

43. Sorrentino, M.; Cirillo, V.; Nappi, L. Development of flexible procedures for co-optimizing design and control of fuel cell hybrid vehicles. *Energy Convers. Manag.* **2019**, *185*, 537–551. [CrossRef]

44. Bizon, N.; Thounthong, P. Real-time strategies to optimize the fueling of the fuel cell hybrid power source: A review of issues, challenges and a new approach. *Renew. Sustain. Energy Rev.* **2018**, *91*, 1089–1102. [CrossRef]

45. Secanell, M.; Wishart, J.; Dobson, P. Computational design and optimization of fuel cells and fuel cell systems: A review. *J. Power Sources* **2011**, *196*, 3690–3704. [CrossRef]

46. Raga, C.; Barrado, A.; Miniguano, H.; Lazaro, A.; Quesada, I.; Martin-Lozano, A. Analysis and Sizing of Power Distribution Architectures Applied to Fuel Cell Based Vehicles. *Energies* **2018**, *11*, 2597. [CrossRef]

47. Hu, Z.; Li, J.; Xu, L.; Song, Z.; Fang, C.; Ouyang, M.; Dou, G.; Kou, G. Multi-objective energy management optimization and parameter sizing for proton exchange membrane hybrid fuel cell vehicles. *Energy Convers. Manag.* **2016**, *129*, 108–121. [CrossRef]

48. Chen, Y.-S.; Lin, S.-M.; Hong, B.-S. Experimental Study on a Passive Fuel Cell/Battery Hybrid Power System. *Energies* **2013**, *6*, 6413–6422. [CrossRef]

49. Kim, M.-J.; Peng, H. Power Management and Design Optimization of Fuel Cell/Battery Hybrid Vehicles. *J. Power Sources* **2007**, *165*, 819–832. [CrossRef]

Energies **2019**, *12*, 1889

50. Koubaa, R.; Krichen, L. Double layer metaheuristic based energy management strategy for a Fuel Cell/Ultra-Capacitor hybrid electric vehicle. *Energy* **2017**, *133*, 1079–1093. [CrossRef]

51. De Lorenzo, G.; Fragiacomo, P. Electrical and thermal analysis of an intermediate temperature IIR-SOFC system fed by biogas. *Energy Sci. Eng.* **2018**, *6*, 60–72. [CrossRef]

52. Fathabadi, H. Novel fast and high accuracy maximum power point tracking method for hybrid photovoltaic/fuel cell energy conversion systems. *Renew. Energy* **2017**, *106*, 232–242. [CrossRef]

53. Bizon, N.; Oproescu, M. Experimental Comparison of Three Real-Time Optimization Strategies Applied to Renewable/FC-Based Hybrid Power Systems Based on Load-Following Control. *Energies* **2018**, *11*, 3537. [CrossRef]

54. Eriksson, E.L.V.; Mac, E.; Gray, A. Optimization and integration of hybrid renewable energy hydrogen fuel cell energy systems—A critical review. *Appl. Energy* **2017**, *202*, 348–364. [CrossRef]

55. Bizon, N.; Lopez-Guede, J.M.; Kurt, E.; Thounthong, P.; Mazare, A.G.; Ionescu, L.M.; Iana, G.V. Hydrogen Economy of the Fuel Cell Hybrid Power System optimized by air flow control to mitigate the effect of the uncertainty about available renewable power and load dynamics. *Energy Convers. Manag.* **2019**, *179*, 152–165. [CrossRef]

56. Pukrushpan, J.T.; Stefanopoulou, A.G.; Peng, H. *Control of Fuel Cell Power Systems*; Springer: New York, NY, USA, 2004.

57. Bizon, N.; Culcer, M.; Iliescu, M.; Mazare, A.G.; Ionescu, L.M.; Beloiu, R. Real-time strategy to optimize the Fuel Flow rate of Fuel Cell Hybrid Power Source under variable load cycle. In Proceedings of the ECAI–9th Edition of International Conference on Electronics, Computers and Artificial Intelligence, Targoviste, Romania, 29 June–1 July 2017. [CrossRef]

58. Bizon, N.; Culcer, M.; Oproescu, M.; Iana, G.; Ionescu, L.M.; Mazare, A.G.; Iliescu, M. Real-time strategy to optimize the Airflow rate of Fuel Cell Hybrid Power Source under variable load cycle. In Proceedings of the 22rd International Conference on Applied Electronics—APPEL 2017, Pilsen, Czech Republic, 5–6 September 2017. [CrossRef]

59. Bizon, N.; Iana, G.; Kurt, E.; Thounthong, P.; Oproescu, M.; Culcer, M.; Iliescu, M. Air Flow Real-Time Optimization Strategy for Fuel Cell Hybrid Power Sources with Fuel Flow Based on Load-Following. *Fuel Cell* **2018**, *18*, 809–823. [CrossRef]

60. Ramos-Paja, C.A.; Spagnuolo, G.; Petrone, G.; Emilio Mamarelis, M. A perturbation strategy for fuel consumption minimization in polymer electrolyte membrane fuel cells: Analysis, Design and FPGA implementation. *Appl. Energy* **2014**, *119*, 21–32. [CrossRef]

61. Bizon, N.; Hoarcă, C.I. Hydrogen saving through optimized control of both fueling flows of the Fuel Cell Hybrid Power System under a variable load demand and an unknown renewable power profile. *Energy Convers. Manag.* **2019**, *184*, 1–14. [CrossRef]

62. *SimPowerSystems TM Reference*; Hydro-Québec and the MathWorks, Inc.: Natick, MA, USA, 2010.

63. Bizon, N.; Thounthong, P.; Raducu, M.; Constantinescu, L.M. Designing and modelling of the asymptotic perturbed extremum seeking control scheme for tracking the global extreme. *Int. J. Hydrogen Energy* **2017**, *42*, 17632–17644. [CrossRef]

64. Bizon, N.; Kurt, E. Performance Analysis of Tracking of the Global Extreme on Multimodal Patterns using the Asymptotic Perturbed Extremum Seeking Control Scheme. *Int. J. Hydrogen Energy* **2017**, *42*, 17645–17654. [CrossRef]

Article

Optimal Capacitor Bank Allocation in Electricity Distribution Networks Using Metaheuristic Algorithms

Ovidiu Ivanov *, Bogdan-Constantin Neagu, Gheorghe Grigoras * and Mihai Gavrilas

Department of Power Engineering, Gheorghe Asachi Technical University of Iasi, 700050 Iasi, Romania; bogdan.neagu@tuiasi.ro (B.-C.N.); mgavrilas@yahoo.com (M.G.)
* Correspondence: ovidiuivanov@tuiasi.ro (O.I.); ggrigor@tuiasi.ro (G.G.)

Received: 16 September 2019; Accepted: 1 November 2019; Published: 6 November 2019

Abstract: Energy losses and bus voltage levels are key parameters in the operation of electricity distribution networks (EDN), in traditional operating conditions or in modern microgrids with renewable and distributed generation sources. Smart grids are set to bring hardware and software tools to improve the operation of electrical networks, using state-of the art demand management at home or system level and advanced network reconfiguration tools. However, for economic reasons, many network operators will still have to resort to low-cost management solutions, such as bus reactive power compensation using optimally placed capacitor banks. This paper approaches the problem of power and energy loss minimization by optimal allocation of capacitor banks (CB) in medium voltage (MV) EDN buses. A comparison is made between five metaheuristic algorithms used for this purpose: the well-established Genetic Algorithm (GA); Particle Swarm Optimization (PSO); and three newer metaheuristics, the Bat Optimization Algorithm (BOA), the Whale Optimization Algorithm (WOA) and the Sperm-Whale Algorithm (SWA). The algorithms are tested on the IEEE 33-bus system and on a real 215-bus EDN from Romania. The newest SWA algorithm gives the best results, for both test systems.

Keywords: electricity distribution networks; optimal capacitor allocation; Genetic Algorithm; Particle Swarm Optimization; Bat Algorithm; Whale Algorithm; Sperm-Whale Algorithm

---

1. Introduction

Distribution Network Operators take into account the implementation of smart solutions to improve both the voltage level in the subordinate networks and the power factor, with the aim to maintain the balance between power generation and consumption while meeting the quality of supply standards and regulations.

In this context, the use of capacitor banks is an easy solution to be implemented with technical and economic benefits to the smart grid, maximizing the long-term return on investment as the network develops. An intelligent control of capacitor banks leads to improved energy efficiency and voltage level in the buses of distribution networks, resulting in an increase in the percentage of energy delivered to consumers [1].

The advantages of integrating capacitor banks in the flexible smart grid communication and control infrastructure are the increase of network energy efficiency and power quality improvement [2]. Thus, the technologies and modern techniques enable today the large-scale integration of capacitor banks managed with smart control algorithms.

In the literature, many methods have been proposed to solve the Optimal Capacitor Banks Allocation (OCBA) in distribution networks as a combinatorial optimization problem. These techniques can be grouped in four main categories: numerical [3]; analytical [4]; heuristic [5–7]; and artificial intelligence, population based (Artificial Neural Networks, metaheuristics) [8,9]. An overview about the metaheuristics used for the problem of capacitor banks allocation is made in the following, highlighting their specific purpose. The OCBA solution for power losses or cost minimization is obtained using a genetic algorithm in [10,11], a fuzzy technique in [12] and an artificial neural network in [9]. Regarding the metaheuristics, a significant number of papers consider the joule loss minimization, voltage bus improvement, and total cost minimization. Thus, in [13–15] a Multi-Objective Particle Swarm Optimization (MOPSO) algorithm is proposed. For active power loss reduction using load flow computation, the branch and bound method is generally preferred, for its reduced computation time. For example, for the minimization of the total annual costs, the Crow Search Algorithm (CSA) is used in [16,17], the Particle Swarm Optimization (PSO) and hybrid PSO algorithm are adapted in [18–21], the Flower Pollination Algorithm (FPA) is preferred in [22,23], and an Improved Harmony Algorithm is chosen in [24]. On the other hand, the OCBA problem based on active power minimization was approached in [25,26] using the Bacterial Foraging Optimization Algorithm, the Intersect Mutation Differential Evolution (IMDE) Algorithm in [27], the Artificial Bee Colony (ABC) in [5,28] and the Ant Lion Optimization Algorithm in [29]. The improvement of the voltage profile carried out using the Symbiotic Organisms Search Algorithm (SOSA) in [30]. Another paper proposes the JAYA optimization algorithm [31] for power factor correction. For voltage profile improvement, the Oppositional Cuckoo Optimization Algorithm (OCOA) was used in [32]. It must be mentioned that the authors' previous approaches regarding the OCBA problem used several metaheuristic algorithms, such as PSO, BOA, Fireworks Algorithm (FWA), and WOA [33].

A brief description of the papers that use metaheuristics in the CBA problem considering both objective functions (OF) and constraints (C) is presented in Table 1. The considered objective functions are: OF1, active power losses minimization; OF2, voltage profile improvement; OF3, voltage deviation minimization; OF4, cost minimization; OF5—net savings maximization; OF6, voltage stability improvement. The main constraints for the OCBA problem are a combination of the following: C1, bus voltage allowable limits; C2, current flow limits on the branches; C3, bus reactive allowable limits; C4, maximum stock of capacitors; C5, bus apparent power balance; C6, maximum number of transformer tap changer steps; C7, the total reactive power injected should not exceed the total reactive power demand; C8, power flow limits on the branches; and C9, bus power factor limits.

This paper is focused on a comparative study of several metaheuristic algorithms adapted for solving the OCBA problem with the objective of energy loss minimization in MV distribution networks. During the analysis, the well-known GA and PSO are tested against two newer metaheuristics that have seen previous uses in power engineering applications, the BOA and WOA, and another recent but much less used method, the SWA. The latter is shown to outperform all its predecessors, when tested on two MV distribution networks with different characteristics: the smaller IEEE 33—bus test network [5,13,25] and a larger 215—bus 20/0.4 kV distribution network from Romania. During the case study, the algorithms use the same initial population and fitness function. Results are shown regarding active power and energy losses and bus voltage levels, for which the best results are obtained with the SWA.

Table 1. Literature review regarding the capacitor allocation problem based on artificial intelligence.

Objective Function	Constraints									Test Network	References
	C1	C2	C3	C4	C5	C6	C7	C8	C9		
OF1	X	X	-	-	-	-	-	-	-	38 bus—Roy–Billinton Test System	[8]
	X	-	X	-	X	X	-	-	-	IEEE 30, 57, 118 and 300 bus	[10,27,29]
	X	-	X	-	X	-	-	-	-	IEEE 33 bus	[13]
	X	X	X	X	-	-	-	-	-	IEEE 33 and 94 bus	[14,15]
	X	X	X	-	X	X	-	-	-	IEEE 30 bus	[21]
	X	X	X	-	X	-	-	-	-	IEEE 33 and 85 bus	[25,28]
	X	X	X	X	X	-	-	-	-	IEEE 33 and 119 bus	[5]
OF2	X	-	-	-	-	-	-	-	-	IEEE 10, 23 and 34 bus	[12]
	X	-	X	-	X	-	-	-	-	IEEE 22, 69, 85 and 141 bus	[32]
OF3	X	-	X	-	X	X	-	-	-	IEEE 30, 57, 118 and 300 bus	[10,27]
	X	X	X	-	X	X	-	-	-	IEEE 30 bus	[21]
OF4	X	-	X	-	-	-	X	-	X	IEEE 10, 33 and 69 bus	[16,17,22,26]
	X	-	X	-	-	-	-	X	-	IEEE 10, 15 and 34 bus	[19]
	X	-	X	-	-	-	-	-	-	IEEE 30 and 85 bus	[20]
	X	-	X	-	-	-	-	-	-	IEEE 33, 34, 69 and 85 bus	[23,27]
	X	-	X	-	X	-	X	-	X	IEEE 85 and 118 bus	[24]
OF5	X	-	-	-	X	-	-	-	-	IEEE 28-bus	[11]
	X	-	-	-	-	-	-	-	-	IEEE 9-bus	[30]
OF6	X	-	X	-	-	X	X	-	-	IEEE 30 bus	[18]
	X	-	X	-	X	X	X	-	-	IEEE 30, 57 and 118 bus	[27]
	X	-	X	-	X	-	-	-	-	IEEE 30, 118 and 300 bus	[29]

2. Metaheuristic Algorithms

Metaheuristics are a special class of algorithms that can be used to solve search and optimization problems. As described in [34], they are approximate, usually non-deterministic methods that aim to search for solutions near the global optimum, exploring this space through a partly guided and partly random search. While the main disadvantage of metaheuristics is the uncertainty of reaching the global optimal solution, their advantages lie in not being problem-specific (allowing the flexibility of applying the same solving principle to several types of problems) and having intuitive mathematical models, borrowing concepts and approaches from the natural world, rather than from theoretical mathematical models. This contributes to their accessibility for a wider range of users. Most modern metaheuristics are population-based, starting from an initial group of solutions, called 'population', generated randomly, and refining it in an iterative process, according to a set of specific steps, until a stopping criterion is met. The performance of each individual from the population is assessed by computing its fitness function. The basic block diagram of a population-based metaheuristic algorithm (PMA) is depicted in Figure 1, where the steps common to all algorithms are represented with white boxes, and the part specific to each algorithm, delimited by symbols (A) and (B) is presented in gray.

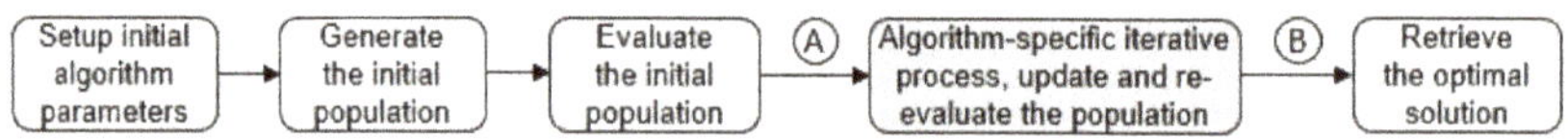

Figure 1. The basic flowchart of a population-based metaheuristic algorithm.

The initial parameters are partially common to all algorithms, such as population size N or maximum number of iterations *maxit*, and partially specific to each algorithm, such as the mutation rate *rmut* for the Genetic Algorithm (GA) or inertia value w for the Particle Swarm Optimization (PSO). An individual from a population with N members, denoted in the following as

$$X_i = [x_1, x_2, \ldots, x_m], \quad i = 1 \ldots N \tag{1}$$

is encoded as a vector with length m, and element types and values dictated by the problem that needs to be solved. It usually represents an input parameter combination or a possible solution for the problem, which must satisfy all the constraints of the optimization model. The fitness evaluation of each population member requires the decoding of the information contained in the solution that it represents, solving the problem and evaluating the results. The optimality degree of the solution is assessed with the fitness function value associated to the respective population member. For a population with N members, X_i, $i = 1, \ldots, N$, N fitness functions will be computed and ranked.

The (A) to (B) section from Figure 1 consists of several steps, which describe each specific metaheuristic algorithm. While in the figures accompanying Sections 2.1–2.5 are presented all the details specific to each algorithm, delimited by (A) to (B), Table 2 summarizes their main steps, emphasizing their particularities.

Table 2. The metaheuristic algorithms used in the paper for solving the OCBA problem.

Algorithm	Main Steps
Genetic Algorithm (GA)	Selection, crossover, mutation, using the entire population
Particle Swarm Optimization (PSO)	Speed and position update, using the entire population, and exploration followed by exploitation of the search space
The Bat Optimization Algorithm (BOA)	Speed and position update, frequency adaptation, and local search in each iteration, using the entire population, exploration followed by exploitation
The Whale Optimization Algorithm (WOA)	Continuous choice between three search methods: exploration, encircling, and spiral attack, using the entire population
The Sperm Whale Algorithm (SWA)	Population divided in subgroups that perform the search independently, using dominant crossover in each subgroup

Among the various metaheuristic algorithms available in the literature, those from Table 2 were chosen taking into account the following reasoning: the genetic algorithm and the particle swarm optimization are the best known and widely used metaheuristics, with numerous applications in power systems, which makes them a valid basis for comparison. The bat algorithm and the whale optimization algorithm are newer algorithms, previously used by the authors in solving similar optimization problems and shown to improve the quality of the results, compared with GA and PSO [35,36]. On the other hand, the sperm whale algorithm is a novelty in solving optimization problems in the power systems field. The results from the case study will show that the SWA outperforms the previous algorithms, making it a viable new alternative for solving optimization problems related to power systems applications.

The best-known PMAs are the genetic algorithm and the particle swarm optimization, which also describe two fundamental search principles used by metaheuristic algorithms: the evolutionary and performance-based patterns.

2.1. Genetic Algorithms

The Genetic Algorithm (GA), proposed in [37], is probably the best-known metaheuristic algorithm. In the GA, population members are named 'chromosomes', and their elements are 'genes'. The search and optimization mechanisms use Darwinist natural evolution, based on perpetuation through genetic material exchange and mutation inside a population of same-species individuals, across a significant number of generations (iterations).

For finding new and improved solutions for an optimization problem, the GA relies on changing the population by using in each iteration the three main genetic operators (Figure 2):

- Selection: From the existing population, whole individuals are selected based on their performance, expressed by the fitness function. The better-adapted individuals are favored for surviving. In the standard GA, the population size is constant. Thus, the lesser adapted individuals, which are discarded, are replaced by clones of the survivors.
- Crossover: Pairs of parent chromosomes exchange a number of genes, the resulting offspring having new characteristics, possibly resulting in better solutions for the problem.
- Mutation: Randomly generated variations on gene values, resulting in chromosomes with minor structural changes, simulating genetic mutations of real living organisms.

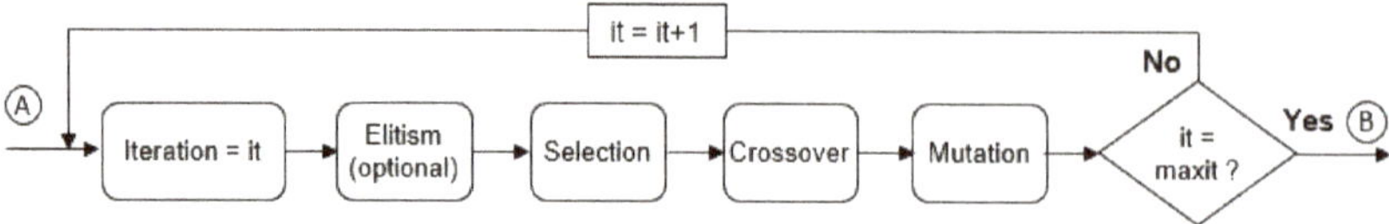

Figure 2. The flowchart of a GA iteration.

Optionally, an elitist procedure can also be incorporated in the GA, which ensures the preservation of the best-found optimal solution and its fitness function across generations.

The literature offers a high variety of selection [38] and crossover [39] types, which together with the user-chosen crossover and mutation rates provide significant customization possibilities, making the GA a flexible problem-solving tool.

In this paper, the tournament selection method was used, which draws randomly p members from the existing population, out of which retains the best q, according to their fitness function. The procedure is repeated until a new population of size N is created.

The method of choice for the crossover operator was the uniform crossover, illustrated in Figure 3. Two parents are randomly chosen from the population and, for each gene, a random number is generated. The parents swap the genes only if the generated random number exceeds a customizable threshold tr.

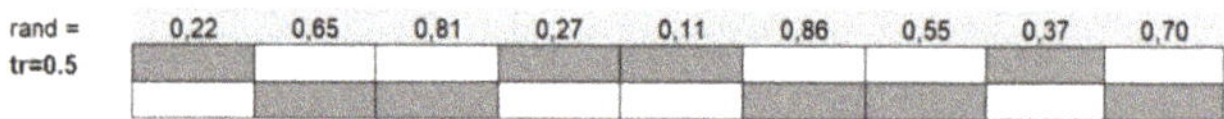

Figure 3. The uniform crossover.

2.2. Particle Swarm Optimization

On the other hand, the PSO algorithm [40] uses a different search method, based on variable travel speeds and position shifting in the search space. Each individual ('particle') from the population ('swarm') changes its speed in each iteration based on its current distance from two reference points: The best solution found so far by the swarm leader and the best position ever achieved by the particle itself. Compared with the GA, the PSO mechanism, presented in Figure 4, is very simple, requiring for each particle j, $j = 1, ..., N$, only the computation of its new speed and position:

$$sp_j^{(it)} = w \cdot sp_j^{(it-1)} + 2 \cdot rnd_1 \cdot (x_{j,best}^{(it)} - x_{j,crt}^{(it)}) + 2 \cdot rnd_2 \cdot (leader^{(it)} - x_{j,crt}^{(it)}) \tag{2}$$

$$x_j^{(it+1)} = x_j^{(it)} + sp_j^{(it)} \tag{3}$$

followed by the update of each particle's best position and the change of the leader position, if better solutions are found. The particle speeds are initialized with low random values, which would not influence the search direction.

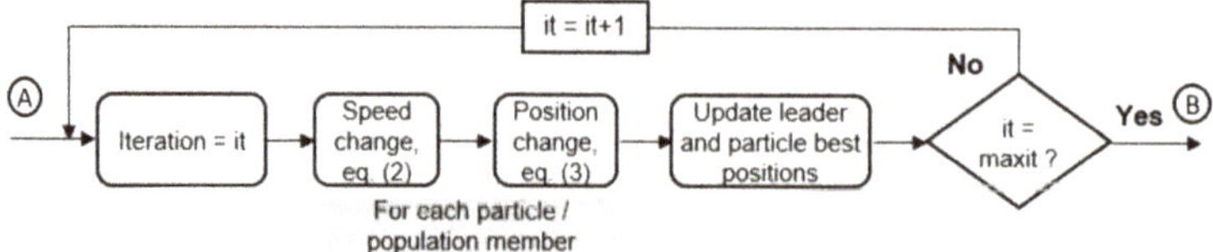

Figure 4. The flowchart of a PSO iteration.

In Equations (2) and (3), $sp_j^{(it)}$ and $sp_j^{(it-1)}$ are the speed of particle x_j ($j = 1 \ldots N$) in the previous ($it - 1$) and current (it) iteration, rnd_1 and rnd_2 are random vectors, $x_{j,best}^{(it)}$ and $x_{j,crt}^{(it)}$ are the best personal and the current position of particle x_j, *leader*$^{(it)}$ is the position of the leader in iteration it, and $x_j^{(it)}$ is the position of particle x in the current iteration. The factor w from Equation (2) is an inertia term, which decreases over the iteration count, larger initial values encouraging exploration, and smaller final values enabling the exploitation or local search around the best-known optimal solution.

It should be noted that while the GA explores the search space using crossover to make random changes of the information that is already present in the population, the mutation probability being much smaller, PSO changes randomly the speed of each particle element, moving it in the direction of the leader and personal best position.

The newer metaheuristic methods used in this paper, while sharing the natural inspiration of GA and PSO, combine elements found in the two algorithms and increase the number of input parameters and the complexity of their mathematical model in order to improve their optimization performance. They are the Bat Optimization Algorithm (BOA), Whale Optimization Algorithm (WOA) and Sperm-Whale Algorithm (SWA).

2.3. The Bat Optimization Algorithm

Bats hunt for prey using echolocation. In the initial search stage, they emit high amplitude/low frequency ultrasound impulses, with low emission rate (10–20 imp/sec), decoding in real time the reflected waves in order to identify the approximate position of the prey. When a potential target is identified, the bat increases the pulse rate up to 200 imp/sec, and the pulse frequency, which enables it to search accurately the space separating it from the prey, identifying the obstacles in its path and precisely locating the victim and its movement pattern.

The bat optimization algorithm [41] uses the PSO principle of changing the speed and position of the population members (here called 'bats'), but the speed update formula is more elaborated, considering the principle of raising the signal frequency and pulse rate as the bats are getting close to the prey, i.e., to the optimal solution. The basic flowchart of a BOA iteration is depicted in Figure 5.

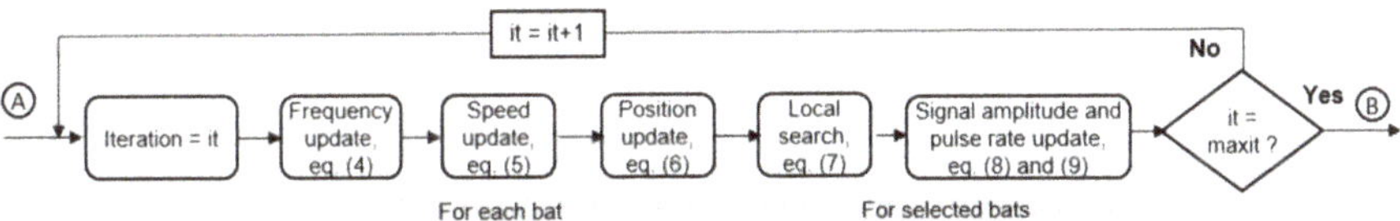

Figure 5. The flowchart of a BOA iteration.

The bats' speeds are initialized in the same manner as in the PSO algorithm but are accompanied by the initial signal amplitude, A_j, maximum pulse rate, $r_{j,max}$ and random pulse frequency $f_j \in [f_{min}, f_{max}]$, $j = 1, ..., N$.

In each iteration it, every bat from the population performs three operations:

- Frequency update:

$$f_j = f_{min} + rnd \cdot (f_{max} - f_{min}) \tag{4}$$

- Speed update, with an equation inspired from (2):

$$sp_j^{(it)} = w \cdot sp_j^{(it-1)} + f_j \cdot rnd_1 \cdot (x_{j,crt}^{(it)} - x_{j,best}^{(it)}) \tag{5}$$

- Position update, identical to the formulation from (3):

$$x_j^{(it+1)} = x_j^{(it)} + sp_j^{(it)} \tag{6}$$

The BOA also includes a local search. The best individuals from the population are randomly moved in the search space, with

$$x_j^{(it+1)} = x_j^{(it+1)} + pp \cdot \overline{A}, \quad j = 1 \ldots M, \quad M < N \tag{7}$$

where $\overline{A}$ is the average bat amplitude for iteration *it* and $pp \in [-1, 1]$.

The new bat positions computed with Equations (4) to (7) are accepted in the population with random probability and only if the newly obtained position is better than the previous.

At the end of each iteration, if a bat improves its position, its signal amplitude is decreased:

$$A_j^{(it+1)} = \alpha \cdot A_j^{(it)} \tag{8}$$

and its pulse emission rate increases:

$$r_j^{(it+1)} = r_j^{(it)} \cdot \left(1 - e^{-\gamma \cdot it}\right) \tag{9}$$

where $\alpha \in (0, 1)$ and $\gamma > 0$.

This behavior, much like the inertia term for PSO, increases the probability of performing local searches when the iteration count is nearing *itmax*.

2.4. The Whale Optimization Algorithm

The hunting behavior of humpback whales is the source of inspiration for the Whale Optimization Algorithm (WOA). The whales hunt in groups, and when they find their prey, consisting of schools of krill or small fish near the water surface, they attack it from below using two maneuvers: encircling and spiraling.

The WOA uses a population of vector solutions ('whales'), which are hunting for prey independently, guiding their search by following a reference individual, usually their leader, i.e., the whale closest to the problem solution ('food'), according to its fitness function.

During the algorithm, whales use initially encircling, then spiral attack, in the same way PSO and BOA use the broad exploration and the exploitation of the search space near the optimal solution.

In each iteration *it*, the encircling performed by each whale *j* from the population is described by [42]:

$$x_j^{(it+1)} = reference^{(it)} - A \cdot D_1 \tag{10}$$

where

$$A = 2 \cdot a \cdot rnd_1 - a \tag{11}$$

$$D_1 = \left| C \cdot reference^{(it)} - x^{(it)} \right|, \quad C = 2 \cdot rnd_2 \tag{12}$$

The coefficient *a* from equation (11) is a scalar value decreasing during the iterative from a positive value to 0. The ($\cdot$) sign denotes the element-by-element multiplying of vectors, and $||$, an absolute value.

For the extreme values of $a = 0$ and $a = 1$, equations (10) to (12) show that position $x_j^{(it+1)}$ will always lie between $x_j^{(it)}$ and $reference^{(it)}$, thus moving any whale towards the reference solution used to guide the population. If values larger than 1 are given to *a*, factor *A* from (11) will also increase, moving the wales beyond the target and exploring a possibly uncharted portion of the search space.

If the reference position is $reference^{(it)} = leader^{(it)}$, the leader from the current iteration, when *A* decreases, whales get closer to the leader, encircling the prey or the optimal solution. If another whale is used as reference, $reference^{(it)} = random(x^{(it)})$, the search will shift towards its path, simulating the exploration of the sea in search for food performed by real whales.

The spiral attack phase is described by an equation that combines oscillatory and exponentially variating components:

$$x_j^{(it+1)} = D_2 \cdot e^{b \cdot l} \cdot \cos(2 \cdot \pi \cdot l) + leader^{(it)} \tag{13}$$

with

$$D_2 = \left| leader^{(it)} - x_j{}^{(it)} \right| \tag{14}$$

In equation (13), b is a constant, and l is a random value from the [−1, 1] interval [42].

The initially large, then gradually decreasing values of a, so that first $|A| > 1$, then $|A| < 1$, $|A| \to 0$, first move the whales away from the leader in exploration, then encourage encircling, followed by spiral attack. If p denotes a random number, the general equation for changing the position of a whale follows as:

$$x_j^{(it+1)} = \begin{cases} reference^{(it)} - A \cdot D_1, & \text{if } \quad p \geq 0.5 \\ D_2 \cdot e^{bl} \cdot \cos(2 \cdot \pi \cdot l) + leader^{(it)} & \text{if } \quad p < 0.5 \end{cases} \tag{15}$$

The flowchart of a WOA iteration is presented in Figure 6.

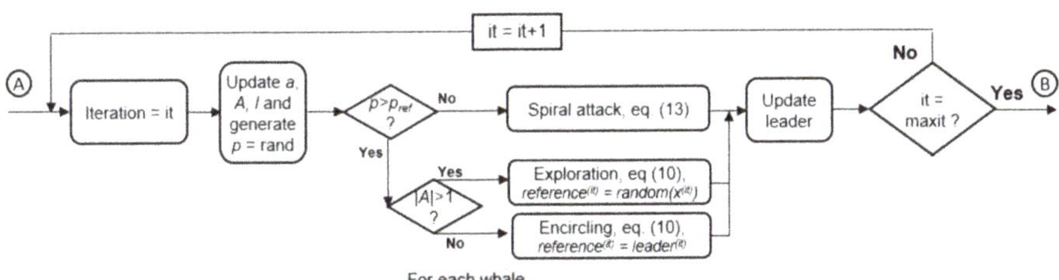

Figure 6. The flowchart of a WOA iteration.

WOA has two specific parameters that can be tuned for better performance: coefficient a from Equation (11) and constant b from (13).

2.5. The Sperm-Whale Algorithm

The search used by the SWA mimics the hunting behavior of sperm whales, which live alone or in small groups at the bottom of the sea and must come to the surface to hunt and breathe [43]. In each iteration, the population of the SWA is split into smaller search groups consisting in uniformly distributed better and worse adapted members ('sperm whales'). Consequently, the search for the optimal solution occurs independently in each group. First, the sperm whales change their position from the bottom of the sea to the surface. This step is simulated only for the worst adapted member of the group, for which the opposite position is computed. The positions of the leader and of the worst individual in a group g, $leader^{(g,it)}$ and $worst^{(g,it)}$, are used to compute an in-between distance $dist^{(g,it)}$:

$$dist^{(g,it)} = worst^{(g,it)} + w \cdot leader^{(g,it)} \tag{16}$$

The reflex position of $worst^{(g,it)}$ is then computed with Equation (17).

$$reflex^{(g,it)} = worst^{(g,it)} + 2 \cdot (dist^{(g,it)} - worst^{(g,it)}) = 2 \cdot dist^{(g,it)} - worst^{(g,it)} \tag{17}$$

The newly computed individual $reflex^{(g,it)}$ will replace $worst^{(g,it)}$ only if its fitness function is better. At the beginning of the iterative process, when the inertia w from Equation (16) is large, the individual will search beyond $leader^{(g,it)}$ (exploration phase, Figure 7a). As w decreases, the search will focus between $worst^{(g,it)}$ and $leader^{(g,it)}$, exploiting the search space around the known optimal solution (Figure 7b).

(a)   (b)

Figure 7. Reflex search in the SWA algorithm: (**a**) exploration, (**b**) exploitation.

In the second stage, a Good Gang is formed within the group, gathering the best gg individuals ranked according to their fitness function. Every Good Gang member performs several local searches in which its elements k are displaced randomly, within a small radius r:

$$x_{j,k} = \pm r \cdot x_{j,k}, \qquad j = 1 \dots gg \tag{18}$$

The original Good Gang members are replaced only if better sperm-whale positions are found during the local search.

Finally, the best Good Gang member from the group (the dominant sperm-whale) performs genetic crossover with all other group individuals. One of the two resulting children is chosen randomly to replace the worst of the two parents.

At the end of the iteration, the groups are reunited in the final population, which will repeat the search process until the stopping criterion of the algorithm is met. The basic flowchart of a SWA iteration is presented in Figure 8.

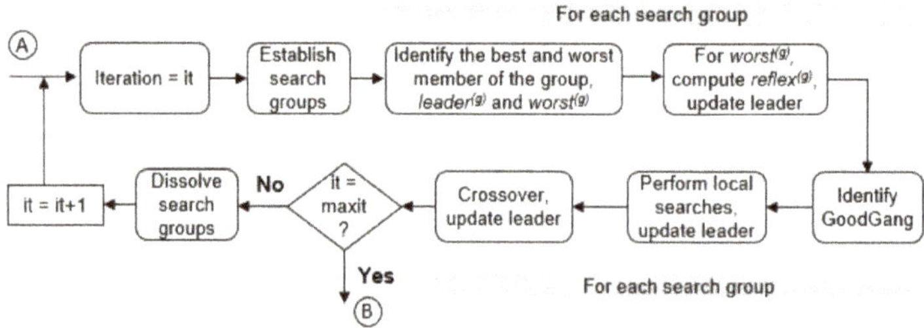

Figure 8. The flowchart of a SWA iteration.

The SWA offers several tuning options for the user. The population size, number of search groups within the population, the inertia w and its decrement, the Good Gang size and number of local searches for its members, the local search radius r, and the crossover method can be adjusted for better performance.

3. The Implementation of the Optimal Reactive Compensation for Loss Minimization Problem

The five metaheuristic algorithms presented in the previous chapter were run in an implementation of the Optimal Capacitor Banks Allocation (OCBA) problem for active energy loss minimization. The approach used in this paper is stated as follows: Find the optimal buses in an EDN where capacitor banks (CB) should be installed and the amount of reactive load compensation in each bus, with the objective of operating the EDN with minimal active power and energy losses for the interval of a typical day. For an EDN with NN buses (nodes) and NB branches, the mathematical expression of the objective function of the OCBA problem was defined as:

$$\Delta P[\%] = \sum_{h=1}^{24} \Delta P_h \Big/ \left(\sum_{h=1}^{24} \sum_{bus=1}^{NN} P_{b,h} + \sum_{h=1}^{24} \Delta P_h \right) * 100 = \min \tag{19}$$

The mathematical model of the fitness function considered the following constraints:

Cr1. The available CB stock cannot be exceeded:

$$\sum_{bus=1}^{NN} NCB_{bus} \leq stock_{CB} \tag{20}$$

Cr2. The compensation level in each bus cannot exceed the reactive bus load (avoid reversed reactive power flows):

$$NCB_{bus} \cdot q_{CB} \leq Q_{bus}, \quad bus = 1 \ldots NN \tag{21}$$

Cr3. Branch current flows after compensation cannot exceed the branch rated current:

$$I_{br} <= I_{max,br}, \quad br = 1 \ldots NB \tag{22}$$

Cr4. Bus voltages after compensation cannot exceed the maximum allowed value:

$$U_{bus} <= U_{max,bus}, \quad bus = 1 \ldots NN \tag{23}$$

In Equations (19) to (22), $\Delta P[\%]$ is the percent active power loss in the EDN over 24 h, ΔP_h is the active power loss in the EDN at hour h, NCB_{bus} is the number of CBs installed in a generic bus; q_{CB} is the reactive power rating of a CB, Q_{bus} is the reactive load of a generic bus, $stock_{CB}$ is the CB stock; I_{br} is the current flow on a generic branch br, $I_{max,br}$ is the rated current of branch br, U_{bus} is the voltage of a generic bus after compensation, and $U_{max,bus}$ is the maximum bus voltage allowed in a generic bus.

All the algorithms tested in the case study used the same solution encoding for their population members. They were generated as vectors of the type described by Equation (1), with integer numbers and length equal to the number of buses in which compensation was possible in the network. The significance of the value of a generic element represented the number of CBs placed in the bus to which it was designated. All the algorithms started in the first iteration with the same population, generated randomly but considering constraint Cr2 of maximum allowed number of CBs in each bus and Cr1, the maximum CB stock (Figure 9).

	bus 1	bus 2	bus 3	...	bus M-1	bus M
max. allowed in bus	5	0	3		2	4
actual compensation	3	0	1		2	2

Figure 9. The encoding of the solutions used by the OCBA problem.

The fitness of the optimal solutions was assessed in all the algorithms using the objective function (19), which also considered the constraints from Equations (20) to (23). The methodology employed for calculating the fitness of each solution is described in Figure 10.

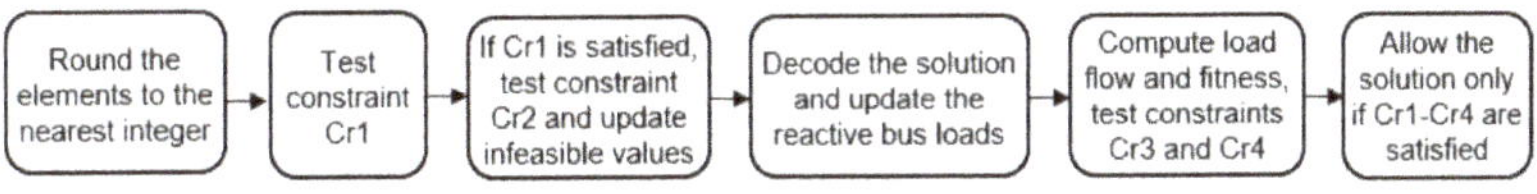

Figure 10. Fitness computation and solution validation for the OCBA problem.

By applying any of the equations (2) to (18) or by genetic crossover and mutation, the changes undergone by population members can result in their invalidation because of

- Non-integer values, leading to invalid number of CBs installed in a bus;
- Values exceeding the interval $[0, NCB_{bus}]$ allowed by the constraint Cr2;
- Violation of constraint Cr1, by exceeding the available CB stock;
- Solutions otherwise valid but which lead to the violation of the constraints Cr3 or Cr4.

Thus, every newly generated population member, for each algorithm, must pass through a validation procedure before being allowed in the population created for the next iteration.

If constraint Cr1 is not satisfied, the solution is always discarded as unfeasible. If the constraint Cr2 is not fulfilled, the solution is modified by setting the unfeasible values to the nearest allowed value, using for each element x_j from Figures 1 and 9 the following correction:

$$x_j = \begin{cases} 0, & \text{if } x_j = 0 \\ NCB_j, & otherwise \end{cases} \tag{24}$$

For each population member, the active power losses used in equation (19) are computed with the Newton–Raphson load flow (LF) algorithm, which is slower, but generally considered more accurate than the branch-and-bound methods preferred in the analysis of distribution networks. The LF algorithm also provides the results required for checking the constraints Cr3 and Cr4, bus voltage and branch current flow limits. Prior to computing the LF, the compensation solution is simulated subtracting from the bus reactive loads the CB injection for each compensated bus, according to the population member/solution being tested.

4. Results

The OCBA problem was solved using the metaheuristics presented in Section 2, with the aim of comparing the performance of the newer algorithm designs, BOA, WOA, and SWA against the two well established methods, GA and PSO. For comparison purposes, all the algorithms had a common representation for the members of the population, illustrated in Figure 9, and the same fitness function, active power and energy loss minimization, computed with Equation (19) and subjected to the constraints given by Equations (20) to (23), Section 3. The initial parameters used for the algorithms are presented in Table 3.

Since for all algorithms there was no improvement for the optimal solution beyond generation 350, the graphical representations of the results will be limited to the first 360 generations.

Two test networks were used to validate the comparison: the smaller sized IEEE 33-bus MV radial voltage distribution system (Figure 11) and a larger 215-bus MV EDN from a residential area of a major city from Romania (Figure 12).

Synthetic data regarding the physical characteristics for the IEEE 33-bus system is given in Table 4. The system does not include MV/LV transformer data; thus, the active power losses computed by the algorithms do not include transformer losses. The extended data regarding branch characteristics (connecting buses, type line or transformer, electrical parameters resistance and reactance, maximum branch current) is provided in Appendix A, Table A1. For the active and reactive bus loads, the study uses a custom representation consisting of daily 24-h profiles, described below.

For this particular test system, the literature provides only a set of instantaneous active and reactive bus loads. In this paper, these values were used as reference, in conjunction with a set of typical load profiles (TLP) provided in the Supplementary Materials attached to the paper, to create 24-h profiles. The TLPs considered several types of loads, residential, industrial, and their distribution in the network is summarized in Table 5.

Table 3. The initial parameters used for the metaheuristic algorithms.

Common Parameters for All the Algorithms:
Population size: 40, Number of generations: 500

GA
Selection method: tournament—keep best 2 from a random draw of 4 Crossover method: uniform at threshold 0.5 Mutation: random Crossover and mutation rates: 0.9/0.1

PSO
Inertia: 0.9 decreasing to 0.4, linear descent

BOA
Initial amplitude: 1 Initial pulse emission rate: 0.3 Amplitude attenuation rate: 0.7 Pulse emission increase rate: 0.3 Pulse frequency domain: [−0.9, 1.2] Number of bats used for local search: dynamic. If a randomly generated number is greater than the bat's pulse emission rate, the bat is selected for local search in the current iteration.

WOA
Coefficient a: decreasing from 3 to 0, linear descent Coefficient l: random Coefficient b: 1

SWA
Number of search groups: 4 Good Gang size: half of the search group size, rounded to the nearest integer Number of searches performed by each Good Gang member: 10 Local search radius: 1 Crossover type: uniform

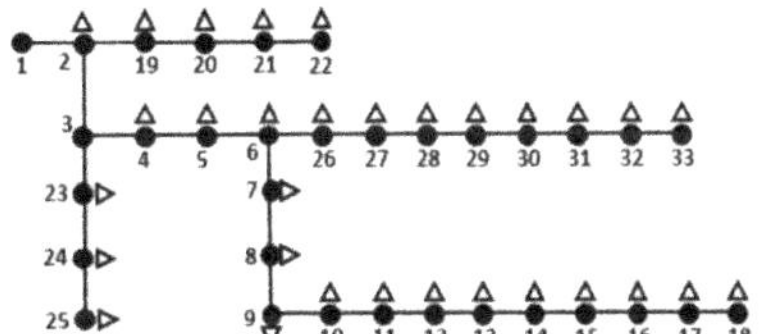

Figure 11. The IEEE 33-bus test system.

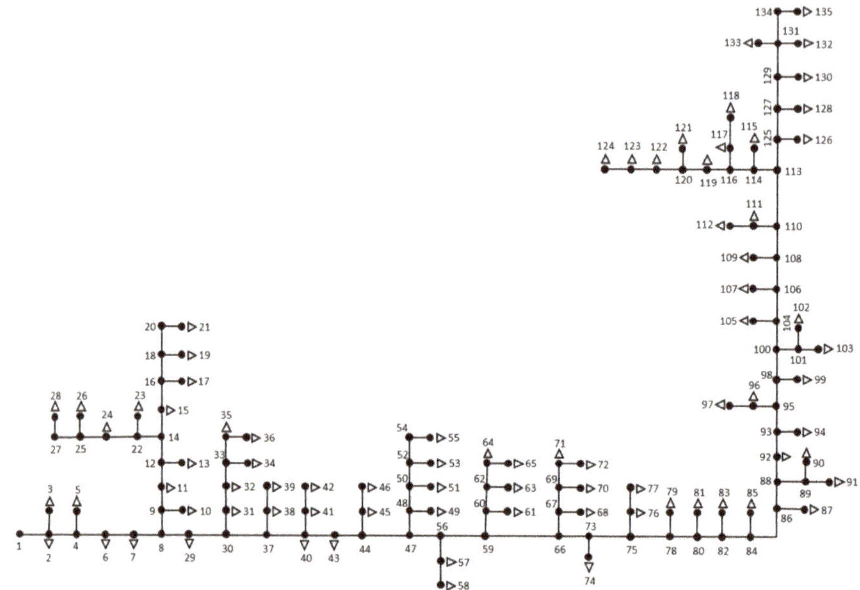

Figure 12. The residential 215-bus EDN.

Table 4. The physical characteristics of the IEEE 33-bus system.

Number of Buses	Transformer Rated Power/Number	Feeder Type	Feeder Cross-Section (mm^2)	Total Length(km)
12.66 kV: 33	None	Unknown[1]	Unknown[1]	Unknown[1]

[1] Only the total resistance and reactance is known for each branch

Table 5. The TLP categories used for the IEEE 33 -bus system and their bus distribution.

TLP Category	Bus
Residential	4, 5, 6, 7, 8, 9, 10, 11, 12, 13 14, 15, 16, 17, 18
Industry	2, 3, 19, 20, 21, 22, 23, 24, 25
Commercial	26, 27, 28, 29, 30, 31, 32, 33

The second test system used in the case study is a much larger EDN, consisting of 135 MV buses to which 80 loads are connected through MV/LV transformers. For simplicity, the transformers are omitted in Figure 12, together with the corresponding 80 bus numbers for the transformer LV busbars (from 136 to 215). A summary of the transformer rated power, together with the feeder and bus general information, is given in Table 6. The electrical parameters of the branches are given in Appendix A, Table A2. For this network, the bus loads were also modelled as 24-h daily profiles, being measured by the smart metering infrastructure installed by the local distribution utility for a typical working day.

Table 6. The physical characteristics of the 215-bus EDN.

Number of Buses	Transformer Rated Power/Number	Feeder Type	Feeder Cross-Section (mm^2)	Total Length (km)
20 kV: 135 (1–135) 0.4 kV: 80 (136–215)	63 kVA: 33; 100 kVA: 14 160 kVA: 15, 250 kVA: 15 400 kVA: 3	Cable	3 × 95 3 × 120 3 × 150	2.5 84.45 3.2

Because a 24-h voltage profile was not available for the 215-bus network, the voltage reference for the slack bus was recommended by the distribution utility at the value of 1.06 pu. For the IEEE 33-bus system, the setpoint was 1 pu (nominal voltage). The slack bus for both networks is bus 1, and all the other buses are modeled as PQ (consumer) buses.

4.1. Results for the IEEE 33-Bus System

The first step of the study was the choice of the maximum CB stock used for optimization. For this purpose, the load profile of the network for 24 h, given in Figure 13, was analyzed. Since the purpose was to test the performance of each algorithm, the CB stock was set at 70 × 7.5 kVar units, which would ensure a maximum of 525 kVar of VAR compensation, about half of the minimal value of the reactive load, occurring at night hours. In this way, the number of possible CB allocation variants is maximized, while reducing the investment cost.

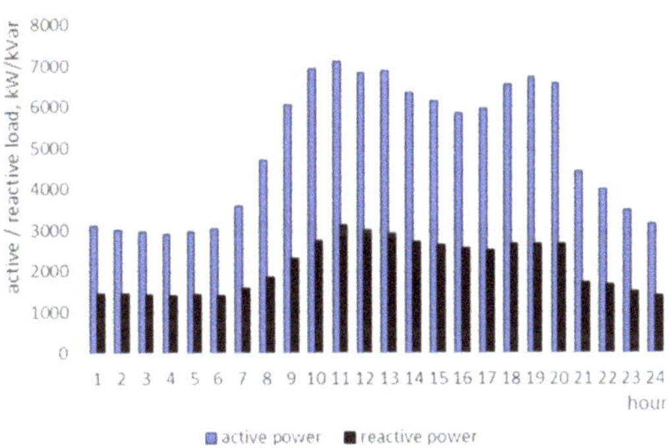

Figure 13. The active and reactive load profiles of the IEEE 33-bus test system—hourly values.

The GA, PSO, BOA, WOA, and SWA were run using this CB stock, the initial parameters from Table 3, and the same initial population. The solution identified by each algorithm, compared with the reference case (no reactive power compensation), and their corresponding fitness functions (percent losses) are presented in Table 7. The first line of Table 7 also presents the maximum number of CBs that can be allocated in each bus, computed according to the minimal reactive power load, so that constraint Cr2 would be always fulfilled. The same results are displayed graphically in Figure 14. In Figure 15, the parallel evolution of the fitness function of each algorithm over the first 350 generations is presented, on a typical run, emphasizing the fact that the SWA and BOA obtain the solutions corresponding to the lowest loss values, followed by the PSO, GA, and BAT.

Table 7. The OCBA solutions found by the metaheuristic algorithms for the IEEE 33-bus system.

Bus	2	3	4	5	6	7	8	9	10	11	12	13	14	15	16	17	18
CB limit	3	2	4	0	0	6	3	1	4	0	6	4	4	0	0	0	14
Reference	0	0	0	0	0	0	0	0	0	0	0	0	0	0	0	0	0
GA	1	1	1	0	0	1	1	1	2	0	5	1	2	0	0	0	13
PSO	0	0	0	0	0	4	0	1	0	0	0	4	4	0	0	0	14
BAT	0	2	4	0	0	3	1	1	1	0	6	4	3	0	0	0	2
WOA	0	0	0	0	0	0	0	1	0	0	1	4	4	0	0	0	14
SWA	0	0	0	0	0	0	0	0	0	0	4	4	4	0	0	0	14

Bus	19	20	21	22	23	24	25	26	27	28	29	30	31	32	33	CB Used	Fitness
CB limit	6	1	17	2	3	7	5	0	0	0	3	39	4	5	0	143	N/A
Reference	0	0	0	0	0	0	0	0	0	0	0	0	0	0	0	0	5.482
GA	1	1	1	1	1	1	1	0	0	0	1	27	4	2	0	70	4.982
PSO	0	1	0	0	0	0	0	0	0	0	3	39	0	0	0	70	4.947
BAT	0	1	0	1	3	7	2	0	0	0	2	24	2	1	0	70	5.038
WOA	0	1	0	0	0	0	0	0	0	0	0	39	1	5	0	70	4.939
SWA	0	0	0	0	0	0	0	0	0	0	0	35	4	5	0	70	4.934

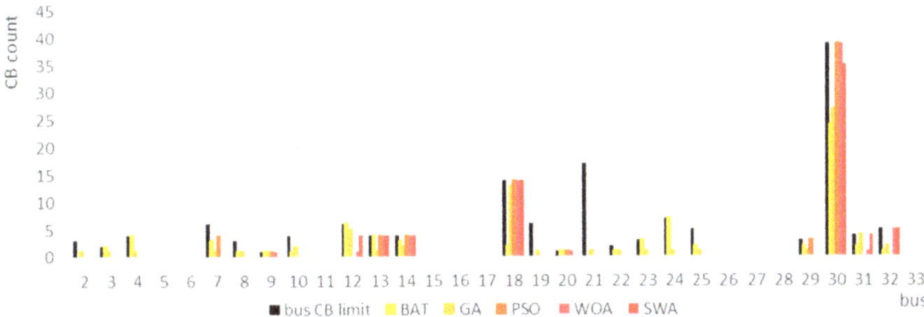

Figure 14. The number of CBs allocated in the buses of the IEEE 33-bus system by each algorithm.

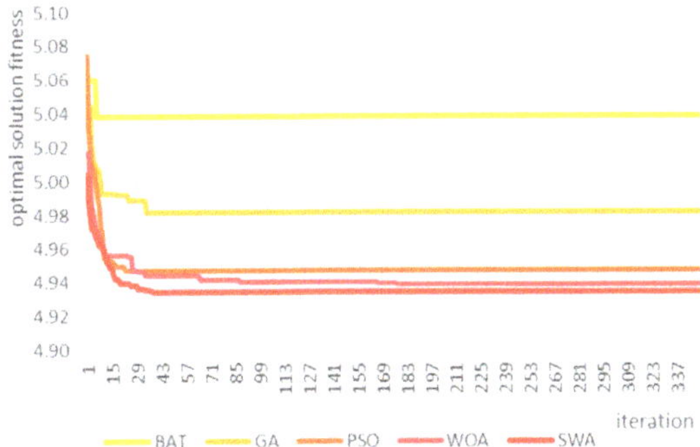

Figure 15. The fitness of the optimal solution found by the metaheuristic algorithms after 360 iterations, for the IEEE 33-bus system.

The results from Table 7 and Figure 14 show that the best fitness function values are obtained when maximum compensation is applied at buses 18 (feeder end), 29–32 (feeder end), and 12–14, while for other buses with compensation potential, such as 21–25, where the reactive load is high, no capacitor banks are allocated. All the algorithms use the entire CB stock available, with differences in the buses chosen for compensation and number of CBs allocated to each bus.

The results regarding the active power losses, for each hour and algorithm, compared with the reference case are plotted in Figure 16 and presented in Table 8. Table 9 gives the loss reduction in percent, against the reference case, for which the total values are represented in Figure 17. The loss reduction ranges between 6.55% and 16.78%, depending on the algorithm and network load. The improvement is higher in off-peak hours, and the best results are obtained with SWA (8.17% to 16.78%), with a global value of 10.51% over 24 h. PSO, WOA, and SWA are the closest to the optimal solution.

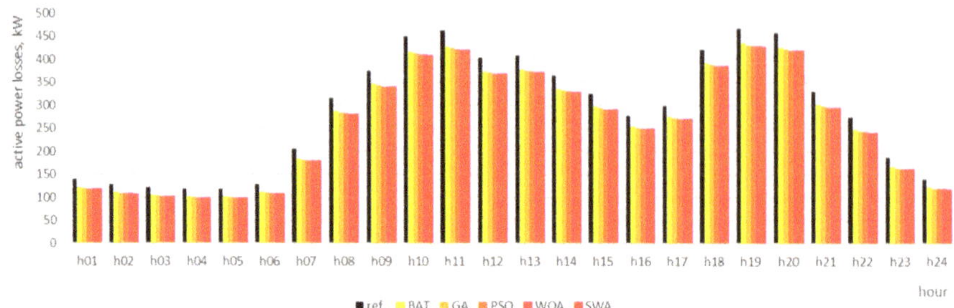

Figure 16. Hourly active power losses in the IEEE 33-bus system, for each algorithm.

Table 8. Hourly and total active power losses in the IEEE 33-bus system, in kW, for each algorithm.

Hour	Ref.	BAT	GA	PSO	WOA	SWA
h01	140.04	122.53	120.22	118.98	118.74	118.51
h02	128.61	111.53	109.32	108.14	107.93	107.70
h03	122.58	105.89	103.74	102.65	102.45	102.22
h04	119.17	102.76	100.65	99.59	99.39	99.17
h05	118.50	102.35	100.31	99.31	99.13	98.90
h06	128.54	112.25	110.25	109.23	109.04	108.82
h07	205.62	184.05	181.30	179.43	178.99	178.75
h08	314.56	287.82	284.37	281.89	281.25	280.98
h09	374.57	345.89	342.50	340.05	339.35	339.12
h10	448.92	416.40	412.92	410.21	409.41	409.15
h11	462.22	427.75	424.10	421.26	420.40	420.17
h12	404.06	373.11	370.05	367.81	367.11	366.87
h13	408.02	377.17	373.94	371.64	370.97	370.72
h14	364.61	335.05	331.31	328.94	328.27	328.03
h15	324.39	296.90	293.33	291.25	290.70	290.45
h16	277.75	254.32	251.42	250.06	249.70	249.46
h17	298.98	275.44	272.52	271.10	270.71	270.50
h18	421.48	391.95	388.02	385.68	385.03	384.78
h19	466.73	436.14	431.94	429.53	428.93	428.60
h20	458.26	427.14	422.83	420.32	419.70	419.36
h21	329.63	302.53	298.48	295.98	295.44	295.03
h22	273.68	248.43	244.68	242.37	241.88	241.51
h23	187.07	167.53	164.73	163.25	162.98	162.67
h24	139.61	123.09	120.90	119.80	119.60	119.37
total	6917.59	6328.03	6253.83	6208.47	6197.11	6190.86

Table 9. Hourly and total power loss reduction in the IEEE 33-bus system, in %, for each algorithm.

Hour	BAT	GA	PSO	WOA	SWA
h01	12.50	14.15	15.03	15.21	15.37
h02	13.28	15.00	15.91	16.08	16.26
h03	13.62	15.37	16.26	16.42	16.61
h04	13.77	15.54	16.43	16.60	16.78
h05	13.62	15.35	16.19	16.35	16.54
h06	12.67	14.22	15.02	15.17	15.34
h07	10.49	11.83	12.74	12.95	13.06
h08	8.50	9.60	10.39	10.59	10.67
h09	7.66	8.56	9.22	9.40	9.47
h10	7.24	8.02	8.62	8.80	8.86
h11	7.46	8.25	8.86	9.05	9.10
h12	7.66	8.42	8.97	9.15	9.20
h13	7.56	8.35	8.92	9.08	9.14
h14	8.11	9.13	9.78	9.97	10.03
h15	8.47	9.57	10.21	10.38	10.46
h16	8.44	9.48	9.97	10.10	10.18
h17	7.87	8.85	9.32	9.46	9.53
h18	7.00	7.94	8.49	8.65	8.71
h19	6.55	7.46	7.97	8.10	8.17
h20	6.79	7.73	8.28	8.41	8.49
h21	8.22	9.45	10.21	10.37	10.50
h22	9.23	10.60	11.44	11.62	11.76
h23	10.45	11.94	12.73	12.88	13.04
h24	11.83	13.40	14.19	14.33	14.49
total	8.52	9.60	10.25	10.42	10.51

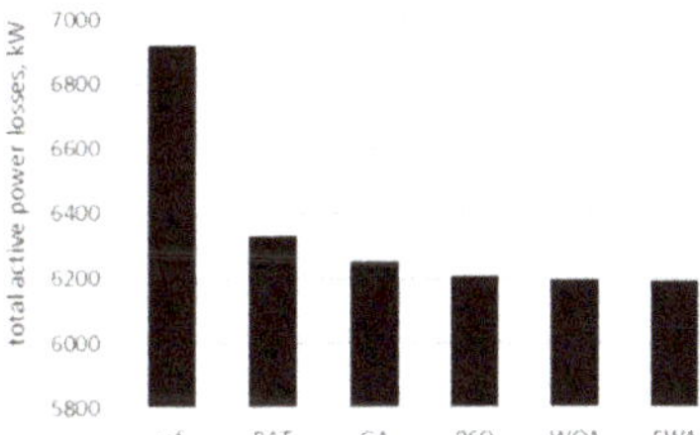

Figure 17. The total active power loss reduction in the IEEE 33-bus system, in kW, for each algorithm.

Compared with the reference case, the best compensation solution found by the SWA leads to a loss reduction of 726.73 kW for the analyzed day, which, if it is extrapolated for a year, amounts to 265.26 MW loss saving. The difference between SWA and the second best result, given by WOA, is of 6.25 kW per day or 2.28 MW for an entire year. As Figure 16 shows, SWA achieves these savings mainly during two hours, at 19.00 and 24.00.

Reactive power compensation with capacitor banks is mainly used in EDN for voltage profile improvement, where specific networks configurations and load patterns lead to high voltage drops along the feeders. In the case of the IEEE 33-bus system, the nominal voltage setting for the slack bus and the load profiles from Appendix A lead, in the reference case without compensation, to the voltage profile described by Figure 18.

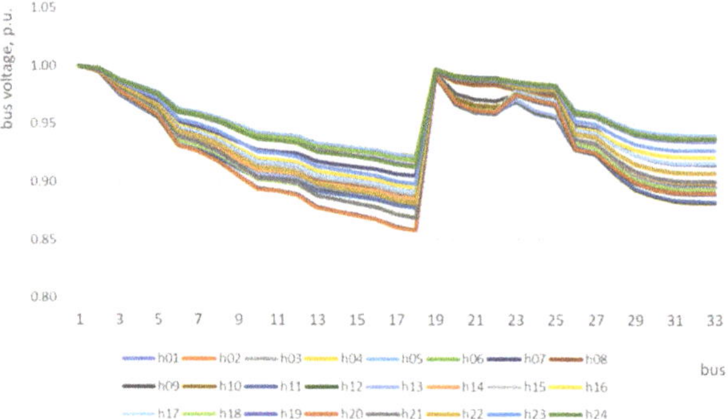

Figure 18. The bus voltages in the IEEE 33-bus system without compensation, for each hour from the analyzed day.

The values show voltage drops that exceed the lower limit of −10% prescribed by the Romanian standards, in several buses located near the end of the main two supply paths, ending in buses 18 and 33. For bus 18, the voltage has the minimum value of 0.858 pu, at hours 19.00 and 20.00, while for bus 33, the minimum voltage is 0.882 pu, at hour 10.00.

The improvement of the voltages obtained hourly with each algorithm is depicted in Figure 19 for bus 18 and in Figure 20 for bus 33. The percent improvements over the reference values (no compensation) are given in Tables 10 and 11, respectively. Again, the SWA gives the best results, with the maximum voltage improvement. The minimum reference voltage value of 0.858 pu in bus 18 (hour 10.00) is raised by 1.65%, to 0.872 pu. However, the voltages remain below the 0.9 pu minimum allowed limit, for 10 h from the 24-h analysis interval. Better voltage regulation can be possible using a larger CB stock or raising the voltage in the reference bus, by changing the HV/MV transformer tap position from the substation at bus 1.

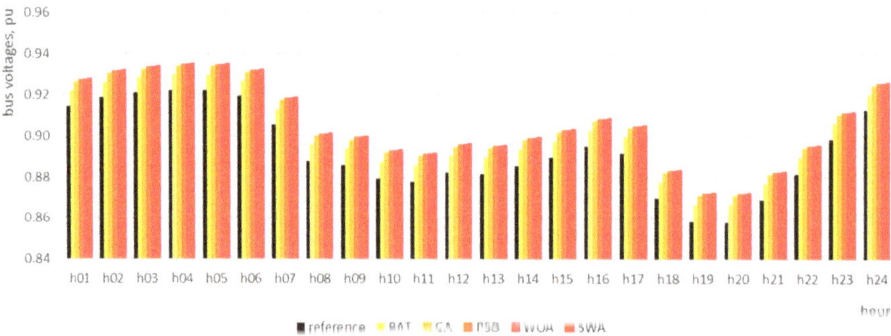

Figure 19. Voltage improvement after compensation for bus 18, the IEEE 33-bus system.

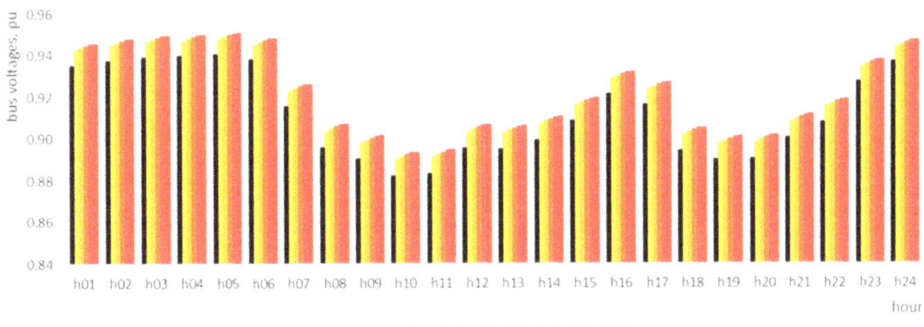

Figure 20. Voltage improvement after compensation for bus 33, the IEEE 33-bus system.

Table 10. Hourly reference voltage values for bus 18 and percent improvements after compensation, the IEEE 33-bus test system.

Hour	Voltage, pu	Improvement, %				
	Ref.	BAT	GA	PSO	WOA	SWA
h01	0.914	0.82	1.26	1.38	1.38	1.42
h02	0.919	0.81	1.25	1.37	1.37	1.41
h03	0.921	0.80	1.24	1.36	1.36	1.40
h04	0.922	0.80	1.23	1.36	1.36	1.40
h05	0.922	0.80	1.23	1.36	1.36	1.40
h06	0.919	0.80	1.24	1.36	1.36	1.40
h07	0.905	0.84	1.29	1.42	1.42	1.46
h08	0.888	0.88	1.35	1.48	1.48	1.53
h09	0.886	0.88	1.36	1.49	1.49	1.54
h10	0.879	0.90	1.38	1.52	1.52	1.56
h11	0.878	0.91	1.39	1.53	1.53	1.57
h12	0.882	0.89	1.37	1.50	1.50	1.55
h13	0.882	0.89	1.37	1.51	1.51	1.55
h14	0.885	0.89	1.36	1.50	1.50	1.54
h15	0.890	0.88	1.35	1.48	1.48	1.52
h16	0.895	0.86	1.32	1.45	1.45	1.50
h17	0.892	0.86	1.33	1.46	1.46	1.51
h18	0.870	0.92	1.41	1.55	1.55	1.60
h19	0.858	0.95	1.45	1.60	1.60	1.65
h20	0.858	0.95	1.45	1.60	1.60	1.65
h21	0.869	0.92	1.42	1.55	1.56	1.60
h22	0.882	0.89	1.37	1.51	1.51	1.55
h23	0.898	0.85	1.31	1.44	1.44	1.48
h24	0.913	0.82	1.26	1.38	1.39	1.43
minimum value, pu	0.858					
max. improvement, %		0.95	1.45	1.60	1.60	1.65

Table 11. Hourly reference voltage values for bus 33 and percent improvements after compensation, the IEEE 33-bus test system.

Hour	Voltage, pu	Improvement, %				
	Reference	BAT	GA	PSO	WOA	SWA
h01	0.935	0.78	0.91	1.02	1.10	1.11
h02	0.937	0.77	0.90	1.01	1.09	1.10
h03	0.939	0.77	0.90	1.01	1.09	1.10
h04	0.939	0.77	0.90	1.01	1.09	1.10
h05	0.940	0.77	0.90	1.01	1.08	1.10
h06	0.938	0.77	0.90	1.01	1.09	1.10
h07	0.915	0.82	0.96	1.08	1.16	1.17
h08	0.896	0.86	1.01	1.14	1.22	1.24
h09	0.890	0.88	1.02	1.15	1.24	1.25
h10	0.882	0.90	1.05	1.18	1.27	1.28
h11	0.883	0.90	1.05	1.18	1.27	1.28
h12	0.895	0.87	1.01	1.14	1.23	1.24
h13	0.895	0.87	1.01	1.14	1.23	1.24
h14	0.899	0.86	1.00	1.13	1.22	1.23
h15	0.908	0.84	0.98	1.10	1.19	1.20
h16	0.921	0.81	0.94	1.06	1.14	1.15
h17	0.916	0.82	0.96	1.07	1.16	1.17
h18	0.894	0.87	1.02	1.14	1.23	1.24
h19	0.890	0.88	1.03	1.15	1.24	1.26
h20	0.890	0.88	1.03	1.15	1.24	1.26
h21	0.900	0.85	1.00	1.12	1.21	1.22
h22	0.908	0.84	0.98	1.10	1.19	1.20
h23	0.927	0.79	0.93	1.04	1.12	1.14
h24	0.937	0.77	0.90	1.01	1.09	1.11
minimum value, p.u.	0.882					
max. improvement, %		0.90	1.05	1.18	1.27	1.28

The voltage improvements are smaller for bus 33, with a maximum of 1.28% with the SWA, but with three algorithms (SWA, PSO and WOA), the voltage levels are raised after compensation above the maximum −10% deviation allowed by the regulations during 7 h (8.00, 9.00, 12.00, 13.00, 18.00, 19.00, and 20.00), only two hours remaining below this threshold (10.00 and 11.00), as it can be seen in Figure 20.

4.2. Results for the 215-Bus Distribution Network

In this case, the CB stock used for compensation was chosen using the 24-h load profile of the network from Figure 21. In comparison with the IEEE 33-bus system, the minimum off-peak reactive load is reduced, while the number of buses available for compensation increases significantly, allowing a higher number of possible solutions. Thus, the CB stock was set at 90 units of 7.5 kVar each, providing a maximum of 675 kVar of reactive power. As the solutions from Figure 22 and Table 12 show, all the algorithms use the entire stock, with different bus allocation. The SWA provides the best solution (6.19% active power losses), followed by the PSO (6.21%). In Table 12, since the VAR compensation is performed at the LV side of the substation transformers, the bus numbers are given for both the MV buses denoted in Figure 12, and for their corresponding LV transformer busbars. In Figure 22, only the LV bus numbers are used, for better readability. Figure 23 presents the evolution of the fitness function of each metaheuristic algorithm over the first 360 generations, on a typical run.

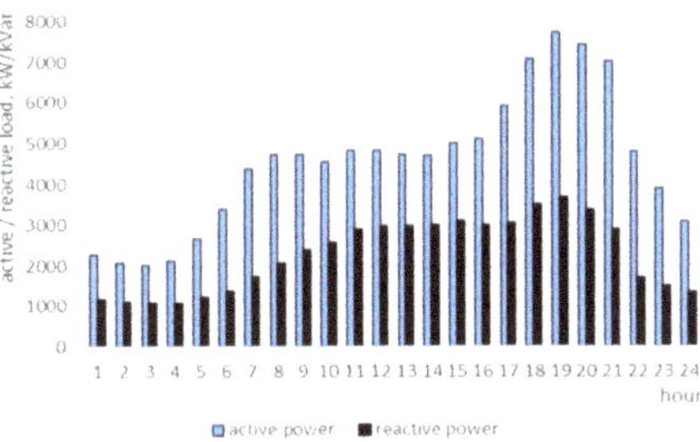

Figure 21. The active and reactive load profiles of the 215-bus network—hourly values.

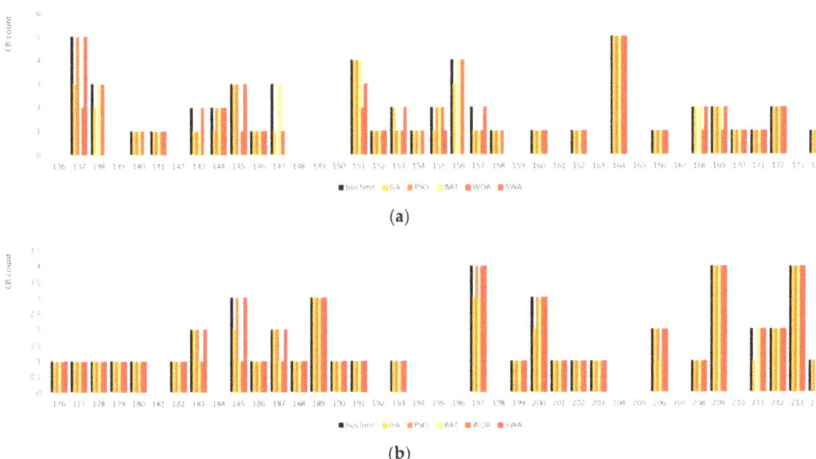

(a)

(b)

Figure 22. The number of CBs allocated in the buses of the 215-bus network by each algorithm: (**a**)—buses 136-175, (**b**)—buses 176-215.

For this network, as Figure 22 shows, the best two solutions (SWA, PSO) are mainly differentiated by the CB allocation at buses 151, LV side (28, MV side); 153 (31); 155 (34); 156 (35); and 157 (36), located at the beginning of the network, and having significant reactive power loads. This behavior is triggered by the use of the 90 CB stock, which is close to the maximum possible number of CBs that can be allocated for compensation, 107, and because of the sufficient stock, most of the buses can use the maximum possible CB allocation.

The comparison between the hourly active power losses computed by the Newton–Raphson algorithm for the reference case (no compensation) and the losses determined for each best compensation solution found by the metaheuristic algorithms is presented in Table 13 and Figure 24. Furthermore, Table 14 gives the percent reduction in losses obtained using the compensation solutions, while Figure 25 allows for an overview of the total losses obtained in each case, based on the values computed in Table 13.

Table 12. The OCBA solutions found by the metaheuristic algorithms for the 215-bus network.

MV Bus	2	3	5	6	7	10	11	13	15	17	19	21	23	24	26	28	29	31	32	34	35
LV Bus	136	137	138	139	140	141	142	143	144	145	146	147	148	149	150	151	152	153	154	155	156
bus limit	0	5	3	0	1	1	0	2	2	3	1	3	0	0	0	4	1	2	1	2	4
reference	0	0	0	0	0	0	0	0	0	0	0	0	0	0	0	0	0	0	0	0	0
GA	0	3	2	0	1	1	0	1	1	3	1	1	0	0	0	4	1	2	1	1	3
PSO	0	5	0	0	1	1	0	1	2	3	1	0	0	0	0	4	1	1	1	2	0
BAT	0	0	3	0	1	1	0	1	2	0	1	3	0	0	0	4	1	1	1	2	4
WOA	0	2	3	0	1	1	0	2	2	1	1	1	0	0	0	2	1	1	1	2	4
SWA	0	5	0	0	0	1	0	0	2	3	1	0	0	0	0	3	1	2	0	1	0

MV Bus	36	38	39	40	41	42	43	45	46	49	51	53	55	57	58	61	63	64	65	68	70
LV Bus	157	158	159	160	161	162	163	164	165	166	167	168	169	170	171	172	173	174	175	176	177
bus limit	2	1	0	1	0	1	0	5	0	1	0	2	2	1	1	2	0	1	1	1	1
reference	0	0	0	0	0	0	0	0	0	0	0	0	0	0	0	0	0	0	0	0	0
GA	1	1	0	1	0	1	0	5	0	1	0	2	2	1	1	2	0	1	1	1	1
PSO	1	1	0	1	0	1	0	5	0	1	0	0	2	1	1	2	0	1	1	1	1
BAT	1	1	0	1	0	1	0	5	0	1	0	2	2	1	1	2	0	1	1	1	1
WOA	1	1	0	1	0	1	0	5	0	1	0	1	1	1	1	2	0	1	1	1	1
SWA	2	0	0	1	0	1	0	5	0	1	0	2	2	1	1	2	0	1	1	1	1

MV Bus	71	72	74	76	77	79	81	83	85	87	90	91	92	94	96	97	99	102	103	105	107
LV Bus	178	179	180	181	182	183	184	185	186	187	188	189	190	191	192	193	194	195	196	197	198
bus limit	1	1	1	0	1	2	0	3	1	2	1	3	1	1	0	1	0	0	0	4	0
reference	0	0	0	0	0	0	0	0	0	0	0	0	0	0	0	0	0	0	0	0	0
GA	1	1	1	0	1	2	0	2	1	2	1	3	1	1	0	1	0	0	0	3	0
PSO	1	1	1	0	1	2	0	3	1	2	1	3	1	1	0	1	0	0	0	4	0
BAT	1	1	1	0	1	1	0	1	1	1	1	3	1	1	0	1	0	0	0	3	0
WOA	1	1	1	0	1	1	0	1	1	1	1	3	1	1	0	1	0	0	0	4	0
SWA	1	1	1	0	1	2	0	3	1	2	1	3	1	1	0	1	0	0	0	4	0

MV Bus	109	111	112	115	117	118	119	121	122	123	124	126	128	130	132	133	135		
LV Bus	199	200	201	202	203	204	205	206	207	208	209	210	211	212	213	214	215	CB Used	Fitness
bus limit	1	3	1	1	1	0	0	2	0	1	4	0	2	2	4	1	3	107	
reference	0	0	0	0	0	0	0	0	0	0	0	0	0	0	0	0	0	0	6.8578
GA	1	2	1	1	1	0	0	2	0	1	4	0	1	2	4	1	0	90	6.2385
PSO	1	3		1	1	0	0	2	0	1	4	0	0	2	4	1	3	90	6.2050
BAT	1	3		1	1	0	0	1	0	1	4	0	2	2	4	1	3	90	6.2278
WOA	1	3		1	1	0	0	2	0	1	4	0	2	2	4	1	3	90	6.2175
SWA	1	3		1	1	0	0	2	0	1	4	0	2	2	4	1	3	90	6.1910

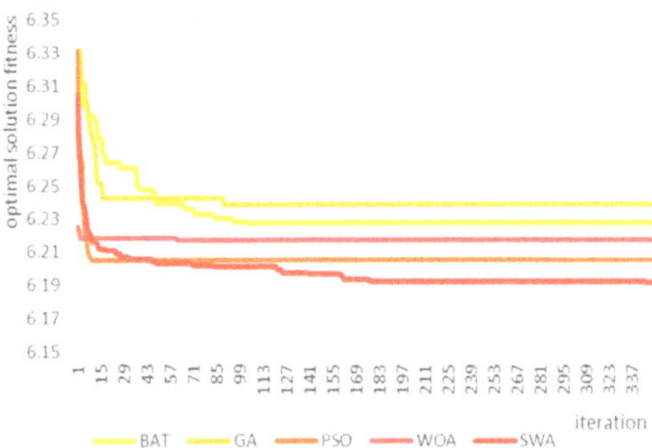

Figure 23. The fitness of the optimal solution found by the metaheuristic algorithms after 360 iterations, for the 215-bus network.

Table 13. Hourly and total active power losses in the 215-bus network, in kW, for each algorithm.

Hour	Ref.	BAT	GA	PSO	WOA	SWA
h01	67.331	55.673	55.609	55.539	55.583	55.305
h02	56.182	45.636	45.575	45.517	45.541	45.319
h03	53.214	43.113	43.053	43.000	43.016	42.817
h04	58.190	47.878	47.818	47.762	47.782	47.572
h05	90.377	77.534	77.467	77.382	77.446	77.109
h06	147.228	131.143	131.068	130.941	131.065	130.562
h07	245.758	223.025	223.059	222.583	222.867	222.029
h08	290.292	262.257	262.479	261.527	261.962	260.864
h09	303.227	270.552	271.077	269.405	270.026	268.696
h10	296.025	261.012	261.713	259.595	260.336	258.874
h11	354.181	312.943	313.922	311.078	312.050	310.247
h12	363.306	320.607	321.658	318.606	319.647	317.753
h13	352.690	310.099	311.147	308.090	309.134	307.242
h14	350.992	308.310	309.373	306.281	307.332	305.437
h15	394.674	349.439	350.570	347.313	348.428	346.400
h16	393.582	349.764	350.799	347.811	348.843	346.908
h17	505.757	458.845	459.845	456.827	457.938	455.824
h18	741.209	680.993	682.464	678.467	679.882	677.100
h19	892.805	825.937	827.555	823.210	824.773	821.637
h20	795.742	736.564	737.748	734.379	735.670	732.945
h21	672.197	623.613	624.385	621.999	622.971	620.808
h22	304.238	280.233	280.128	279.887	280.171	279.245
h23	194.076	175.399	175.315	175.153	175.327	174.688
h24	122.607	107.163	107.100	106.949	107.055	106.594
total	8045.88	7257.73	7270.93	7229.30	7244.84	7211.97

Figure 24. Hourly active power losses in the IEEE 33-bus system, for each algorithm.

Table 14. Hourly and total power loss reduction in the 215-bus network, in %, for each algorithm.

Hour	BAT	GA	PSO	WOA	SWA
h01	17.31	17.41	17.51	17.45	17.86
h02	18.77	18.88	18.98	18.94	19.33
h03	18.98	19.09	19.19	19.16	19.54
h04	17.72	17.82	17.92	17.89	18.25
h05	14.21	14.28	14.38	14.31	14.68
h06	10.92	10.98	11.06	10.98	11.32
h07	9.25	9.24	9.43	9.31	9.66
h08	9.66	9.58	9.91	9.76	10.14
h09	10.78	10.60	11.15	10.95	11.39
h10	11.83	11.59	12.31	12.06	12.55
h11	11.64	11.37	12.17	11.90	12.40
h12	11.75	11.46	12.30	12.02	12.54
h13	12.08	11.78	12.65	12.35	12.89
h14	12.16	11.86	12.74	12.44	12.98
h15	11.46	11.17	12.00	11.72	12.23
h16	11.13	10.87	11.63	11.37	11.86
h17	9.28	9.08	9.67	9.45	9.87
h18	8.12	7.93	8.46	8.27	8.65
h19	7.49	7.31	7.80	7.62	7.97
h20	7.44	7.29	7.71	7.55	7.89
h21	7.23	7.11	7.47	7.32	7.64
h22	7.89	7.92	8.00	7.91	8.22
h23	9.62	9.67	9.75	9.66	9.99
h24	12.60	12.65	12.77	12.68	13.06
total	9.80	9.63	10.15	9.96	10.36

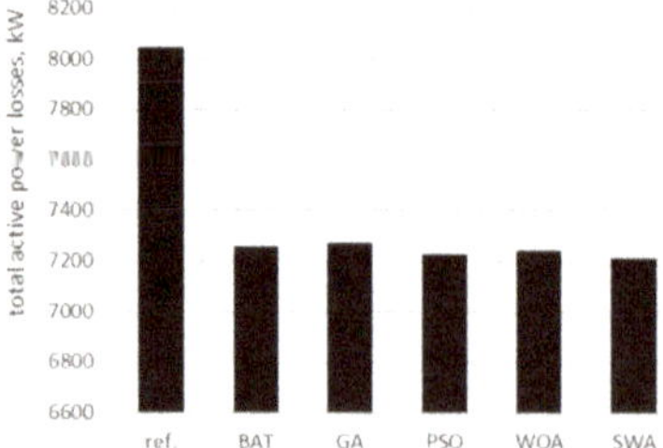

Figure 25. The total active power loss reduction in the 215-bus network, in kW, for each algorithm.

The best CB allocation solution, obtained with the SWA, achieves a loss reduction of 833.91 kW or 10.36% for the entire network, in 24 h, which amounts to 304.38 MW in an entire year. The next best solution, found with the PSO algorithm, achieves only 816.58 kW (10.15%). The difference between the two solutions is of 17.33 kW in the analyzed day, or 6.32 MW in a year. The improvement over the IEEE 33-bus system regarding the loss reduction can be attributed to the presence of the MV/LV transformers.

The bus voltage levels for all 24 h and buses from the 215-bus network are presented in Figure 26. The length of the main feeder and the bus loadings lead to low voltage levels at the last buses on the main supply path, with values below the 0.9 pu limit at the LV side. At the MV side, the voltages are inside the allowed range, varying from 1.060 pu in the slack bus to 0.940 pu.

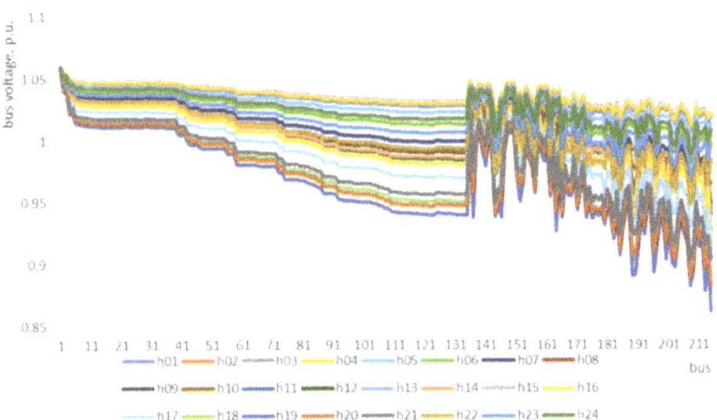

Figure 26. The bus voltages in the 215-bus network without compensation, for each hour from the analyzed day.

Figures 27 and 28 depict the effect of VAR compensation on the voltages at bus 135 (MV) and 215 (LV). By allocating the available CB stock according to the solutions found by the five metaheuristic algorithms, the bus voltages increase with maximum 1.36%, as shown in Table 15 for bus 215 (LV). This increase is not sufficient for raising the lowest voltage values above the desired limit of 0.9 pu. Since the CB stock is near the maximum allowed reactive load compensation which fulfills constraint Cr2 specified by the optimization model, an alternative solution is to change the MV/LV transformer tap settings in the affected buses.

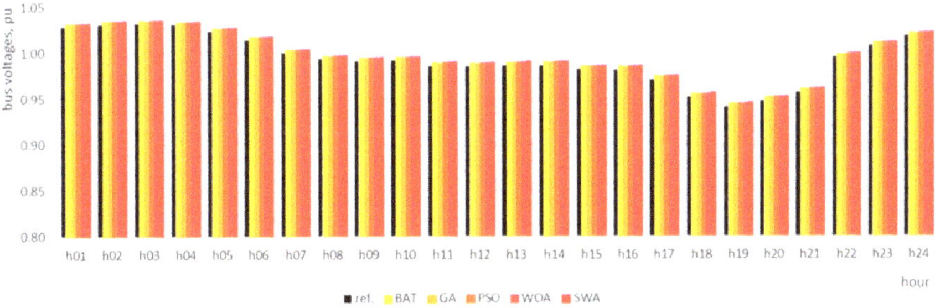

Figure 27. Voltage improvement after compensation for bus 135, medium voltage, the 215-bus network.

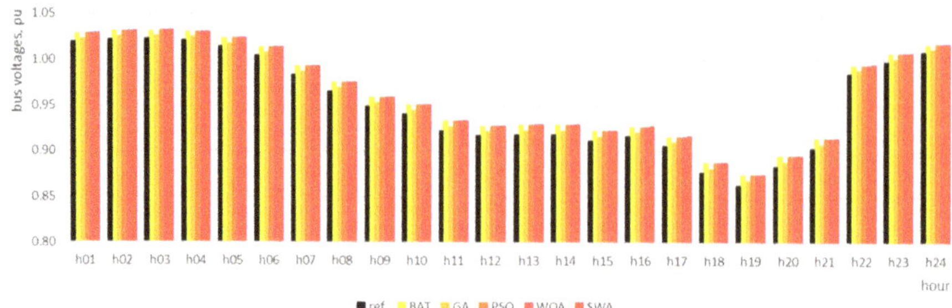

Figure 28. Voltage improvement after compensation for bus 215, low voltage, the 215-bus network.

Table 15. Hourly reference voltage values for bus 215 and percent improvements after compensation, the 215-bus network.

Hour	Voltage, pu	Improvement, %				
	Ref.	BAT	GA	PSO	WOA	SWA
h01	1.019	0.86	0.36	0.86	0.87	0.88
h02	1.022	0.85	0.35	0.86	0.86	0.87
h03	1.022	0.85	0.35	0.86	0.86	0.87
h04	1.021	0.86	0.35	0.86	0.86	0.87
h05	1.014	0.87	0.36	0.87	0.88	0.89
h06	1.004	0.89	0.37	0.89	0.90	0.91
h07	0.983	0.94	0.39	0.94	0.95	0.96
h08	0.965	0.99	0.41	0.99	0.99	1.01
h09	0.948	1.04	0.43	1.04	1.04	1.06
h10	0.940	1.06	0.45	1.07	1.07	1.09
h11	0.922	1.12	0.47	1.13	1.13	1.15
h12	0.917	1.14	0.48	1.15	1.15	1.16
h13	0.918	1.14	0.48	1.14	1.15	1.16
h14	0.918	1.14	0.48	1.14	1.15	1.16
h15	0.911	1.16	0.49	1.17	1.17	1.18
h16	0.916	1.14	0.48	1.15	1.15	1.17
h17	0.905	1.17	0.49	1.18	1.18	1.20
h18	0.876	1.28	0.54	1.28	1.29	1.30
h19	0.863	1.33	0.57	1.33	1.34	1.36
h20	0.884	1.24	0.53	1.25	1.25	1.27
h21	0.903	1.17	0.50	1.17	1.18	1.19
h22	0.985	0.93	0.39	0.94	0.94	0.95
h23	0.998	0.90	0.38	0.91	0.91	0.92
h24	1.008	0.88	0.37	0.89	0.89	0.90
minimum value, pu	0.863					
max. improvement, %		1.33	0.57	1.33	1.34	1.36

5. Discussions and Conclusions

The reactive power flow in the active electricity distribution networks has an important influence on the bus voltage level and the active power losses. Therefore, in order to control the reactive power absorbed by consumers, their consumption must be characterized by a power factor approximately equal to the neutral value (0.9 in Romania). Optimal allocation of reactive sources in the electricity distribution networks is made for power losses reduction, power factor correction and/or voltage profile improvement.

The optimization model considered in the paper has as main objective the optimal allocation of capacitor banks (CBs) in the medium voltage networks to minimize the power/energy losses, taking into account the technical restrictions imposed by the available CB stock, the compensation level in each bus, branch current flows, and bus voltages. This is very useful for distribution network operators that install now large amounts of capacitor banks (CB) in the distribution networks. In order to optimize the location of these CBs, the used test networks (the IEEE 33-bus system and a real 215-bus EDN from Romania) were modelled considering the MV lines, the MV/LV transformers from the electric distribution substations, where available, and the MV and LV buses. The different algorithms (Genetic Algorithm (GA), Particle Swarm Optimization (PSO), Bat Optimization Algorithm (BOA), Whale Optimization Algorithm (WOA) and Sperm-Whale Algorithm (SWA)) were tested to see which would be the best to solve the problem of capacitor bank allocation. The study, made using the IEEE 33-bus system, highlighted the fact that the SWA leads to the best results compared to the other algorithms. Compared with the reference case, the best compensation solution found by the SWA leads to a loss reduction of 726.73 kW for the analyzed day, which, if it is extrapolated for a year, amounts to 265.26 MW loss saving. The difference between SWA and the second best result, given by WOA, is of 6.25 kW per day or 2.28 MW for an entire year. In the case of the voltage level, an improvement was observed on the entire electricity distribution network, in all nodes, also obtained with the help of SWA. The minimum reference voltage value in bus 18 (the farthest node), at hour 10.00, was increased by 1.65%.

Moreover, the algorithms were tested in a real electricity distribution network (215-bus EDN) from Romania. The best CB allocation solution, obtained again with the SWA, achieves a loss reduction of 833.91 kW or 10.36% for the entire network, in 24 h, which amounts to 304.38 MW in an entire year. The next best solution, found with the PSO algorithm, achieves only 816.58 kW (10.15%). The difference between the two solutions is of 17.33 kW in the analyzed day, or 6.32 MW in a year. The solutions found led to an increase of the voltage in the farthest node (215) with maximum 1.36%.

Based on the obtained results, it can be affirmed that the use of capacitor banks is an easy solution to be implemented with technical and economic benefits to the electricity distribution networks that maximizes long-term return on investment as the network develops. An intelligent control of capacitor banks leads to improved energy efficiency and voltage level in the buses of electricity distribution networks, resulting in an increase in the percentage of energy delivered to consumers. Amongst the tested algorithms, the SWA finds the best compensation solutions, which can lead to significant additional loss savings and shorter investment recovery times.

Supplementary Materials: The following are available online at http://www.mdpi.com/1996-1073/12/22/4239/s1, file IEEE33_load_data.xls, active and reactive load profiles for the IEEE-33bus test system, and file EDN215_load_data.xls, active and reactive load profiles for the 215-bus distribution network.

Author Contributions: Conceptualization, O.I., B.-C.N. and G.G.; methodology, O.I. and G.G.; software, O.I. and B.-C.N.; validation, B.-C.N. and G.G.; formal analysis, M.G.; investigation, O.I. and B.-C.N.; data curation, B.-C.N.; writing—original draft preparation, O.I. and B.-C.N.; writing—B.-C.N., G.G. and M.G.; supervision, M.G.

Funding: This research received no external funding.

Conflicts of Interest: The authors declare no conflict of interest.

Appendix A

Table A1. Branch parameters for the IEEE 33-bus system.

Bus 1	Bus 2	Imax, A	Type[1]	Resistance, Ω	Reactance, Ω
1	2	420	1	0.092	0.047
2	3	420	1	0.493	0.251
3	4	420	1	0.366	0.186
4	5	420	1	0.381	0.194
5	6	420	1	0.819	0.707
6	7	420	1	0.187	0.619
7	8	420	1	0.711	0.235
8	9	420	1	1.03	0.74

Table A1. *Cont.*

Bus 1	Bus 2	Imax, A	Type[1]	Resistance, Ω	Reactance, Ω
9	10	420	1	1.044	0.74
10	11	420	1	0.197	0.065
11	12	420	1	0.374	0.124
12	13	420	1	1.468	1.155
13	14	420	1	0.542	0.713
14	15	420	1	0.591	0.526
15	16	420	1	0.746	0.545
16	17	420	1	1.289	1.721
17	18	420	1	0.732	0.574
19	20	420	1	1.504	1.356
2	19	420	1	0.164	0.157
20	21	420	1	0.41	0.478
21	22	420	1	0.709	0.937
23	24	420	1	0.898	0.709
24	25	420	1	0.896	0.701
26	27	420	1	0.284	0.145
27	28	420	1	1.059	0.934
28	29	420	1	0.804	0.701
29	30	420	1	0.508	0.259
3	23	420	1	0.451	0.308
30	31	420	1	0.975	0.963
31	32	420	1	0.311	0.362
32	33	420	1	0.341	0.53
6	26	420	1	0.203	0.103

[1] Branch type can be 1—line; 2—transformer.

Table A2. Branch parameters for the 215-bus distribution network.

Bus 1	Bus 2	Imax, A	Type	Resistance, Ω	Reactance, Ω
1	2	235	1	0.80625	0.314159
2	3	315	1	0.1051	0.061104
2	4	295	1	0.7701	0.358142
4	5	295	1	0.00154	0.000716
4	6	295	1	0.30804	0.143257
6	7	295	1	0.021306	0.009909
7	8	295	1	0.021306	0.009909
8	9	295	1	0.272102	0.126543
8	29	295	1	0.056474	0.026264
9	10	295	1	0.02567	0.011938
9	11	295	1	0.213061	0.099086
11	12	295	1	0.115515	0.053721
101	102	295	1	0.141185	0.065659
101	103	295	1	0.02567	0.011938
104	105	295	1	0.02567	0.011938
104	106	295	1	0.361947	0.168327
106	107	295	1	0.02567	0.011938
106	108	295	1	0.2567	0.119381
108	109	295	1	0.05134	0.023876
108	110	295	1	0.467194	0.217273

Table A2. *Cont.*

Bus 1	Bus 2	Imax, A	Type	Resistance, Ω	Reactance, Ω
110	111	295	1	0.115515	0.053721
110	113	295	1	0.12835	0.05969
12	13	295	1	0.07701	0.035814
12	14	295	1	0.019253	0.008954
111	112	295	1	0.041072	0.019101
113	114	295	1	0.2567	0.119381
113	125	295	1	0.07701	0.035814
114	115	295	1	0.173273	0.080582
114	116	295	1	0.035938	0.016713
116	117	295	1	0.369648	0.171908
116	119	295	1	0.07701	0.035814
117	118	295	1	0.202793	0.094311
119	120	295	1	0.562173	0.261443
120	121	295	1	0.15402	0.071628
120	122	295	1	0.110381	0.051334
122	123	295	1	0.187391	0.087148
123	124	295	1	0.187391	0.087148
125	126	295	1	0.038505	0.017907
125	127	295	1	0.169422	0.078791
127	128	295	1	0.02567	0.011938
127	129	295	1	0.064175	0.029845
129	130	295	1	0.20536	0.095504
129	131	295	1	0.361947	0.168327
14	15	295	1	0.097546	0.045365
14	22	295	1	0.035938	0.016713
131	132	295	1	0.120649	0.056109
131	133	295	1	0.17969	0.083566
131	134	295	1	0.2567	0.119381
134	135	295	1	0.033371	0.015519
15	16	295	1	0.15402	0.071628
16	17	295	1	0.15402	0.071628
16	18	295	1	0.019253	0.008954
18	19	295	1	0.17969	0.083566
18	20	295	1	0.019253	0.008954
20	21	295	1	0.02567	0.011938
22	23	295	1	0.351679	0.163551
22	24	295	1	0.248999	0.115799
24	25	295	1	0.23103	0.107442
25	26	295	1	0.071876	0.033427
25	27	295	1	0.392751	0.182652
27	28	295	1	0.10268	0.047752
29	30	295	1	0.038505	0.017907
30	31	295	1	0.071876	0.033427
30	37	295	1	0.056474	0.026264
31	32	295	1	0.123216	0.057303
32	33	295	1	0.087278	0.040589
33	34	295	1	0.192525	0.089535
33	35	295	1	0.071876	0.033427
35	36	295	1	0.351679	0.163551
37	38	295	1	0.10268	0.047752
37	40	295	1	0.318308	0.148032
39	38	309	1	0.46244	0.268858
40	41	295	1	0.41072	0.191009
40	43	295	1	0.403019	0.187427
42	41	295	1	0.12835	0.05969

Table A2. *Cont.*

Bus 1	Bus 2	Imax, A	Type	Resistance, Ω	Reactance, Ω
43	44	295	1	0.084711	0.039396
44	45	295	1	0.33371	0.155195
44	47	295	1	0.189958	0.088342
45	46	295	1	0.02567	0.011938
47	48	295	1	0.238731	0.111024
47	56	295	1	0.338844	0.157582
48	49	295	1	0.17969	0.083566
48	50	295	1	0.238731	0.111024
50	51	295	1	0.288788	0.134303
50	52	295	1	0.040045	0.018623
52	53	295	1	0.07701	0.035814
52	54	295	1	0.10268	0.047752
54	55	295	1	0.23103	0.107442
56	57	295	1	0.05134	0.023876
56	59	295	1	0.659719	0.306808
57	58	295	1	0.10268	0.047752
59	60	295	1	0.043639	0.020295
59	66	295	1	0.084711	0.039396
60	61	295	1	0.02567	0.011938
60	62	295	1	0.074443	0.03462
62	63	295	1	0.441524	0.205335
62	64	295	1	0.028237	0.013132
64	65	295	1	0.17969	0.083566
66	67	295	1	0.17969	0.083566
66	73	295	1	0.467194	0.217273
67	68	295	1	0.10268	0.047752
67	69	295	1	0.120649	0.056109
69	70	295	1	0.02567	0.011938
69	71	295	1	0.210494	0.097892
71	72	295	1	0.02567	0.011938
73	74	295	1	0.02567	0.011938
73	75	295	1	0.467194	0.217273
75	76	295	1	0.64175	0.298451
75	78	295	1	0.021306	0.009909
76	77	295	1	0.30804	0.143257
78	79	295	1	0.15402	0.071628
78	80	295	1	0.084711	0.039396
80	81	295	1	0.02567	0.011938
80	82	295	1	0.12835	0.05969
82	83	295	1	0.10268	0.047752
82	84	295	1	0.189958	0.088342
84	85	295	1	0.23103	0.107442
84	86	295	1	0.2567	0.119381
86	87	295	1	0.05134	0.023876
86	88	295	1	0.5134	0.238761
88	89	295	1	0.07701	0.035814
88	92	295	1	0.084711	0.039396
89	90	295	1	0.02567	0.011938
89	91	295	1	0.02567	0.011938
92	93	295	1	0.5134	0.238761
93	94	295	1	0.02567	0.011938
93	95	295	1	0.084711	0.039396
95	96	295	1	0.07701	0.035814
95	98	295	1	0.12835	0.05969
96	97	295	1	0.053907	0.02507

Table A2. *Cont.*

Bus 1	Bus 2	Imax, A	Type	Resistance, Ω	Reactance, Ω
98	99	295	1	0.084711	0.039396
98	100	295	1	0.064175	0.029845
100	101	295	1	0.07701	0.035814
100	104	295	1	0.12835	0.05969
2	136	1000	2	0.06	0.082
3	137	1000	2	0.011	0.037
5	138	1000	2	0.011	0.037
6	139	1000	2	0.011	0.037
7	140	1000	2	0.011	0.037
10	141	1000	2	0.019	0.036
11	142	1000	2	0.06	0.082
102	195	1000	2	0.04	0.05
103	196	1000	2	0.06	0.082
105	197	1000	2	0.011	0.037
107	198	1000	2	0.06	0.082
109	199	1000	2	0.011	0.037
111	200	1000	2	0.006	0.023
112	201	1000	2	0.011	0.037
115	202	1000	2	0.06	0.082
117	203	1000	2	0.06	0.082
118	204	1000	2	0.06	0.082
119	205	1000	2	0.06	0.082
13	143	1000	2	0.019	0.036
121	206	1000	2	0.011	0.037
122	207	1000	2	0.06	0.082
123	208	1000	2	0.06	0.082
124	209	1000	2	0.019	0.036
126	210	1000	2	0.06	0.082
128	211	1000	2	0.019	0.036
130	212	1000	2	0.019	0.036
132	213	1000	2	0.006	0.023
133	214	1000	2	0.06	0.082
135	215	1000	2	0.019	0.036
15	144	1000	2	0.06	0.082
17	145	1000	2	0.04	0.05
19	146	1000	2	0.06	0.082
21	147	1000	2	0.011	0.037
23	148	1000	2	0.06	0.082
24	149	1000	2	0.06	0.082
26	150	1000	2	0.06	0.082
28	151	1000	2	0.019	0.036
29	152	1000	2	0.06	0.082
31	153	1000	2	0.04	0.05
32	154	1000	2	0.019	0.036
34	155	1000	2	0.019	0.036
35	156	1000	2	0.011	0.037
36	157	1000	2	0.04	0.05
38	158	1000	2	0.011	0.037
39	159	1000	2	0.019	0.036
40	160	1000	2	0.019	0.036
41	161	1000	2	0.04	0.05
42	162	1000	2	0.06	0.082
43	163	1000	2	0.019	0.036
45	164	1000	2	0.006	0.023
46	165	1000	2	0.06	0.082
49	166	1000	2	0.06	0.082
51	167	1000	2	0.06	0.082

Table A2. *Cont.*

Bus 1	Bus 2	Imax, A	Type	Resistance, Ω	Reactance, Ω
53	168	1000	2	0.04	0.05
55	169	1000	2	0.04	0.05
57	170	1000	2	0.019	0.036
58	171	1000	2	0.06	0.082
61	172	1000	2	0.011	0.037
63	173	1000	2	0.06	0.082
64	174	1000	2	0.06	0.082
65	175	1000	2	0.06	0.082
68	176	1000	2	0.06	0.082
70	177	1000	2	0.04	0.05
71	178	1000	2	0.06	0.082
72	179	1000	2	0.06	0.082
74	180	1000	2	0.06	0.082
76	181	1000	2	0.06	0.082
77	182	1000	2	0.04	0.05
79	183	1000	2	0.04	0.05
81	184	1000	2	0.06	0.082
83	185	1000	2	0.04	0.05
85	186	1000	2	0.04	0.05
87	187	1000	2	0.011	0.037
90	188	1000	2	0.011	0.037
91	189	1000	2	0.011	0.037
92	190	1000	2	0.06	0.082
94	191	1000	2	0.04	0.05
96	192	1000	2	0.019	0.036
97	193	1000	2	0.04	0.05
99	194	1000	2	0.019	0.036

References

1. T&D World Magazine. Capacitor Bank Control Adapts to Evolving Challenges of Smart Grid. Available online: https://www.tdworld.com/test-monitor-amp-control/capacitor-bank-control-adapts-evolving-challenges-smart-grid (accessed on 15 July 2019).
2. Cooper Power Series. Smart Grid Ready Capacitor Bank Control Delivers Automation and Efficiency. Available online: https://www.eaton.com/content/dam/eaton/products/utility-and-grid-solutions/grid-automation-systems/capacitor-bank-control/cbc-8000-capacitor-bank-control/cbc-8000-capacitor-bank-pa916001en.pdf (accessed on 15 July 2019).
3. Hogan, P.M.; Rettkowski, J.D.; Bala, J.L. Optimal capacitor placement using branch and bound. In Proceedings of the 37th Annual North American Power Symposium, Ames, IA, USA, 25 October 2005; pp. 84–89.
4. Grainger, J.; Lee, S. Optimum Size and Location of Shunt Capacitors for Reduction of Losses on Distribution Feeders. *IEEE Trans. Power Appar. Syst.* **1981**, *3*, 1105–1118. [CrossRef]
5. Muthukumar, K.; Jayalalitha, S. Optimal placement and sizing of distributed generators and shunt capacitors for power loss minimization in radial distribution networks using hybrid heuristic search optimization technique. *Int. J. Electr. Power Energy Syst.* **2016**, *78*, 299–319. [CrossRef]
6. Segura, S.; Romero, R.; Rider, M.J. Efficient heuristic algorithm used for optimal capacitor placement in distribution systems. *Int. J. Electr. Power Energy Syst.* **2010**, *32*, 71–78. [CrossRef]
7. Biagio, M.A.; Coelho, M.A.; Franco, P.E.C. Heuristic for solving capacitor allocation problems in electric energy radial distribution networks. *Pesquisa Oper.* **2012**, *32*, 121–138. [CrossRef]
8. Bakker, E.; Deljou, V.; Rahmani, J. Optimal Placement of Capacitor Bank in Reorganized Distribution Networks Using Genetic Algorithm. *Int. J. Comp. App. Techn. Res. (IJCATR)* **2019**, *8*, 2319–8656. [CrossRef]
9. Ozgonenel, O.; Karagol, S. A novel generation and capacitor integration technique for today's distribution systems, Turk. *J. Elec. Eng. Comp. Sci.* **2017**, *25*, 2434–2443. [CrossRef]
10. Villa-Acevedo, W.; López-Lezama, J.; Valencia-Velásquez, J. A novel constraint handling approach for the optimal reactive power dispatch problem. *Energies* **2018**, *11*, 2352. [CrossRef]

11. Rahmani-andebili, M. Simultaneous placement of DG and capacitor in distribution network. *Electr. Power Syst. Res.* **2016**, *131*, 1–10. [CrossRef]
12. Bhattacharya, S.K.; Goswami, S.K. A new fuzzy based solution of the capacitor placement problem in radial distribution system. *Expert Syst. Appl.* **2009**, *36*, 4207–4212. [CrossRef]
13. Kumar, M.; Nallagownden, P.; Elamvazuthi, I. Optimal Placement and Sizing of Renewable Distributed Generations and Capacitor Banks into Radial Distribution Systems. *Energies* **2017**, *10*, 811. [CrossRef]
14. Zeinalzadeh, A.; Mohammadi, Y.; Moradi, M.H. Optimal multi objective placement and sizing of multiple DGs and shunt capacitor banks simultaneously considering load uncertainty via MOPSO approach. *Int. J. Electr. Power Energy Syst.* **2015**, *67*, 336–349. [CrossRef]
15. Sharaf, A.M.; El-Gammal, A.A. Optimal selection of capacitors in distribution networks for voltage stabilization and loss reduction. In Proceedings of the 2010 IEEE Electrical Power & Energy Conference, Halifax, NS, Canada, 25–27 August 2010; pp. 1–5.
16. Díaz, P.; Pérez-Cisneros, M.; Cuevas, E.; Avalos, O.; Gálvez, J.; Hinojosa, S.; Zaldivar, D. An improved crow search algorithm applied to energy problems. *Energies* **2018**, *11*, 571. [CrossRef]
17. Montazeri, M.; Askarzadeh, A. Capacitor placement in radial distribution networks based on identification of high potential busses. *Int. Trans. Elec. Energy Syst.* **2019**, *29*, e2754. [CrossRef]
18. Jiang, F.; Zhang, Y.; Zhang, Y.; Liu, X.; Chen, C. An Adaptive Particle Swarm Optimization Algorithm Based on Guiding Strategy and Its Application in Reactive Power Optimization. *Energies* **2019**, *12*, 1690. [CrossRef]
19. Elsheikh, A.; Helmy, Y.; Abouelseoud, Y.; Elsherif, A. Optimal capacitor placement and sizing in radial electric power systems. *Alex. Eng. J.* **2014**, *53*, 809–816. [CrossRef]
20. Arulraj, R.; Kumarappan, N. Optimal economic-driven planning of multiple DG and capacitor in distribution network considering different compensation coefficients in feeder's failure rate evaluation. *Eng. Sci. Technol. Int. J.* **2019**, *22*, 67–77. [CrossRef]
21. Sahli, Z.; Hamouda, A.; Bekrar, A.; Trentesaux, D. Reactive Power Dispatch Optimization with Voltage Profile Improvement Using an Efficient Hybrid Algorithm. *Energies* **2018**, *11*, 2134. [CrossRef]
22. Abdelaziz, A.Y.; Ali, E.S.; Elazim, S.A. Flower pollination algorithm and loss sensitivity factors for optimal sizing and placement of capacitors in radial distribution systems. *Int. J. Electr. Power Energy Syst.* **2016**, *78*, 207–214. [CrossRef]
23. Tamilselvan, V.; Jayabarathi, T.; Raghunathan, T.; Yang, X.S. Optimal capacitor placement in radial distribution systems using flower pollination algorithm. *Alex. Eng. J.* **2018**, *57*, 2775–2786. [CrossRef]
24. Ali, E.S.; Elazim, S.A.; Abdelaziz, A.Y. Improved Harmony Algorithm for optimal locations and sizing of capacitors in radial distribution systems. *Int. J. Electr. Power Energy Syst.* **2016**, *79*, 275–284. [CrossRef]
25. Devabalaji, K.R.; Ravi, K.; Kothari, D.P. Optimal location and sizing of capacitor placement in radial distribution system using bacterial foraging optimization algorithm. *Int. J. Electr. Power Energy Syst.* **2015**, *71*, 383–390. [CrossRef]
26. Kishore, C.; Ghosh, S.; Karar, V. Symmetric fuzzy logic and IBFOA solutions for optimal position and rating of capacitors allocated to radial distribution networks. *Energies* **2018**, *11*, 766. [CrossRef]
27. Khodabakhshian, A.; Andishgar, M.H. Simultaneous placement and sizing of DGs and shunt capacitors in distribution systems by using IMDE algorithm. *Int. J. Electr. Power Energy Syst.* **2016**, *82*, 599–607. [CrossRef]
28. Dixit, M.; Kundu, P.; Jariwala, H.R. Incorporation of distributed generation and shunt capacitor in radial distribution system for techno-economic benefits. *Eng. Sci. Technol. Int. J.* **2017**, *20*, 482–493. [CrossRef]
29. Mouassa, S.; Bouktir, T.; Salhi, A. Ant lion optimizer for solving optimal reactive power dispatch problem in power systems. *Eng. Sci. Technol. Int. J.* **2017**, *20*, 885–895. [CrossRef]
30. Baysal, Y.A.; Altas, İ.H. Power quality improvement via optimal capacitor placement in electrical distribution systems using symbiotic organisms search algorithm. *Mugla J. Sci. Technol.* **2017**, *3*, 64–68. [CrossRef]
31. Helmy, W.; Abbas, M.A.E. Optimal sizing of capacitor-bank types in the low voltage distribution networks using JAYA optimization. In Proceedings of the 9th International Renewable Energy Congress (IREC), Hammamet, Tunisia, 20–22 March 2018; pp. 1–5.
32. Sultana, S.; Roy, P.K. Capacitor Placement in Radial Distribution System Using Oppositional Cuckoo Optimization Algorithm. *Int. J. Swarm Intel. Res. (IJSIR)* **2018**, *9*, 64–95. [CrossRef]
33. Neagu, B.C.; Ivanov, O.; Gavrilaş, M. A comprehensive solution for optimal capacitor allocation problem in real distribution networks. In Proceedings of the Conference on Electromechanical and Power System (SIELMEN), Iaşi, Romania, 11–13 October 2017; pp. 565–570.

34. Blum, C.; Roli, A. Metaheuristics in Combinatorial Optimization: Overview and Conceptual Comparison. *ACM Comp. Surv.* **2003**, *35*, 268–308. [CrossRef]
35. Neagu, B.C.; Ivanov, O.; Georgescu, G. Reactive Power Compensation in Distribution Networks Using the Bat Algorithm. In Proceedings of the International Conference and Exposition on Electrical and Power Engineering, Iasi, Romania, 20–22 October 2016; pp. 711–714.
36. Neagu, B.C.; Ivanov, O.; Gavrilas, M. Voltage profile improvement in distribution networks using the whale optimization algorithm. In Proceedings of the International Conference on Electronics, Computers and Artificial Intelligence (ECAI 2017), Targoviste, Romania, 29 June–1 July 2017.
37. Holland, J.H. *Adaptation in Natural and Artificial Systems: An Introductory Analysis with Applications to Biology, Control and Artificial Systems*; University of Michigan Press: Ann Arbor, MI, USA, 1975.
38. Jebari, K. Selection Methods for Genetic Algorithms. *Int. J. Emerg. Sci.* **2013**, *3*, 333–344.
39. Umbarkar, A.J.; Sheth, P.D. Crossover Operators in Genetic Algorithms: A Review. *ICTACT J. Soft Comput.* **2015**, *6*, 1083–1092.
40. Kennedy, J.; Eberhart, R.C. Particle Swarm Optimization. In Proceedings of the ICNN'95—International Conference on Neural Networks, Perth, WA, Australia, 27 November–1 December 1995; IEEE: Piscataway, NJ, USA, 1995.
41. Yang, X.S. A New Metaheuristic Bat-Inspired Algorithm. In *Nature Inspired Cooperative Strategies for Optimization (NISCO 2010)*; Cruz, C., González, J.R., Krasnogor, N., Pelta, D.A., Terrazas, G., Eds.; Springer: Berlin, Germany, 2010; pp. 65–74.
42. Mirjalili, S.; Lewis, A. The Whale Optimization Algorithm. *Adv. Eng. Soft.* **2016**, *95*, 51–67. [CrossRef]
43. Ebrahimi, A.; Khamehchi, E. Sperm whale algorithm: An effective metaheuristic algorithm for production optimization problems. *J. Nat. Gas. Sci. Eng.* **2016**, *29*, 211–222. [CrossRef]

Article

Better Fuel Economy by Optimizing Airflow of the Fuel Cell Hybrid Power Systems Using Fuel Flow-Based Load-Following Control

Nicu Bizon [1,2,*], Alin Gheorghita Mazare [1], Laurentiu Mihai Ionescu [1], Phatiphat Thounthong [3], Erol Kurt [4], Mihai Oproescu [1], Gheorghe Serban [1] and Ioan Lita [1]

[1] Faculty of Electronics, Communication and Computers, University of Pitesti, 1 Targu din Vale, 110040 Pitesti, Romania
[2] University Politehnica of Bucharest, 313 Splaiul Independentei, 060042 Bucharest, Romania
[3] Renewable Energy Research Centre (RERC), King Mongkut's University of Technology North Bangkok, Bangkok 10800, Thailand
[4] Gazi University, Faculty of Technology, Department of Electrical and Electronics Engineering, 06500 Teknikokullar, Ankara, Turkey
* Correspondence: nicu.bizon@upit.ro

Received: 23 June 2019; Accepted: 16 July 2019; Published: 19 July 2019

Abstract: In this paper, the results of the sensitivity analysis applied to a fuel cell hybrid power system using a fuel economy strategy is analyzed in order to select the best values of the parameters involved in fuel consumption optimization. The fuel economy strategy uses the fuel and air flow rates to efficiently operate the proton-exchange membrane (PEM) fuel cell (FC) system based on the load-following control and the global extremum seeking (GES) algorithm. The load-following control will ensure the charge-sustained mode for the batteries' stack, improving its lifetime. The optimization function's optimum, which is defined to improve the fuel economy, will be tracked in real-time by two GES algorithms that will generate the references for the controller of the boost DC-DC converter and air regulator. The optimization function and performance indicators (such as FC net power, FC electrical efficiency, fuel efficiency, and fuel economy) have a multimodal behavior in dithers' frequency. Furthermore, the optimum in the considered range of frequencies depends on the load level. So, the best value could be selected as the frequency where the optimum is obtained for the most load levels. Considering a dither frequency of 100 Hz selected as the best value, the sensitivity analysis of the fuel economy is further analyzed for different values of the weighting parameter k_{eff}, highlighting the multimodal feature in the parameters for the optimization function and fuel economy as well. A k_{eff} value around of 20 lpm/W seems to give the best fuel economy in the full range of load.

Keywords: proton exchange membrane fuel cell; hydrogen economy; fueling flows control; global extremum seeking; load following; optimization

---

1. Introduction

Instead of a diesel generator [1], the proton-exchange membrane fuel cell (PEM FC) system could be used as a backup green energy source [2] for an FC hybrid power system (FC HPS) to mitigate the load variability by the load-following (LF) control [3–5]. A hybrid energy storage system (ESS) using batteries and ultracapacitors is mandatory to dynamically compensate the power flow balance [6,7]. The most used ESS topologies are the active and semi-active topologies, using two and one bidirectional DC-DC converters integrated into a multiport topology, respectively [8–10]. An active topology with two bidirectional DC-DC converters is more flexible as a control structure compared to a semi-active topology [9]. The two control references will be generated by the energy management strategies

(EMS), usually to regulate the DC voltage and mitigate the load pulses via the bidirectional DC-DC converters for the batteries and ultracapacitors stacks [5,7,11,12]. The EMS has an important role in the optimal and safe operation of the FC system [13,14]. The control objectives for PEM FC system are as follows [15–17]: (1) minimization of the fuel consumption; (2) supplying the dynamic loads with energy, such as FC vehicles; (3) safe operation by using appropriate control loops to mitigate the load pulses, to limit the FC current slope, and to avoid fuel starvation.

The first objective can be ensured using a real-time optimization (RTO) strategy based on different optimization functions, which integrate performance indicators related to fuel consumption such as FC net power, FC lifetime, and cost [18–20].

The equivalent consumption minimization strategy (ECMS) is a well-known strategy applied to FC vehicles, which converts the energy difference in battery charging (at the start and end of a load cycle) into additional fuel consumption, to compensate this loss of energy by discharging the battery [21]. In last decade, several algorithms searching for the global optimum using different optimization functions have been proposed in the literature [22–24]. Intelligent concepts are usually involved in these algorithms [25–27]. In this study, the global extremum seeking (GES) algorithm is used due to the good performance reported in the previous work for FC systems [28–31], photovoltaic (PV) systems [22,23,32], and wind turbine (WT) systems [33,34].

The objective of this paper is to use the sensitivity approach to identify the best value of the parameters used by the optimization function and control loops. Except the tuning parameters of the GES algorithm, which will be designed to ensure the imposed performance and stability of the tracking loops, the dither's frequency f_d is the most important parameter that could dynamically affect the performance of the GES algorithm.

The GES algorithm must search the optimization function's optimum in real-time, which is defined in this study as a weighted function of the FC net power and the fuel consumption efficiency using the weighting parameters k_{net} and k_{eff}. So, if $k_{net} = 1$, then it is important to know the value of the weighting parameter k_{eff}, where the best fuel economy is obtained. So, the sensitivity approach in this study will be performed using the parameters f_d and k_{eff}.

The structure of the paper is as follows. The FC HPS architecture and LF control and optimization loops are briefly presented in Section 2. The EMS based on LF control and optimization loops is detailed in Section 3. The obtained results are presented and discussed in Section 4. Section 5 concludes this study.

2. FC Hybrid Power System

The synoptic architectures of the FC HPS and ESS are presented in Figure 1a. The FC power delivered on the DC can be controlled via the boost converter using the switching (SW) command generated by the boost controller under the EMS proposed in this paper. Also, the generated FC power depends on the fuel and air regulators controlled by the input references $I_{refFuel}$ and I_{refAir}.

The Matlab-Simulink®diagram of the FC HPS architecture is presented in Figure 1b. The FC system is the main energy source to supply the load demand on the DC bus, which is modeled by an equivalent DC load for the inverter and an AC load in order to speed up the simulation.

The LF control will set the current reference I_{refLF} for the fuel flow rate (*FuelFr*) regulator to comply with the power flow balance on the DC bus (1), with the battery operating in charge-sustaining mode (2):

$$C_{DC}u_{dc}du_{dc}/dt = \eta_{boost}p_{FCgen} + p_{ESS} - p_{load};$$ (1)

$$P_{ESS(AV)} \cong 0.$$ (2)

So, considering (2), (1) will be rewritten in the average value (AV) during a load cycle (LC) as (3):

$$0 = \eta_{boost}P_{FCgen(AV)} - P_{load(AV)} \Rightarrow P_{FCgen(AV)} = P_{load(AV)}/\eta_{boost}.$$ (3)

Thus, the FC current generated by the FC system will be given by (4):

$$I_{FC(AV)} = P_{load(AV)} / \left(\eta_{boost} V_{FC(AV)} \right). \tag{4}$$

Consequently, the current reference I_{refLF} will be estimated by the LF control using the AV filtering method, based on the mean value (MV) technique of the FC current during a dithers' period:

$$I_{refLF} \cong I_{FC(AV)} = P_{load(AV)} / \left(\eta_{boost} V_{FC(AV)} \right) \tag{5}$$

The MV technique or another filtering technique can be used to smooth the values used in (5) for the safe operation of the FC system, thus avoiding fuel starvation due to sharp changes in load demand. The current reference I_{refLF} will set the fuel flow rate *FuelFr* value using the *FuelFr* regulator's relationship (6):

$$FuelFr = \frac{60000 \cdot R \cdot (273 + \theta) \cdot N_C \cdot I_{LFref}}{2F \cdot (101325 \cdot P_{f(H2)}) \cdot (U_{f(H2)}/100) \cdot (x_{H2}/100)}. \tag{6}$$

Because of (5),

$$I_{FC} \cong I_{FC(AV)} \cong I_{refLF}. \tag{7}$$

So, the current reference I_{ref2} will optimally adjust the air flow rate (*AirFr*) value using the *AirFr* regulator's relationship (8):

$$AirFr = \frac{60000 \cdot R \cdot (273 + \theta) \cdot N_C \cdot \left(I_{FC} + I_{ref2} \right)}{4F \cdot (101325 \cdot P_{f(O2)}) \cdot (U_{f(O2)}/100) \cdot (y_{O2}/100)}. \tag{8}$$

Thus, the input references of the fueling regulators are set as follows: $I_{refFuel} = I_{refLF}$ and $I_{refAir} = I_{FC} + I_{ref2}$.

The optimization function's optimum (computed in the block called "optimization function", in Figure 1) will be searched using the current reference I_{ref1} that will set the FC current based on the 0.1 A hysteresis controller for the boost DC-DC converter.

A semi-active 100 Ah battery/100 F ultracapacitors ESS topology was chosen to dynamically compensate power in (1) and stabilize the DC voltage at 200 V ($u_{DC} \cong 200$ V). The battery will operate in charge-sustaining mode due to the LF control applied to the fuel regulator (6), which will set the FC power close to the requested FC power on the DC bus. The FC power is $P_{FCgen(AV)}$ given by (3) if (2) is considered. Thus, the battery will compensate the minor imbalance in energy flow on the DC bus (1) due to the changes in load profile or the small difference between the FC current set by the LF control (5) $I_{FC} \cong I_{LFref} \cong I_{FC(AV)} = P_{load(AV)} / \left(\eta_{boost} V_{FC(AV)} \right)$ and the value $I_{FC} + I_{ref2}$ set by the GES-based optimization loop (which is applied to the air regulator (7)). So, a 100 Ah capacity of the battery has been obtained for a 1 kW imbalance in energy flow during a 12 s load cycle using the design relationships from [5,7]. This capacity is more than sufficient, considering the sizing design presented in [2]. The sharp changes in load profile, such as pulses, can produce an imbalance in power flow on the DC bus (1) due to the large response time of the FC system and the battery, which is in the order of hundreds of milliseconds and tens of minutes, respectively. During the transitory regimes of the FC system and battery, the ultracapacitors' stack will dynamically compensate the power flow on the DC bus (1) via the bidirectional DC-DC power converter. The 100 F capacitance was obtained for a 1 kW/100 ms pulse using the design relationships from [5,7]. The DC voltage regulation was implemented on the ultracapacitors' stack side due to the slow response of the FC system, but also because the controlled inputs *AirFr* and *FuelFr* (via the fueling regulator) of the FC system and the controller of the FC boost converter are involved in the LF and optimization loops. Thus, if the DC voltage regulation will be implemented on the FC system's side using a proportional–integral–derivative (PID) compensation technique of one of the aforementioned loops, then a degradation of the performance indicators will be obtained compared with the cases when the DC voltage regulation is implemented on the ultracapacitors' stack side or on the battery's side if an active ESS topology is selected instead of a semi-active type. If an active mitigation of the pulses on the DC bus must be implemented, then an active control of the bidirectional DC-DC power converter should be implemented in order to generate an anti-pulse from the ultracapacitors' stack [5,7]. In this case, the DC voltage regulation remains to be implemented on the battery side using a PID compensation technique of the control designed to compensate the minor imbalance in energy flow on the DC bus (1). The high frequency ripple on the DC voltage will be mitigated in all cases by a 100 µF capacitor (C_{DC}) designed using the design relationships from [5,7].

The FC parameters (N_C, θ, $U_{f(H2)}$, $U_{f(O2)}$, $P_{f(H2)}$, $P_{f(O2)}$, x_{H2}, y_{O2}) were set for a 6 kW FC system, and $R = 8.3145$ J/(mol K) and $F = 96485$ As/mol are two well-known constants. The energy efficiency of the boost DC-DC converter may be considered constant in this study ($\eta_{boost} = 0.9$), because the boost controller operated in continuous current mode in the load range higher than 1 kW (so the case of light load was not considered), where the energy efficiency may vary in the range of 88% to 92%, but without significantly modifying the results obtained for a constant efficiency (as will be explained in the Discussion section).

The current references I_{ref1} and I_{ref2} were generated by the GES controller shown in Figure 2, as will be explained in next section.

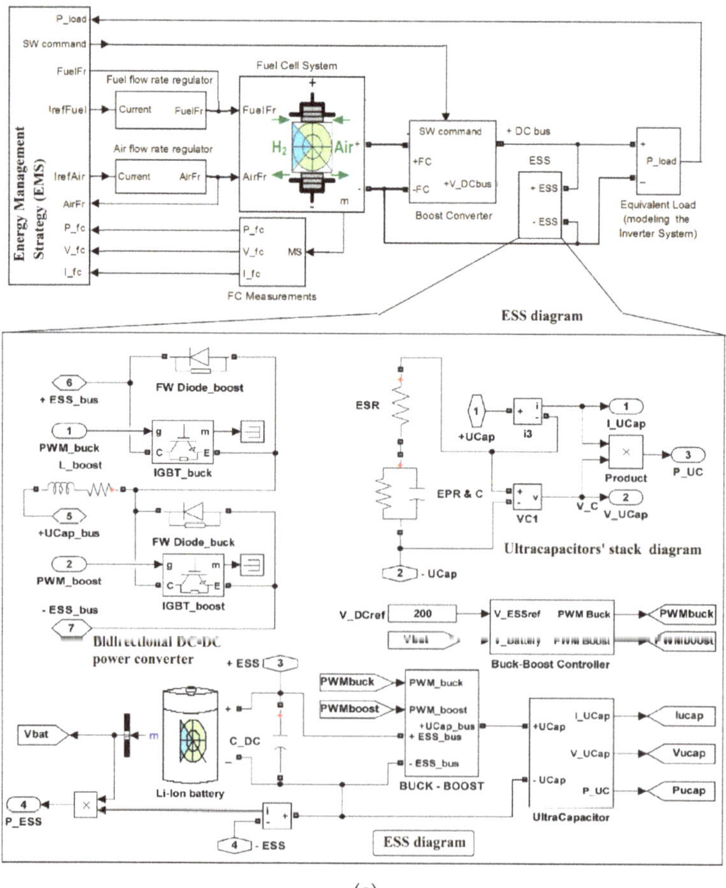

(a)

Figure 1. *Cont.*

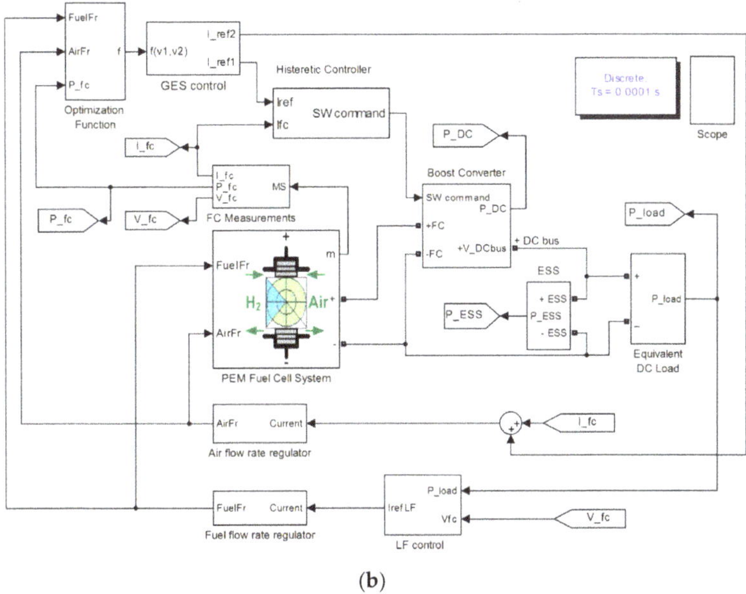

(b)

Figure 1. (**a**) The synoptic architectures of the fuel cell hybrid power system (FC HPS) and energy storage system (ESS). (**b**) The diagram of the FC HPS.

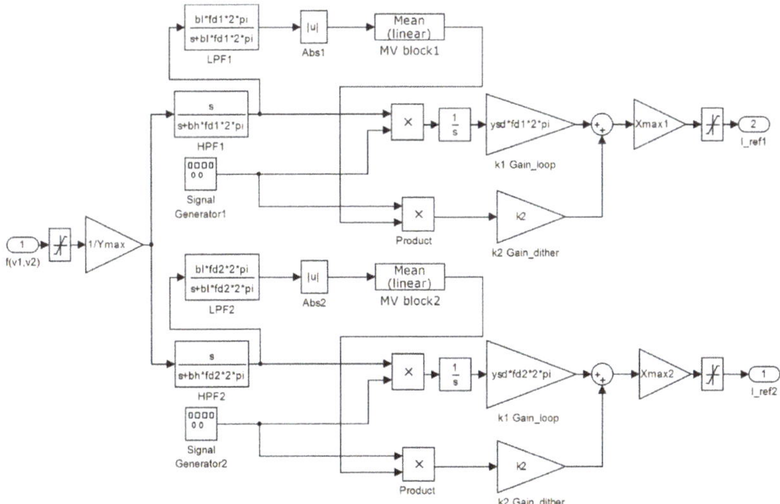

Figure 2. The diagram of the global extremum seeking (GES) control.

3. Energy Management Strategy

Each GES control was implemented (see Figure 2) based on the relationships from [30]. The dither's frequencies were chosen to be $f_{d1} = 100$ Hz and $f_{d2} = 2f_{d1} = 200$ Hz in order to improve the dithers' persistence by interference of the harmonics due to the nonlinear characteristics of the FC system. The cut off frequencies of the low-pass filter (LPF1 and LPF2 in Figure 2) and the high-pass filter (HPF1 and HPF2 in Figure 2) were set by the parameters $b_l = 1.5$ and $b_h = 0.1$. The tuning parameters were designed based on relationships presented in [30], as follows: $k_1 = 1$, $k_2 = 2$. The normalization values of the input (the optimization function $y = f(v_1, v_2)$, where v_1 and v_2 are the search variables) and the outputs (the reference currents I_{ref1} and I_{ref2} for the boost controller and the air regulator) were set to $k_{Nf} = 1/Y_{max} = 1/1000$, $k_{Nv1} = X_{max1} = 50$ and $k_{Nv2} = X_{max2} = 20$. The normalization values were not strict, because the GES control was of the adaptive type, but choosing the right value would improve the searching speed. Readers interested in analyzing and designing a GES control for FC systems or PV, WT, and PV/WT FC HPSs may read [14,17] or [22,23,32], [33], and [35], respectively.

The searching speed was limited to 100 A/s by the slope limiters included in the *FuelFr* and *AirFr* regulators in order to ensure the safe operation of the PEM FC system. So, the FC current could increase up to 20 A during the FC time constant (T_{FC}) of 0.2 s. The GES algorithm needed about 10 periods of dither (T_d) to find the Global Maximum Power Point (GMPP) of the PV system for different steps in irradiance (because there was no limitation related to searching slope in this case, excepting the safe values given by the devices used in the boost the boost DC-DC power converter, which were very high compared to 100 A/s) [22,23]. Considering 10 dither periods to find the maximum efficiency point (MEP) of the FC system or the optimization function's optimum (which is defined in relation to the FC system's performance indicators), to avoid limitation due to FC response time, it was recommended to choose:

$$T_{d(max)} \cong T_{FC}/10. \tag{9}$$

This means $T_{d(max)} \cong 20$ ms, so $f_{d(min)} \cong 50$ Hz. The maximum frequency of the dither would be considered $f_{d(max)} = 200$ Hz in order to have an acceptable increment per dithers' period (about 0.5 A per $T_{d(min)} = 5$ ms). So, the frequency range of the dither was 70 Hz to 220 Hz, with a 30 Hz step in evaluation of the fuel economy based on the optimization function f defined by (10) [35]:

$$f(x, AirFr, FuelFr, P_{Load}) = k_{net} \cdot P_{FCnet} + k_{fuel} \cdot Fuel_{eff}; \tag{10}$$

$$\dot{x} = g(x, AirFr, FuelFr, P_{Load}), \; x \in X; \tag{11}$$

where (11) models the dynamic part of the FC HPS [36], x is the state vector, and P_{Load} is the disturbance. The weighting coefficients k_{net} (1/W) and k_{fuel} (liters per minute (lpm)/W) were defined in accordance with the chosen EMS objective. For example, the FC system energy efficiency ($\eta_{sys} = P_{FCnet}/P_{FC}$) was maximized if $k_{net} = 1$ and $k_{fuel} = 0$ (in this case $f = P_{FCnet}$). If $k_{net} = 1$ and $k_{fuel} \neq 0$, then the fuel consumption efficiency ($Fuel_{eff} \cong P_{FCnet}/FuelFr$) was also considered in the optimization function. So, it was possible to improve the fuel economy for a value of the k_{fuel} weighting coefficient in range 5 lpm/W to 50 lpm/W, with 5 lpm/W step in evaluation of the fuel economy.

The LF control was implemented based on (5) in order to set the value of the FC net power requested by power flow balance (3). So,

$$P_{FCgen} = P_{FCnet} \cong P_{load}/\eta_{boost}, \tag{12}$$

where,

$$P_{FCnet} \cong P_{FC} - P_{cm}; \tag{13}$$

$$P_{cm} = I_{cm} \cdot V_{cm} = \left(a_2 \cdot AirFr^2 + a_1 \cdot AirFr + a_0\right) \cdot \left(b_1 \cdot I_{FC} + b_0\right). \tag{14}$$

The air compressor power (P_{cm}) was estimated with (14), considering the coefficients [37]: $a_0 = 0.6$, $a_1 = 0.04$, $a_2 = -0.00003231$, $b_0 = 0.9987$, and $b_1 = 46.02$.

In order to not exceed the maximum FC power, the range of the load demand was considered from 2 kW to 8 kW, with a 1 kW step in the evaluation of the fuel economy.

The total fuel consumption, $Fuel_T = \int FuelFr(t)dt$, will be estimated in the next section for different load levels in order to evaluate the fuel economy, measured in liters (L).

4. Results

The values of the performance indicators P_{FCnet}, η_{sys}, $Fuel_{eff}$, and $Fuel_T$ using different dither's frequencies and constant load levels are recorded in Tables 1–4. The values obtained by simulation using the static feed-forward (sFF) strategy [36], which is considered in this study as a reference strategy because it is the most known strategy implemented in commercial FC systems, are mentioned in the first column of these Tables. The differences in the performance indicators will be defined compared to the sFF strategy by using (15):

$$\Delta P_{FCnet} \cong P_{FCnet} - P_{FCnet0}; \tag{15a}$$

Table 1. FC net power for different dithers' frequencies and constant load levels.

Load Level	sFF Strategy	Dithers' Frequency f_d (Hz)					
P_{load} (kW)	P_{FCnet0} (kW)	70	100	130	160	190	220
2	1942	1895	1903	1889	1922	1900	1916
3	2880	2843	2848	2836	2805	2837	2817
4	3773	3723	3680	3711	3673	3710	3704
5	4638	4526	4601	4580	4566	4584	4557
6	5437	5315	5323	5337	5385	5346	5333
7	6188	6107	6130	6081	6116	6147	6125
8	6841	6813	6807	6805	6840	6827	6820

Table 2. FC electrical efficiency for different dithers' frequencies and constant load levels.

Load Level	sFF Strategy	Dithers' Frequency f_d (Hz)					
P_{load} (kW)	η_{sys0} (%)	70	100	130	160	190	220
2	93.17	90.66	90.47	90.93	92.24	92.77	92.34
3	91.45	90.25	91.44	89.89	90.67	91.62	91.22
4	90.24	90.24	88.77	89.77	88.82	90.42	89.9
5	88.52	89.73	89.18	88.84	88.9	88.92	88.77
6	86.55	86.75	85.51	86.41	87.15	87.62	87.3
7	84.37	85.33	85.6	86.39	86.21	87.36	85.81
8	82	84.02	83.86	83.85	83.85	83.66	83.7

Table 3. Fuel efficiency for different dithers' frequencies and constant load levels.

Load Level	sFF Strategy	Dithers' Frequency f_d (Hz)					
P_{load} (kW)	$Fuel_{eff0}$ (W/lpm)	70	100	130	160	190	220
2	136.1	134.3	133.4	133.8	136.2	136.7	136.3
3	128.3	128.9	129.7	129.9	128.9	129.6	129.5
4	119.5	122	121.1	122.1	121.3	122.9	122.4
5	111.6	116.8	115.4	115.1	115.4	115.2	115.5
6	102.6	108.1	107.5	107.1	107.5	108.7	108.4
7	92.65	99.94	100	101.9	101.3	102.5	100.5
8	81	92.3	92.27	92.21	91.31	91.45	91.67

Table 4. Total fuel consumption for different dithers' frequencies and constant load levels.

Load Level	sFF Strategy	Dithers' Frequency f_d (Hz)					
P_{load} (kW)	$Fuel_{T0}$ (L)	70	100	130	160	190	220
2	34	33.46	33.53	33.12	33.44	32.95	33.3
3	54.75	51.15	51.6	51.41	51.25	51.46	51.21
4	76.88	72.2	71.24	71.28	70.72	70.51	70.59
5	98.57	86.4	92.47	90.87	90.67	91.05	90.91
6	125.5	112.2	114.2	114.7	115.2	110.5	111.2
7	152.5	136.7	138	127.4	127.9	127.6	131.3
8	193	167	160.1	161.3	163.6	165.7	165.7

$$\Delta\eta_{sys} = \eta_{sys} - \eta_{sys0}; \tag{15b}$$

$$\Delta Fuel_{eff} = Fuel_{eff} - Fuel_{eff0}; \tag{15c}$$

$$\Delta Fuel_T = Fuel_T - Fuel_{T0}. \tag{15d}$$

The differences are recorded in Tables 5–8 and represented in Figures 3–6. Note the multimodal behavior in dithers' frequency for all performance indicators. Also, it is worth mentioning that the optimum's position (maximum of ΔP_{FCnet}, $\Delta\eta_{sys}$, $\Delta Fuel_{eff}$, and minimum of $\Delta Fuel_T$) depends on the load level. So, the best value in the frequencies' range could be selected as the frequency where the optimum is obtained for most of load levels, and this seems to be the dither frequency of 100 Hz.

Table 5. Differences in FC net power compared to reference.

Load Level	Dithers' Frequency f_d (Hz)					
P_{load} (kW)	70	100	130	160	190	220
2	−47	−39	−53	−20	−42	−26
3	−37	−32	−44	−75	−43	−63
4	−50	−93	−62	−100	−63	−69
5	−112	−37	−58	−72	−54	−81
6	−122	−114	−100	−52	−91	−104
7	−81	−58	−107	−72	−41	−63
8	−28	−34	−36	−1	−14	−21

Table 6. Differences in FC electrical efficiency compared to reference.

Load Level	Dithers' Frequency f_d (Hz)					
P_{load} (kW)	70	100	130	160	190	220
2	−2.51	−2.7	−2.24	−0.93	−0.4	−0.83
3	−1.2	−0.01	−1.56	−0.78	0.17	−0.23
4	0	−1.47	−0.47	−1.42	0.18	−0.34
5	1.21	0.66	0.32	0.38	0.4	0.25
6	0.2	−1.04	−0.14	0.6	1.07	0.75
7	0.96	1.23	2.02	1.84	2.99	1.44
8	2.02	1.86	1.85	1.85	1.66	1.7

Table 7. Differences in fuel efficiency compared to reference.

Load Level	Dithers' Frequency f_d (Hz)					
P_{load} (kW)	−1.8	−2.7	−2.3	0.1	0.6	0.2
2	0.6	1.4	1.6	0.6	1.3	1.2
3	2.5	1.6	2.6	1.8	3.4	2.9
4	5.2	3.8	3.5	3.8	3.6	3.9
5	5.5	4.9	4.5	4.9	6.1	5.8
6	7.29	7.35	9.25	8.65	9.85	7.85
7	11.3	11.27	11.21	10.31	10.45	10.67
8	2.02	1.86	1.85	1.85	1.66	1.7

Table 8. Differences in total fuel consumption compared to reference.

Load Level	Dithers' Frequency f_d (Hz)					
P_{load} (kW)	−0.54	−0.47	−0.88	−0.56	−1.05	−0.7
2	−3.6	−3.15	−3.34	−3.5	−3.29	−3.54
3	−4.68	−5.64	−5.6	−6.16	−6.37	−6.29
4	−12.17	−6.1	−7.7	−7.9	−7.52	−7.66
5	−13.3	−11.3	−10.8	−10.3	−15	−14.3
6	−15.8	−14.5	−25.1	−24.6	−24.9	−21.2
7	−26	−32.9	−31.7	−29.4	−27.3	−27.3
8	−0.54	−0.47	−0.88	−0.56	−1.05	−0.7

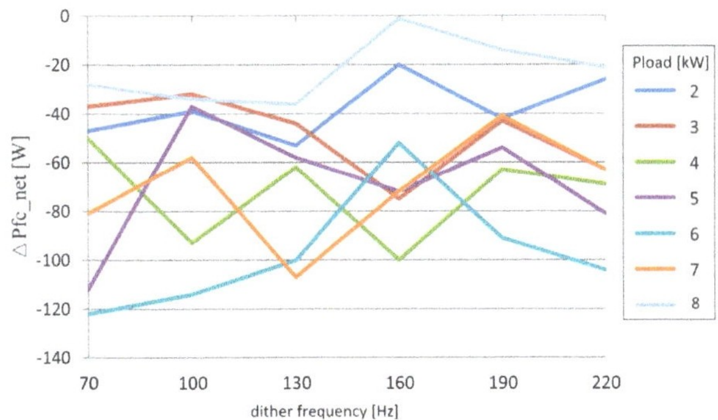

Figure 3. Differences in FC net power.

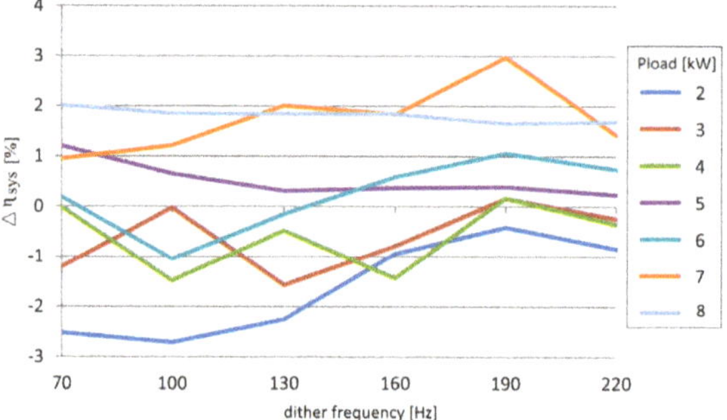

Figure 4. Differences in FC electrical efficiency.

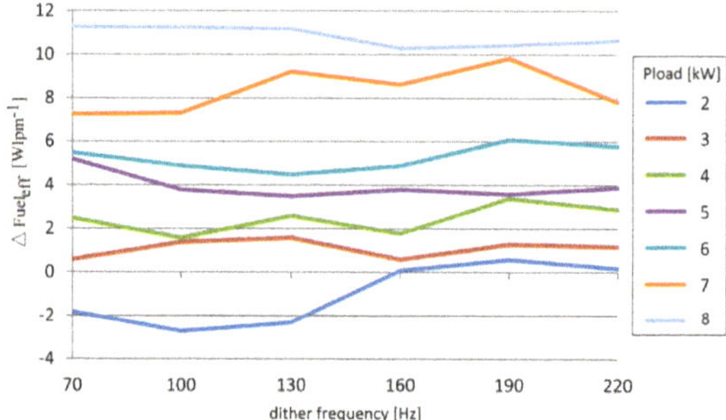

Figure 5. Differences in fuel efficiency.

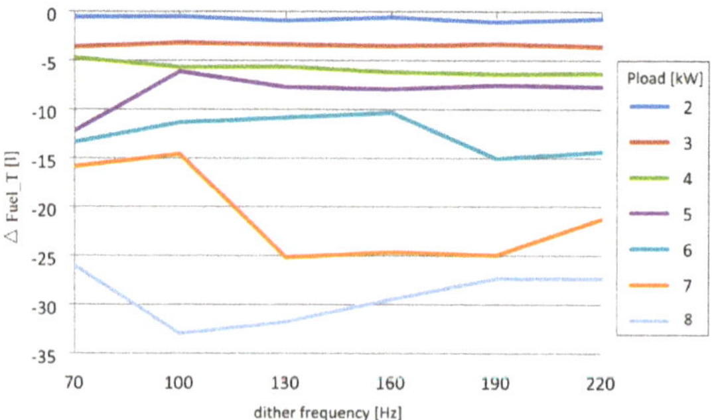

Figure 6. Differences in total fuel consumption.

Considering a dither frequency of 100 Hz, the total fuel consumption ($Fuel_T$) for different values of the parameters k_{eff} and P_{load} is recorded in Table 9. The values for $k_{eff} = 0$ (mentioned in the first column of the Table 9) are used as reference values. So, the differences in total fuel consumption ($\Delta Fuel_T$) are estimated in Table 10 and represented in Figure 7.

Table 9. Total fuel consumption for different values of the parameters k_{eff} and P_{load}.

Load Level	Weighting Parameter k_{eff} (lpm/W)									
P_{load} (kW)	0	5	10	15	20	25	30	35	40	45	50
2	34	33.74	33.72	33.53	33.51	33.53	33.54	33.91	33.45	33.49	33.76
3	54.75	51.68	51.67	52.55	51.61	51.6	51.69	51.56	51.56	51.54	51.76
4	76.88	71.5	71.47	71.37	71.25	71.24	71.1	71.11	70.98	71.37	71.63
5	99.7	92.39	92.38	92.65	92.13	92.47	92.71	92.7	92.5	92.59	92.42
6	125.5	113.8	114.3	114.2	114	114.2	114.6	114.7	114.8	114.6	114.8
7	152.5	132.1	136.7	137	137.5	138.1	138	138.7	138.8	139	139.4
8	193	155.8	156.6	158	158.9	160.1	160	161	161	162	163.4

Table 10. Differences in total fuel consumption compared to $k_{eff} = 0$.

Load Level	Weighting Parameter k_{eff} (lpm/W)								
P_{load} (kW)	5	10	15	20	25	30	35	40	45	50
2	−0.26	−0.28	−0.47	−0.49	−0.47	−0.46	−0.09	−0.55	−0.51	−0.24
3	−3.07	−3.08	−2.2	−3.14	−3.15	−3.06	−3.19	−3.19	−3.21	−2.99
4	−5.38	−5.41	−5.51	−5.63	−5.64	−5.78	−5.77	−5.9	−5.51	−5.25
5	−7.31	−7.32	−7.05	−7.57	−7.23	−6.99	−7	−7.2	−7.11	−7.28
6	−11.7	−11.2	−11.3	−11.5	−11.3	−10.9	−10.8	−10.7	−10.9	−10.7
7	−20.4	−15.8	−15.5	−15	−14.4	−14.5	−13.8	−13.7	−13.5	−13.1
8	−37.2	−36.4	−35	−34.1	−32.9	−33	−32	−32	−31	−29.6

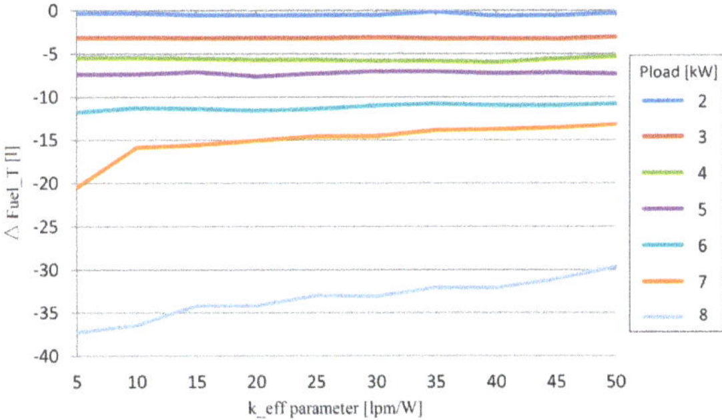

Figure 7. Differences in total fuel consumption for different values of the weighting parameter k_{eff}.

The sensitivity analysis of the fuel economy ($\Delta Fuel_T$) highlights the better fuel economy with increase in load level, and this is normal. Also, note the multimodal behavior in the weighting parameter k_{eff}. This is better shown in Figure 8 (where the high values of fuel economy for a load of 7 kW and 8 kW are canceled). Looking to Table 10 (where the optimum, local minimums, and the minimums at $k_{eff} = 5$ and $k_{eff} = 50$ are highlighted in different colors: yellow, blue, and gray, respectively), a k_{eff} value in the range of 20 lpm/W to 30 lpm/W seems to give the best fuel economy in the load range of 2 kW to 5 kW. However, note the decrease in fuel economy with the increase in k_{eff} value. So, the recommended value for the entire load range is $k_{eff} = 20$ lpm/W.

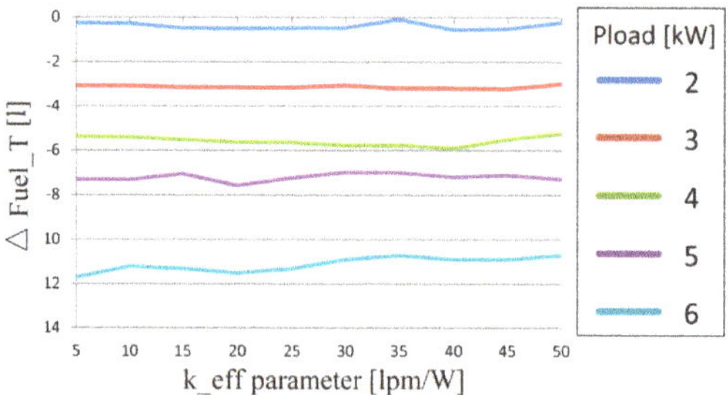

Figure 8. The multimodal behavior of the fuel economy $\Delta Fuel_T$ in weighting parameter k_{eff}.

The effect of a variable energy efficiency of the boost converter on the results obtained at constant energy efficiency will be analyzed and discussed in the next section.

5. Discussion

The dependence of the power loss (P_{loss}) and energy efficiency for a DC-DC power converter were analyzed for low and medium power applications in [38,39] and [40–42], respectively. The main findings of the aforementioned studies are as follows:

1)   The energy efficiency characteristic related to the load current (I_{load}) is dependent on the control mode used, the switching frequency, and coil inductance [42];

2)   For low [38] and high [39] power applications using the control mode based on the pulse width modulation technique and optimal control, respectively, the energy efficiency can be approximated by (16):

$$\eta_{converter} = \eta_{max} - \frac{\eta_{max} - \eta_{min}}{4} \cdot \lg^2\left(\frac{I_{load}}{I_{load(opt)}}\right), \tag{16}$$

where η_{max} is the maximum of the energy efficiency (obtained at the optimal load current $I_{load(opt)}$) and η_{min} is the minimum of the energy efficiency (obtained in the considered load range).

3)   For low-power applications using the control mode based on the pulse frequency modulation technique, the energy efficiency can be approximated by (17) [38,39]:

$$\eta_{converter} = \eta_{max} - \frac{\zeta_{\eta}}{I_{load}}, \tag{17}$$

where η_{max} is the maximum energy efficiency (obtained at the maximum load current $I_{load(max)}$) and ζ_{η} is a parameter (that must be determined using the experimental values in the considered range of load). Relationship (17) highlights the nonlinear increase in energy efficiency in the range of light loads and the saturation that appears in the rest of the load range;

4)   For most types of control used in medium and high-power applications, the energy efficiency in the normal load range (therefore, except for light loads), where the converter operates in continuous current mode [43,44], can be considered as constant or linearly increasing (18):

$$\eta_{converter} = \eta_{min} + \chi_{\eta} \cdot \frac{I_{load}}{I_{load(max)}}, \tag{18}$$

where η_{min} is the energy efficiency obtained at the load current $I_{load(min)}$, which is the upper limit of light loads, and χ_{η} is a parameter to be determined using the experimental values in the considered load range (except the light loads).

The assumption that the energy efficiency linearly increases is valid for the medium-power FC HPS analyzed in this paper, because the load range was higher than 1 kW (so the case of light load was not considered). For different values of the load current, the LF control and optimization loops set the values of the FC current, I_{FC0}, and I_{FC1}, using the sFF strategy and the fuel economy strategy analyzed in this paper. So, (18) can be rewritten using as a variable the FC current as (19):

$$\eta_{boost} = \eta_{\min} + K_\eta \cdot \frac{I_{FC}}{I_{FC(\max)}}, \tag{19}$$

where $\eta_{min} \cong 88.5\%$ is the energy efficiency obtained at the nominal FC current, $I_{FC(min)} \cong 30$ A, and $K_\eta = 4$ is a parameter which has been determined using the experimental values of the energy efficiency $\eta_{max} \cong 92\%$ obtained at the maximum FC current, $I_{FC(max)} \cong 240$ A. Note that the energy efficiency obtained at the nominal FC current, $I_{FC(nom)} \cong 130$ A, was $\eta_{nom} \cong 90.17\%$ (which is very close to the constant value considered in simulation).

The power loss of the boost converter (P_{loss}) was estimated using (20):

$$P_{loss} = P_{load} \cdot \left(\frac{1}{\eta_{boost}} - 1 \right). \tag{20}$$

The FC current, I_{FC0}, and I_{FC1}, using the sFF strategy and the fuel economy strategy, are registered in the second and third columns of Table 11 for different load levels and f_d =100 Hz. The energy efficiency (η_{boost0} and η_{boost1}) and the power loss of the boost converter (P_{loss0} and P_{loss1}) were estimated for the sFF strategy and the fuel economy strategy with (19) and (20), and are registered in Table 11. The difference in power loss of the boost converter (ΔP_{loss}) is registered in the eighth column of Table 11. The influence of ΔP_{loss} on ΔP_{FCnet} for f_d =100 Hz was estimated as $\Delta P_{loss}/\Delta P_{FCnet}$ (%) and is registered in the last column of Table 11. As expected, the biggest error of 0.10485% was obtained at the maximum load.

Table 11. Influence of variable energy efficiency on the FC net power estimated for $f_d = 100$ Hz.

Load Level	sFF Strategy	Fuel Economy Strategy	Dithers' Frequency f_d = 100 Hz					
P_{load} (kW)	I_{FC0} (A)	I_{FC1} (A)	η_{boost0} (%)	η_{boost0} (%)	P_{loss0} (kW)	P_{loss1} (kW)	ΔP_{loss} (kW)	$\Delta P_{loss}/\Delta P_{FCnet}$ (%)
2	36.62	36.59	89.1103	89.1098	0.24441	0.24442	0.00001	0.00003
3	58.95	58.29	89.4825	89.4715	0.35261	0.35302	0.00041	0.00129
4	82.62	77.78	89.8770	89.7963	0.45053	0.45452	0.00400	0.00430
5	108.1	105.2	90.3017	90.2533	0.53700	0.53996	0.00297	0.00801
6	138.9	126	90.8150	90.6000	0.60684	0.62252	0.01568	0.01375
7	173	149.1	91.3833	90.9850	0.66004	0.69358	0.03354	0.05782
8	193	170.6	91.7167	91.3433	0.72251	0.75817	0.03565	0.10485

It is worth mentioning that the biggest differences mentioned in Table 4, Table 6, and Table 8 were less than 0.2% for variable energy efficiency compared to the constant efficiency and also, this value was obtained at maximum load. So, the conclusions of this study are valid for both constant and variable energy efficiency.

6. Conclusion

The sensitivity analysis of the dither's frequency f_d and weighting parameter k_{eff} was performed in this study in order to identify the best value of these parameters, which can be used to improve the fuel economy of an FC HPS. For this, the FC HPS was modeled, and the optimization and control loops of the considered strategy were designed.

The main findings of this study are as follows:

- Firstly, the sensitivity analysis of the dither's frequency f_d revealed that a value of 100 Hz is recommended to improve the performance indicators, such as P_{FCnet}, η_{sys}, $Fuel_{eff}$, and $Fuel_T$;

-   Secondly, the sensitivity analysis of the fuel economy ($\Delta Fuel_T$) in weighting parameter k_{eff} was performed for a 100 Hz dither; a k_{eff} value in the range of 20 lpm/W to 30 lpm/W gave the best fuel economy in the load range of 2–6 kW, but for a load of 6–8 kW, the fuel economy was better with a decrease in k_{eff}. So, k_{eff} = 20 lpm/W is recommended to improve the fuel economy in the full range of load.
-   Thirdly, a better fuel economy with an increase in load level has been highlighted.

Subsequent works will focus on comparing the performance of this strategy (using the load-following for the fuel regulator and the air optimization) with other strategies (for example, with the strategy which considers the fuel optimization and the load-following mode for the air regulator). But first, a sensitivity analysis for both f_d and k_{eff} parameters will be performed for the new strategies, to validate the recommended values of 100 Hz and 20 lpm/W obtained in this study.

Experimental tests have been performed for the first strategies (such as [3,4]) proposed in the research grant mentioned in the Acknowledgments section, but these will continue for recently proposed advanced strategies [14,45–47], including the strategy detailed in this paper.

Author Contributions: Methodology and Writing—Original Draft Preparation: B.N.; Validation and Supervision: N.B., P.T., E.K.; Formal analysis: N.B., G.S., I.L.; Writing—Review and Editing: N.B., A.G.M., L.M.I., M.O.

Funding: This research was funded by Ministry of National Education and Scientific Research, Romania, grant numbers PN-III P1-1.2-PCCDI2017-0332 and PN-III P2-2.1-PED-2016-1223.

Acknowledgments: This work was supported by a grant of the Ministry of National Education and Scientific Research, Romania, CNCS/CCCDI-UEFISCDI, within PNCDI III, code PN-III P1-1.2-PCCDI2017-0332 and title "Increasing the institutional capacity of bioeconomic research for the innovative exploitation of the indigenous vegetal resources in order to obtain horticultural products with high added value", and within PNCDI III, code PN-III P2-2.1-PED-2016-1223, number #53PED and title "Experimental validation of a propulsion system with hydrogen fuel cell for a light vehicle - Mobility with Hydrogen Demonstrator". The sponsors had no role in the design, execution, interpretation, or writing of the study. The research was performed in the Research Center "Modeling and Simulation of the Systems and Processes".

Conflicts of Interest: The authors declare no conflict of interest

Nomenclature

AirFr	Air Flow rate
AV	Average value
f_d	Dither frequency
EMS	Energy Management Strategy
ECMS	Equivalent Consumption Minimization Strategy
ES	Extremum Seeking
ESS	Energy Storage System
FuelFr	Fuel Flow rate
FC	Fuel cell
P_{FC}	FC stack power
P_{Fcne}	FC net power
P_{cm}	Air compressor power
η_{sy}	FC electrical efficiency
FCHPS	Fuel Cell Hybrid Power System
$Fuel_T$	Total Fuel Consumption
$Fuel_{eff}$	Fuel Consumption Efficiency
GES	Global Extremum Seeking

HPS	Hybrid Power System
k_{Nv1} and k_{Nv1}	Output normalization gains
k_{Nf}	Input normalization gain
I_{LFref}	Load-following reference
I_{ref1} and I_{ref2}	GES references
LF	Load-following
LC	Load cycle
LPF	Low-pass filter
HPF	High-pass filter
MEP	Maximum Efficiency Point
MPP	Maximum Power Point
GMPP	Global Maximum Power Point
MV	Mean Value
PV	Photovoltaic
PEMFC	Proton Exchange Membrane Fuel
P_{load}	Stationary load power (constant power demand)
p_{load}	Dynamic load power (variable power demand)
RTO	Real-Time Optimization
sFF	Static Feed-Forward
k_{fuel}	Weighting coefficient of the fuel consumption efficiency
WT	Wind Turbine

References

1. Jamshidi, M.; Askarzadeh, A. Techno-economic analysis and size optimization of an off-grid hybrid photovoltaic, fuel cell and diesel generator system. *Sustain. Cities Soc.* **2019**, *44*, 310–320. [CrossRef]
2. Sarma, U.; Ganguly, S. Determination of the component sizing for the PEM fuel cell-battery hybrid energy system for locomotive application using particle swarm optimization. *J. Energy Storage* **2018**, *18*, 247–259. [CrossRef]
3. Bizon, N. Real-time optimization strategy for fuel cell hybrid power sources with load-following control of the fuel or air flow. *Energy Convers. Manag.* **2018**, *157*, 13–27. [CrossRef]
4. Bizon, N.; Iana, G.; Kurt, E.; Thounthong, P.; Oproescu, M.; Culcer, M.; Iliescu, M. Air Flow Real-Time Optimization Strategy for Fuel Cell Hybrid Power Sources with Fuel Flow Based on Load-Following. *Fuel Cell* **2018**, *18*, 809–823. [CrossRef]
5. Bizon, N. Effective Mitigation of the Load Pulses by Controlling the Battery/SMES Hybrid Energy Storage System. *Appl. Energy* **2018**, *229*, 459–473. [CrossRef]
6. Bendjedia, B.; Rizoug, N.; Boukhnifer, M.; Bouchafaa, F.; Benbouzid, M. Influence of secondary source technologies and energy management strategies on Energy Storage System sizing for fuel cell electric vehicles. *Int. J. Hydrogen Energy* **2018**, *43*, 11614–11628. [CrossRef]
7. Bizon, N. Hybrid power sources (HPS) for space applications: Analysis of PEMFC/Battery/SMES HPS under unknown load containing pulses. *Renew. Sustain. Energy Rev.* **2019**, *105*, 14–37. [CrossRef]
8. Bizon, N. Energy efficiency for the multiport power converters architectures of series and parallel hybrid power source type used in plug-in/V2G fuel cell vehicles. *Appl. Energy* **2013**, *102*, 726–734. [CrossRef]
9. Mihaescu, M. Applications of multiport converters. *J. Electr. Eng. Electron. Control Comput. Sci.* **2016**, *21*, 13–18.
10. Bizon, N. Energy Efficiency of Multiport Power Converters used in Plug-In/V2G Fuel Cell Vehicles. *Appl. Energy* **2012**, *96*, 431–443. [CrossRef]
11. Hou, J.; Song, Z.; Park, H.; Hofmann, H.; Sun, J. Implementation and evaluation of real-time model predictive control for load fluctuations mitigation in all-electric ship propulsion systems. *Appl. Energy* **2018**, *230*, 62–77. [CrossRef]
12. Bizon, N.; Lopez-Guede, J.M.; Kurt, E.; Thounthong, P.; Mazare, A.G.; Ionescu, L.M.; Iana, G. Hydrogen Economy of the Fuel Cell Hybrid Power System optimized by air flow control to mitigate the effect of the uncertainty about available renewable power and load dynamics. *Energy Convers. Manag.* **2019**, *179*, 152–165. [CrossRef]

13. Kaya, K.; Hames, Y. Two new control strategies: For hydrogen fuel saving and extend the life cycle in the hydrogen fuel cell vehicles. *Int. J. Hydrogen Energy* **2019**, *44*, 18967–18980. [CrossRef]

14. Bizon, N. Real-time optimization strategies of FC Hybrid Power Systems based on Load-following control: A new strategy, and a comparative study of topologies and fuel economy obtained. *Appl. Energy* **2019**, *241*, 444–460. [CrossRef]

15. Ahmadi, S.; Bathaee, S.M.T.; Hosseinpour, A.H. Improving fuel economy and performance of a fuel-cell hybrid electric vehicle (fuel-cell battery and ultra-capacitor) using optimized energy management strategy. *Energy Convers. Manag.* **2018**, *160*, 74–84. [CrossRef]

16. Ou, K.; Yuan, W.W.; Choi, M.; Yang, S. Optimized power management based on adaptive-PMP algorithm for a stationary PEM fuel cell/battery hybrid system. *Int. J. Hydrogen Energy* **2018**, *43*, 15433–15444. [CrossRef]

17. Bizon, N.; Thounthong, P. Fuel Economy using the Global Optimization of the Fuel Cell Hybrid Power Systems. *Energy Convers. Manag.* **2018**, *173*, 665–678. [CrossRef]

18. Yuan, J.; Yang, L.; Chen, Q. Intelligent energy management strategy based on hierarchical approximate global optimization for plug-in fuel cell hybrid electric vehicles. *Int. J. Hydrogen Energy* **2018**, *43*, 8063–8078. [CrossRef]

19. Zhao, J.; Ramadan, H.S.; Becherif, M. Metaheuristic-based energy management strategies for fuel cell emergency power unit in electrical aircraft. *Int. J. Hydrogen Energy* **2019**, *44*, 2390–2406. [CrossRef]

20. Bizon, N.; Hoarca, C.I. Hydrogen saving through optimized control of both fueling flows of the Fuel Cell Hybrid Power System under a variable load demand and an unknown renewable power profile. *Energy Convers. Manag.* **2019**, *184*, 1–14. [CrossRef]

21. Zhang, W.; Li, J.; Xu, L.; Ouyang, M. Optimization for a fuel cell/battery/capacity tram with equivalent consumption minimization strategy. *Energy Convers. Manag.* **2017**, *134*, 59–69. [CrossRef]

22. Bizon, N. Global Maximum Power Point Tracking (GMPPT) of Photovoltaic array using the Extremum Seeking Control (ESC): A review and a new GMPPT ESC scheme. *Renew. Sustain. Energy Rev.* **2016**, *57*, 524–539. [CrossRef]

23. Bizon, N. Global Maximum Power Point Tracking based on new Extremum Seeking Control scheme. *Prog. Photovolt. Res. Appl.* **2016**, *24*, 600–622. [CrossRef]

24. Olatomiwa, L.; Mekhilef, S.; Ismail, M.S.; Moghavvemi, M. Energy management strategies in hybrid renewable energy systems: A review. *Renew. Sustain. Energy Rev.* **2016**, *62*, 821–835. [CrossRef]

25. Mehne, H.H. Evaluation of Parallelism in Ant Colony Optimization Method for Numerical Solution of Optimal Control Problems. *J. Electr. Eng. Electron. Control Comput. Sci.* **2015**, *1*, 15–20.

26. Hoarca, I.C.; Raducu, M. On the micro-inverter performance based on three MPPT controllers. *J. Electr. Eng. Electron. Control Comput. Sci.* **2015**, *1*, 7–14.

27. Lopez-Guede, J.M.; Ramos-Hernanz, J.A.; Larrañaga, J.M.; Garmendia, A.; Ionescu, V.M. Study on the influence of Lambda parameter on several performance indexes in Dynamic Matrix Control. *J. Electr. Eng. Electron. Control Comput. Sci.* **2015**, *2*, 1–8.

28. Bizon, N.; Kurt, E. Performance Analysis of Tracking of the Global Extreme on Multimodal Patterns using the Asymptotic Perturbed Extremum Seeking Control Scheme. *Int. J. Hydrogen Energy* **2017**, *42*, 17645–17654. [CrossRef]

29. Ettihir, K.; Cano, M.H.; Boulon, L.; Agbossou, K. Design of an adaptive EMS for fuel cell vehicles. *Int. J. Hydrogen Energy* **2017**, *42*, 1481–1489. [CrossRef]

30. Bizon, N.; Thounthong, P.; Raducu, M.; Constantinescu, L.M. Designing and Modelling of the Asymptotic Perturbed Extremum Seeking Control Scheme for Tracking the Global Extreme. *Int. J. Hydrogen Energy* **2017**, *42*, 17632–17644. [CrossRef]

31. Ettihir, K.; Boulon, L.; Agbossou, K. Optimization-based energy management strategy for a fuel cell/battery hybrid power system. *Appl. Energy* **2016**, *163*, 142–153. [CrossRef]

32. Bizon, N. Global Extremum Seeking Control of the Power Generated by a Photovoltaic Array under Partially Shaded Conditions. *Energy Convers. Manag.* **2016**, *109*, 71–85. [CrossRef]

33. Bizon, N. Optimal Operation of Fuel Cell Wind Turbine Hybrid Power System under Turbulent Wind and Variable Load. *Appl. Energy* **2018**, *212*, 196–209. [CrossRef]

34. Valenciaga, F.; Puleston, P.F.; Battaiotto, P.E. Supervisor Control for a Stand-Alone Hybrid Generation System Using Wind and Photovoltaic Energy. *IEEE Trans. Energy Convers.* **2005**, *20*, 398–405. [CrossRef]

35. Bizon, N.; Oproescu, M. Experimental Comparison of Three Real-Time Optimization Strategies Applied to Renewable/FC-Based Hybrid Power Systems Based on Load-Following Control. *Energies* **2018**, *11*, 3537. [CrossRef]

36. Pukrushpan, J.T.; Stefanopoulou, A.G.; Peng, H. *Control of Fuel Cell Power Systems*; Springer: New York, NY, USA, 2004.

37. Ramos-Paja, C.A.; Spagnuolo, G.; Petrone, G.; Emilio Mamarelis, M. A perturbation strategy for fuel consumption minimization in polymer electrolyte membrane fuel cells: Analysis, Design and FPGA implementation. *Appl. Energy* **2014**, *119*, 21–32. [CrossRef]

38. Shrivastava, A.; Calhoun, B.H. A DC-DC Converter Efficiency Model for System Level Analysis in Ultra Low Power Applications. *J. Low Power Electron. Appl.* **2013**, *3*, 215–232. [CrossRef]

39. Shin, J.; Shin, S.; Kim, Y.; Lee, S.; Song, B.; Jung, G. Optimal current control of a wireless power transfer system for high power efficiency. In Proceedings of the Conference on Electrical Systems for Aircraft, Railway and Ship Propulsion, Bologna, Italy, 16–18 October 2012.

40. Fahem, K.; Chariag, D.E.; Sbita, L. Performance evaluation of continuous and discontinuous pulse width modulation techniques for grid-connected PWM converter. *Int. Trans. Electr. Energy Syst.* **2018**, *28*, 1–14. [CrossRef]

41. Kazimierczuk, M.K. *Pulse-Width Modulated DC–DC Power Converters*, 2nd ed.; John Wiley & Sons: Hoboken, NJ, USA, 2016.

42. Çelebi, M. Efficiency optimization of a conventional boost DC/DC converter. *Electr. Eng.* **2018**, *100*, 803–809. [CrossRef]

43. Hauke, B. Basic Calculation of a Boost Converter's Power Stage. Applications Note. Texas Instruments. Available online: http://www.ti.com/lit/an/slva372c/slva372c.pdf (accessed on 1 July 2019).

44. Raj, A. Calculating Efficiency. Applications note. Texas Instruments. Available online: http://www.ti.com/lit/an/slva390/slva390.pdf (accessed on 1 July 2019).

45. Bizon, N.; Stan, V.A.; Cormos, A.C. Optimization of the Fuel Cell Renewable Hybrid Power System using the control mode of the required load power on the DC bus. *Energies* **2019**, *12*, 1889. [CrossRef]

46. Bizon, N. Efficient fuel economy strategies for the Fuel Cell Hybrid Power Systems under variable renewable/load power profile. *Appl. Energy* **2019**, *251*, 113400–113518. [CrossRef]

47. Bizon, N. Fuel saving strategy using real-time switching of the fueling regulators in the Proton Exchange Membrane Fuel Cell System. *Appl. Energy* **2019**, *252*, 113449–113453. [CrossRef]

Article

Optimization of Component Sizing for a Fuel Cell-Powered Truck to Minimize Ownership Cost

Kyuhyun Sim [1,*], **Ram Vijayagopal** [2], **Namdoo Kim** [2] **and Aymeric Rousseau** [2]

[1]   Sungkyunkwan University, 2066 Seobu-ro, Jangan-gu, Suwon-si, Gyeonggi-do 16419, Korea
[2]   Argonne National Laboratory, 9700 S. Cass Ave, Lemont, IL 60439, USA; rvijayagopal@anl.gov (R.V.);
      nkim@anl.gov (N.K.); arousseau@anl.gov (A.R.)
*   Correspondence: dodo1064@naver.com; Tel.: +82-31-290-7912

Received: 27 February 2019; Accepted: 20 March 2019; Published: 22 March 2019

Abstract: In this study, we consider fuel cell-powered electric trucks (FCETs) as an alternative to conventional medium- and heavy-duty vehicles. FCETs use a battery combined with onboard hydrogen storage for energy storage. The additional battery provides regenerative braking and better fuel economy, but it will also increase the initial cost of the vehicle. Heavier reliance on stored hydrogen might be cheaper initially, but operational costs will be higher because hydrogen is more expensive than electricity. Achieving the right tradeoff between these power and energy choices is necessary to reduce the ownership cost of the vehicle. This paper develops an optimum component sizing algorithm for FCETs. The truck vehicle model was developed in Autonomie, a platform for modelling vehicle energy consumption and performance. The algorithm optimizes component sizes to minimize overall ownership cost, while ensuring that the FCET matches or exceeds the performance and cargo capacity of a conventional vehicle. Class 4 delivery truck and class 8 linehaul trucks are shown as examples. We estimate the ownership cost for various hydrogen costs, powertrain components, ownership periods, and annual vehicle miles travelled.

Keywords: fuel cell powered vehicle; medium- and heavy-duty trucks; component sizing; ownership cost; optimization

---

1. Introduction

Automotive powertrains are being electrified to achieve lower emissions and higher fuel efficiency. Along with battery-powered trucks, fuel cell-powered electric trucks (FCETs) are a promising candidate to replace conventional vehicles [1]. Researches done in California have shown feasibility of this technology for certain types of medium- and heavy-duty trucks [2]. Interest in FCETs for medium-duty vocational trucks such as parcel or package delivery has increased. United Parcel Service (UPS) has unveiled its first prototype for a fuel cell-powered van, and FedEx Express has started delivery using FCETs, which in this case are fuel cell range extenders (FCRExs). FCRExs rely primarily on a battery for power and energy, and they carry a fuel cell to extend the range of the battery pack [3]. The range extender enables the vehicle to cover longer distances that would be difficult with the battery pack alone. In addition, the battery is also used as a buffer for high power and for collecting regenerative energy like a hybrid electric vehicle.

A great deal of work has been done on fuel cell electric vehicle (FCEV) control strategies to improve overall fuel efficiency. In [4], the authors proposed control strategies ruled by fuel cell power. Hames et al. [5] compared different control strategies. However, hybrid powertrains with two or more power sources should optimize powertrain component sizes before developing their energy management control strategy. Fuel efficiency and vehicle performance are always dependent on the sizes of vehicle components. Moreover, improper component sizes can increase vehicle cost, which makes the vehicle unattractive for the consumer to purchase. It is important to define the component system. Fauvel et al. [6] and Kim et al. [7] proposed and validated component sizing processes for hybrid electric vehicle (HEV) powertrain configurations. Marcinkoski et al. [8] optimized component sizes to replace diesel trucks with FCETs while ensuring equivalent performance. Bendjedia et al. [9] presented a methodology to size the energy storage system for different batteries by considering weight, cost, and battery volume. Unlike the presented rule-based sizing processes, Lee [10] proposed a sizing process to minimize fuel consumption with an optimization algorithm to search for an optimum value. It is called POUNDERS (practical optimization using no derivatives for sums of squares), and was developed by Argonne National Laboratory [11]. Using this process, we focused on vehicle ownership costs, including the cost of fuel consumption. Eren et al. [12] sized FCEVs based on, but they did not consider costs that occur when a vehicle is purchased and is being used.

Consumers want to know how much a vehicle will cost throughout the period of ownership; knowing this will help them decide which vehicle to purchase because they know how much money they need to own, purchase, sustain, and operate each vehicle. Total cost of ownership (TCO) is all costs for these. TCO has been used for the estimation of market penetration and the market forecasting [13]. Some researchers have attempted to quantify this cost. The TCO varies in the definition. Mock et al. [14] specify a measure of relevant ownership cost (RCO). Simeu et al. [15] and Rousseau et al. [16] analyzed energy consumption and the costs of plug-in electric vehicles using RCO.

FCEVs use two different energy sources: a battery pack and a hydrogen tank. Properly sizing the fuel cell and the battery can reduce the overall cost of the vehicle while sustaining vehicle performance. The objective of component sizing is to help minimize the overall ownership cost of the FCEV. This paper seeks the optimal component size to minimize RCO for fuel cell hybrid vehicles. We use FCEV models developed in Autonomie and an optimization algorithm from POUNDERS to simulate and optimize a medium- and heavy-duty trucks. The paper is organized as follows. In Section 2, we review rules based on design assumptions from an earlier study. In Section 3, we propose a sizing process based on the cost of ownership and on performance. Sections 4 and 5 describe case studies for a class 4 delivery van and a class 8 linehaul truck, respectively. We analyzed how the sizing result that minimizes the ownership cost changes when some assumptions change. Finally, Section 6 provides a conclusion.

2. Rule-Based Design Process Assumptions

Here we review rule-based design logic from a prior study [8] for fuel cell-powered trucks. In the U.S. Department of Energy, all vehicles are classified based on the gross vehicle weight rating (GVWR), ranging from class 1 to class 8 [17]. We focus on a class 4 delivery van and a class 8 linehaul truck as shown in Table 1 for conventional vehicle simulations. Control algorithms and component sizes determine whether the vehicles are charge-sustaining or range-extended hybrids. There are two powertrain architectures. One is a battery-powered electric vehicle with a fuel cell range extender, called a FCREx. The other is a fuel cell-dominant system with a battery for peak acceleration events, which is known as a fuel cell hybrid electric vehicle (FCHEV). In FCRExs, the electric machine is sized to match baseline vehicle performance. The fuel cell meets the demands of continuous loads. The battery can assist on a road with a 6% grade for 18 km and is sized to drive 50% of daily driving range in electric vehicle (EV) mode, using only electric power. The hydrogen storage is sized to extend this range. The electric range assumption may not be optimum in this case, but it serves as the FCREx baseline for this analysis.

Table 1. Conventional vehicle simulations.

	Properties	Class 4 Delivery Van	Class 8 Linehaul Truck
Summary	Daily driving range	321.9 km	643.7 km
	Baseline power	149 kW	336 kW
Performance	Cargo mass	2395 kg	19,908 kg
	Cruising speed	112.7 km/h	96.6 km/h
	6% grade speed	66.0 km/h	49.9 km/h
	0–48 km/h accel. time	8.0 s	17.1 s
	0–97 km/h accel. time	34.2 s	61.1 s

Likewise, FCHEVs are also evaluated for performance. Fuel cells are sized to meet continuous loads for cruise and grade. The electric machine is sized for performance. The battery is sized for both performance and regenerative braking. Hydrogen storage is sized to satisfy daily driving requirements. We see that this sizing approach results in fuel cell-powered trucks that can match or outperform conventional vehicles, with no sacrifices in payload. More information is found in [8]. One limitation of this study was that it did not consider the cost of building or operating the truck. In this paper, we propose an optimum sizing process for fuel cell-powered trucks that will minimize ownership cost while ensuring that performance goals are met.

3. Sizing Process Based on Cost of Ownership and Performance

3.1. Optimization Algorithm for Sizing Components

We used Autonomie to simulate the vehicle performance of a fuel cell-powered truck. Autonomie is a vehicle system simulation tool for the energy consumption and performance [18]. We optimized its onboard hydrogen storage and battery pack size to minimize the ownership costs. This ensures that all performance requirements are met within a 2% tolerance. Figure 1 shows a sizing process. Input variables to be optimized are hydrogen tank, battery capacity, and fuel cell power. When the three input variables change, the vehicle model developed performs the three-vehicle performance test while checking their constraints. Through the optimization process, the input parameters are modified using previous results. Then the feedback process runs until the algorithm finds the optimal value for the objective. It will find the tradeoff relationship between hydrogen storage and battery power. We propose optimization problem as follows:

$$\min_{r_i}\left\{rco(r_i) : l_i \leq r_i \leq u_i, \ c_j(r_i) \leq 0, \ i, j = 1, 2, 3\right\}, \tag{1}$$

where *rco* is the RCO. It is minimized over r_i whose range is from lower limit l_i to upper limit u_i. The value of c_j represents the vehicle performance constraints. The subscript i represents different components and the j represents additional performance constraints.

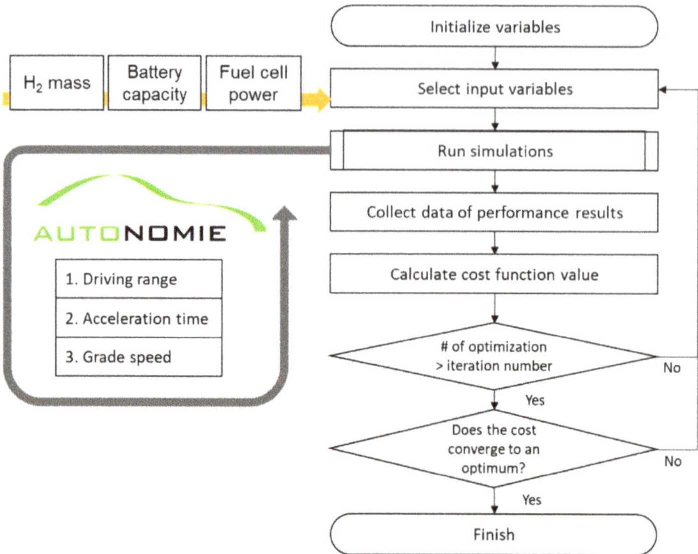

Figure 1. Sizing process with vehicle simulation and POUNDERS (practical optimization using no derivatives for sums of squares).

We used the optimization algorithm for components as the POUNDERS. The POUNDERS is a derivative-free optimization to seek local solutions to a potentially multimodal problem, which is a bound-constrained augmented Lagrangian problem [10]:

$$\min_{r_i}\left\{h(r, s) = rco(r) - \sum_{j\in J} \lambda_j\left(c_j(r) + s_j\right) + \frac{\mu}{2}\sum_{j\in J}\left(c_j(r) + s_j\right)^2 : s \geq 0; 0 \leq r \leq u\right\}, \tag{2}$$

where h is a cost function value, s represents slack variables, λ_j is an estimate of Lagrangian multipliers, and μ is the penalty parameter for the constraints. If the constraints are not satisfied, the slack variables increase the cost function value, h.

The optimization process is conducted based on Autonomie including the POUNDERS, which developed using MATLAB/Simulink. We update the component cost estimate depending on the component sizes, and define the cost function, performance tests and constraints, as explained further in more detail. The cost is calculated by running the performance tests for each vehicle according to the powertrain components. Because the vehicle weight is different and the component performances change themselves, different powertrain components can affect the vehicle performances through the optimization process. Although the algorithm is one objective optimization, it is difficult to optimize powertrain components because each constraint of the vehicle performance has its own objective. The process optimizes the component sizes to minimize the cost, while ensuring the performance. In this study, the process iterates more 31 times, sufficient to find an optimal value.

3.2. Relevant Cost of Ownership

RCO is the net present cost to own and operate the vehicle. It includes the investment cost with the purchase price and any fees, taxes, and incentive or disincentives. It also includes all operating costs, maintenance costs, and a resale or residual value. The following equation is a way to calculate the RCO. For more detail, see reference [16]:

$$rco = cost_{inv} + cost_{pv\_energy} + cost_{pv\_maint} + cost_{pv\_batt\_replace} - cost_{residual}, \tag{3}$$

where $cost_{inv}$ is total investment cost for purchase, initial registration, home electric vehicle service equipment, and vehicle incentive. The $cost_{pv\_energy}$ value is the present value energy cost while considering vehicle fuel efficiency and the current costs of fossil fuel and hydrogen. The $cost_{pv\_maint}$ is the present value of maintenance costs, repairs, and so on. The $cost_{pv\_batt\_replace}$ is the present cost of battery replacement. The $cost_{residual}$ is the residual value of function with initial cost, the vehicle's annual vehicle miles travelled (VMT), and a discount rate. Purchase price and fuel (or energy) costs are the primary variables for RCO. All other factors are either constants or a function of the purchase price.

We assume that the depreciation is 5%, and vehicle life is 15 years. We assume that the ownership period is 5 years but the actual value will depend on class and vocation. The Federal Highway Administration states that the average delivery truck travels an average of 21,108 km annually [19]. We assume that VMT is 22,531 km for class 4. However, class 8 drives more, so we assume 160,934 km per year.

To calculate the energy cost, we assume that the fuel price is $3 per gallon and the hydrogen price is $4 or $12 per gasoline gallon equivalent (gge). We estimate the manufacturing cost based on component costs, which is based on 2017 FCTO/VTO Benefit Analysis Assumptions [20]. The purchase price is set at 1.5 times the cost of manufacturing. Battery cost is estimated using energy and power, which is $243 per kilowatt-hour (kWh) and $20 per watt (W). The fuel cell system is calculated using simplified cost calculations based on the peak efficiency of the stack and weight ratio for the tank. The values are based on the assumptions used for the Fiscal Year 2016 fuel cell technology analysis [21]. The cost of the fuel cell tank is estimated to be $595 per kilogram of usable hydrogen at a 4.4% storage weight ratio. We estimate that the cost of the fuel cell system is $50.69 per kilowatt (kW) at 59.5% efficiency.

Component size also affects vehicle weight, which affects specific power. We assume that the specific power of the fuel cell system is 659 W/kg [22,23]. Motor specific power is 1.9 kW per kilogram (kW/kg), based on U.S. Department of Energy (DOE) estimates [24].

3.3. Optimization Conditions and Vehicle Performance Requirements

There are three performance requirements by checking the car's driving range with all provided energy, an acceleration time, and a final vehicle speed on a specific grade. The car's range should be higher than 241 km on Air Resources Board (ARB) Transient cycle used by the U.S. Environmental Protection Agency (EPA) for class 4 vehicles. Likewise, the driving range for class 8 linehaul truck is 483 km at EPA's 105 km/h rating because the truck drives more highway over long distances. Next, the acceleration time includes the period from stationary to 97 km/h; this may be lower than 34 s for class 4 and 64 s for class 8. In the grade speed test, the vehicle runs on a road with a 6% grade; the end speed should be 66 km/h and 48 km/h for class 4 and class 8, respectively, which is an approximation for the Davis Dam test [25].

4. Case Study 1: Optimizing a Class 4 Delivery Van

4.1. Fuel Cell Range Extenders

The first type of truck we decided to target is a medium-duty or class 4 pickup or delivery van. It has a cargo mass of 2800 kg, just like the conventional baseline. As mentioned above, FCREx is battery-dominant and similar to a series hybrid electric vehicle that uses a fuel cell instead of an internal combustion engine. Figure 2 shows the configuration of the FCEV in Autonomie. This fuel cell connects to the battery for charging. As the battery becomes fully charged, an electric motor runs the vehicle using electrical energy. When it runs on the electrical energy and uses hydrogen with the fuel cell system while its battery sustains a certain charge state, this is called charge sustaining mode. FCRExs often use power from both the battery and the fuel cell when they need the large amounts of wheel power demanded for energy management control strategies. Table 2 shows vehicle specifications. We do not expect much change in fuel cell power or motor power, because the vehicle should still meet all performance requirements.

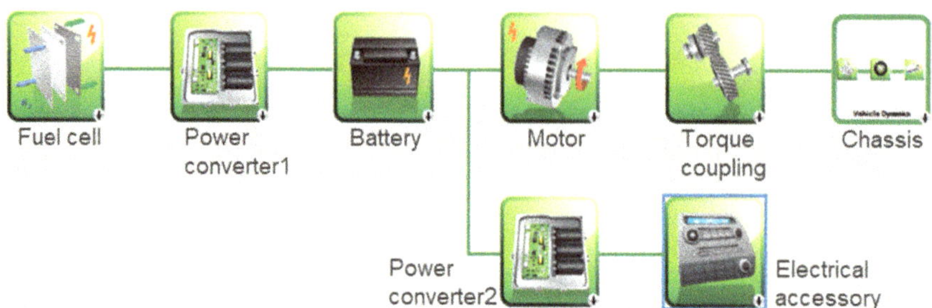

Figure 2. Configuration of fuel cell electric vehicle (FCEV) in Autonomie.

Table 2. Vehicle specifications of a class 4 van and class 8 line haul truck.

Medium Duty	Vehicle Mass	Frontal Area	Drag Coefficient	Electric Motor	Fuel Cell Power	Battery Type	Battery Energy	On Board H_2 Storage
Class 4	7317 kg	7.50 m^2	0.70	211 kW	100 kW	Li-ion	59 kWh	4 kg
Class 8	41,723 kg	10 m^2	0.55	870 kW	340 kW	Li-ion	770 kWh	46.4 kg

Above all, the optimization technique needs to be verified against parameter sweeps for the test cases. For example, there are class 4 FCREx optimization results. The optimization is 4 times faster than parameter sweeps and yields better results. Figure 3a,b show feasible points for grade and range tests, respectively. Figure 3c shows sums the feasible ranges of Figure 3a,b. The right square magnified shows the related results. Blue circles are feasible points and red crosses are surrounding points. Red and blue stars are the estimated and POUNDERS results, respectively. Squares are POUNDERS tracking points. Each number next to a point is an RCO value. Some variables do not coincide exactly with the optimal value because the POUNDERS results can be one of the local optimums. We found the estimated optimal value by using the minimum value from the fixed grid data, which can reveal the optimal battery capacity, hydrogen mass, and fuel cell power. However, the estimated value missed the optimal point because of low resolution, 7 by 7. The lowest RCO is located on the left side and bottom of Figure 3c. Therefore, the bottom left values within the feasible section are optimal, closer to the POUNDERS result. If the resolution for the estimated results rise to explore more points the current estimate did not identify, the new optimal point may be closer to the POUNDERS result. POUNDER uses fewer iterations and smaller step sizes to arrive at a better solution.

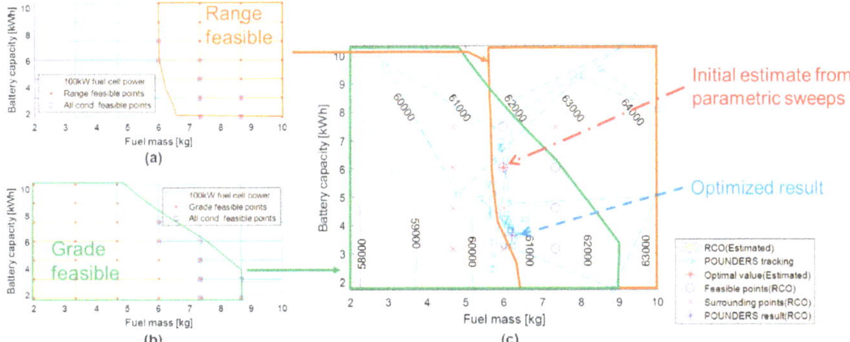

Figure 3. Feasible values of (**a**) range and (**b**) grade for 100 kW of fuel cell power, and (**c**) relevant ownership cost (RCO) results for the estimate and POUNDERS.

Optimized FCRExs have component size that are similar to FCHEVs, as shown in Figure 4. The FCREx can have the same hydrogen tank as FCHEV, but the different driving strategies of both vehicles increase the FCREx battery to achieve the same performance as FCHEV. Table 3 summaries optimization results and compares them with rule-based component sizes. There is tradeoff between fuel mass and battery capacity. The optimized vehicle has lower battery capacity and higher hydrogen mass. Some performance results are lower than those for the rule-based sizing, but they still meet performance requirements. The 0–97 km/h acceleration time increases, but it is still better than that of the conventional vehicle. Grade speed drops, because the battery is no longer assisting the fuel cell during grades. The total vehicle mass remains largely unchanged, although there is a small reduction of 76 kg. These results are independent of fuel cell cost, because fuel cell power remains the same in both cases. In this case, 100 kW of power is necessary to cruise at highway speeds. Therefore, the fuel cell cannot be further downsized.

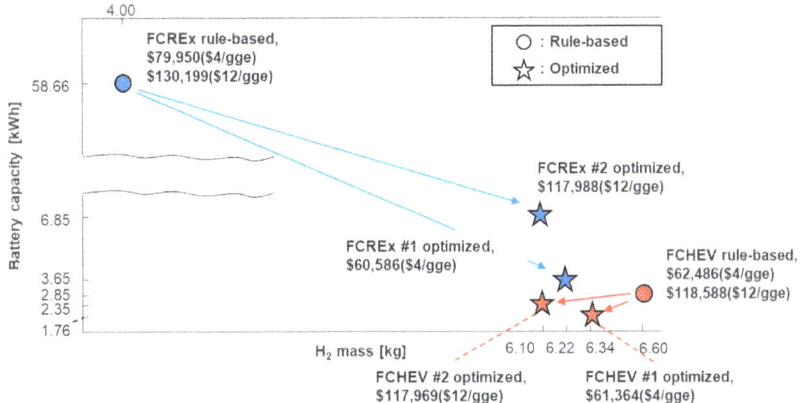

Figure 4. Summary of optimized hydrogen tank and battery pack for class 4 fuel cell vehicles; the #1 case assumes a hydrogen cost of $4/gge (gasoline gallon equivalent) and #2 assumes a hydrogen cost of $12/gge.

Table 3. Comparison between the rule-based and optimized results. FCREx: fuel cell range extenders.

Vehicle Type	Fuel Mass	Battery Capacity	Fuel Cell Power	RCO	Acc. Time	Range	Grade Speed
FCREx rule-based	4.0 kg	58.7 kWh	100.0 kW	$79,588	17.3 s	237.2 km	74.7 km/h
FCREx optimized	6.2 kg	3.6 kWh	101.5 kW	$60,586	21.9 s	241.4 km	64.2 km/h
Difference	55.0%	−93.9%	1.5%	−23.9%	26.6%	1.8%	−14.0%

In addition, the optimum solution changes as hydrogen cost increases. We assume $12 per gge of hydrogen, not $4 per gge. Table 4 shows the optimized results, comparing hydrogen costs of $4 and $12. We observe larger battery capacity and reduced hydrogen fuel use as hydrogen cost increases. Figure 5 shows the RCOs for different hydrogen costs. The manufacturing cost rises with a larger battery and the energy cost increases. The RCO increased by about $20,000 because of the increased energy cost. The overall component design remains fuel cell dominant.

Table 4. Optimized results according to hydrogen price.

H_2 Cost	Fuel Mass	Battery Capacity	Fuel Cell Power	RCO	Range
$4	6.217 kg	3.649 kWh	101.5 kW	$60,586	241.4 km
$12	6.095 kg	4.447 kWh	101.2 kW	$80,231	241.2 km
Difference	−2.0%	21.9%	−0.3%	33.4%	−0.1%

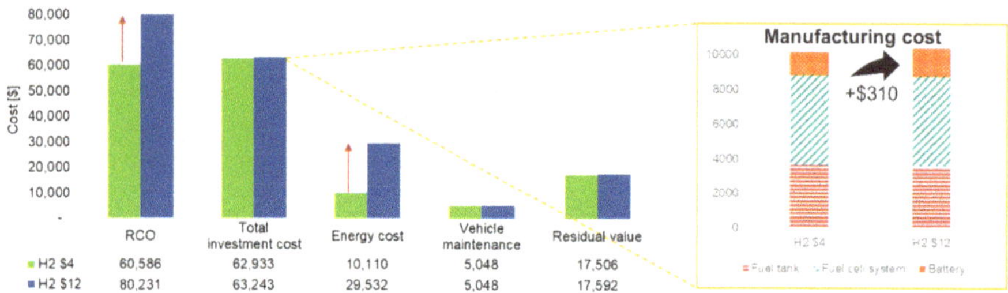

Figure 5. Comparison of relevant cost of ownership (RCO) according to hydrogen cost.

4.2. Fuel Cell Hybrid Electric Vehicles

For a class 4 truck, the FCHEV is optimized to minimize its RCO. When we optimize components, the fuel cell power increases more than that of an FCREx. Because an FCHEV battery has less specific energy than that an FCREx, an FCHEV has a smaller capacity but its mass is higher. Therefore, an FCHEV needs more fuel cell power to reach the same performance as an FCREx. Figure 4 summarizes the optimal fuel tank and battery pack for an FCHEV and an FCREx. Specifically, optimizing battery size for FCHEV is important. Using the rule-based process in Section 2, the battery is sized for maximum regenerative braking. This helps improve overall vehicle fuel economy. Otherwise, optimization to minimize RCO shows that a 38% smaller battery is a better choice. This results in higher fuel consumption and higher operating cost. A smaller battery pack reduces fuel economy in the vehicle optimized for ownership cost. Table 5 shows how battery pack size affects fuel economy. Figure 6a shows battery power operating points and the optimized reduced operating range. Figure 6b,c are motor power and fuel cell power, respectively. The optimized battery reduces regenerative braking torque. Motor size remains unchanged due to performance requirements. As mentioned above, in the rule-based results, the battery is sized to maximize regenerative braking. In the optimized design, however, only the RCO value is considered for component sizing. By optimizing FCHEV components, the fuel mass and battery capacity decrease and the fuel cell power increases while satisfying vehicle performances. Although both processes ensure the same performance, the optimized design reduces RCO. Optimization strikes the right balance between higher initial cost and higher operating cost. A large battery results in higher initial cost but increases fuel economy through additional regenerative braking. A small battery results in higher operating costs but reduces initial cost because the smaller battery is less expensive.

Table 5. Comparison of results for rule-based and optimized FCHEVs.

FCHEV	Rule-Based	Optimized
Fuel economy gasoline equivalent	10.7 km/L	10.4 km/L
Percent regenerative braking at battery	74.1%	63.0%
Battery capacity	2.85 kWh	1.76 kWh

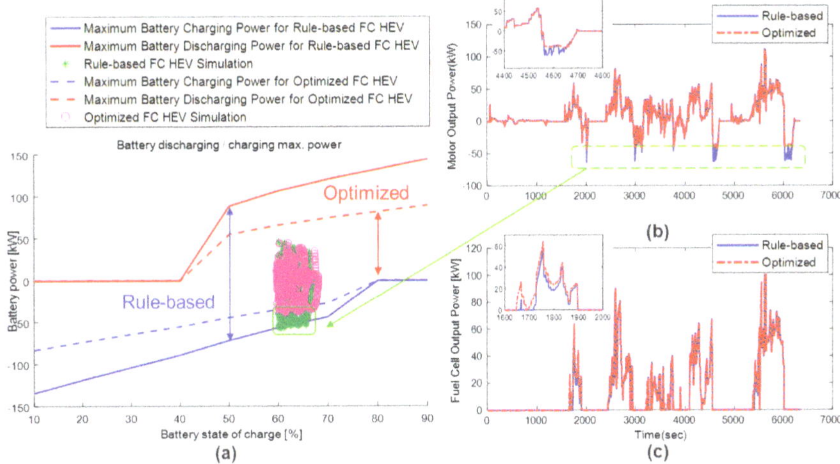

Figure 6. Simulation results for rule-based and optimized FCHEVs: (**a**) battery power operating points, (**b**) motor power, and (**c**) fuel cell power.

When the ownership period and hydrogen cost increase, the optimum solution also changes. Next we consider an ownership period of 10 years and a hydrogen cost of $12 per gge. Table 6 summarizes the results for an FCHEV. Fuel economy increases in importance due to the higher cost of fuel, making a larger battery feasible. This gets closer to the battery size necessary to minimize fuel consumption. The optimization led to a 0.7% drop in fuel economy, reducing the RCO by about $1,000. The optimum solution is sensitive to ownership periods and hydrogen cost assumptions. We compare both rule-based and optimized RCOs for class 4 trucks, as shown in Figure 7. Case #1 results when the ownership period is 5 years and the hydrogen cost is $4/gge. Case #2 results when the ownership period is 10 years and the hydrogen cost is $12/gge. All other assumptions remain the same. In this case, the optimized FCREx is cheaper than the optimized FCHEV by about $800. When the hydrogen cost is $4 over a 5-year ownership period, FCHEV and FCREx RCOs decrease by 1.8% and 24.2%, respectively. On the other hand, when the hydrogen cost is $12 over a 10-year ownership period, FCHEV and FCREx RCOs decrease by 0.5% and 9.4%, respectively. Infrastructure cost is not considered in this analysis. There is no cost assigned to downtime associated with charging.

Table 6. Summary of results for class 4 fuel cell hybrid electric vehicle (FCHEV) with various assumptions.

FCHEV	Rule-Based		Optimized #1	Optimized #2
Ownership period	5 years	10 years	5 years	10 years
H_2 cost	$4/gge	$12/gge	$4/gge	$12/gge
H_2 mass	6.60 kg		6.34 kg	6.10 kg
Battery capacity	2.9 kWh		1.8 kWh	2.35 kWh
Fuel economy gasoline equivalent	10.7 km/L		10.4 km/L	10.6 km/L
RCO	$62,486	$118,588	$61,364	$117,969

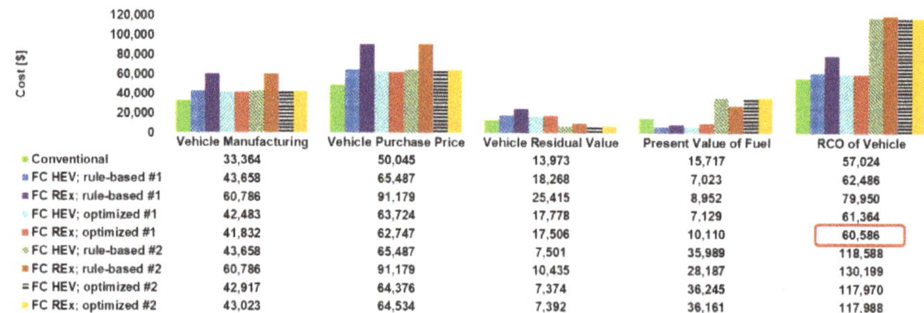

	Vehicle Manufacturing	Vehicle Purchase Price	Vehicle Residual Value	Present Value of Fuel	RCO of Vehicle
Conventional	33,364	50,045	13,973	15,717	57,024
FC HEV; rule-based #1	43,658	65,487	18,268	7,023	62,486
FC REx; rule-based #1	60,786	91,179	25,415	8,952	79,950
FC HEV; optimized #1	42,483	63,724	17,778	7,129	61,364
FC REx; optimized #1	41,832	62,747	17,506	10,110	60,586
FC HEV; rule-based #2	43,658	65,487	7,501	35,989	118,588
FC REx; rule-based #2	60,786	91,179	10,435	28,187	130,199
FC HEV; optimized #2	42,917	64,376	7,374	36,245	117,970
FC REx; optimized #2	43,023	64,534	7,392	36,161	117,988

Figure 7. Comparison of relevant cost of ownership (RCO) for class 4 delivery van.

5. Case Study 2: Optimizing a Class 8 Linehaul Truck

Likewise, for a class 4 delivery van, we apply the optimum sizing process to FCREx and FCHEV class 8 linehaul trucks. The truck requires 483 km of range, less than 64 s of acceleration time, and 48 km/h of grade speed. For a class 8 linehaul truck, we modify this assumption: its VMT is 160,934 km per year and the fuel cell cost is $200 per kW. A class 8 linehaul truck usually operates on the highway, so we simulate it on an EPA 65 driving cycle. Vehicle specifications appear in Table 2. Its objective is also to minimize RCO.

When components are optimized, an FCHEV is cheaper than an FCREx. The optimized FCREx relies primarily on onboard hydrogen storage. Table 7 summarizes the results for a class 8 linehaul truck. Figure 8 compares hydrogen mass and battery capacity sizes for an FCHEV and an FCREx. Battery size decreases from 770 kWh to 24 kWh, and the vehicle runs mostly in charge sustaining mode. Arbitrarily sizing the battery power to supply half the daily driving is not optimum. Fuel cell power and hydrogen storage compensates for the reduction in battery size. The fuel cell power and hydrogen storage increase by 50 kW and 22 kg, respectively. FCHEV sizing remains largely unchanged from the rule-based approach. The result chose a slightly smaller fuel cell and battery. This is likely because the optimization utilized the 2% tolerance allowed in grade speed and acceleration.

Table 7. Summary of optimization results for a class 8 linehaul truck driving 482.8 km.

Vehicle Type	Fuel Mass	Battery Capacity	Fuel Cell Power	RCO	Acc. Time	Range	Grade Speed
FCHEV initial	60.0 kg	4.56 kWh	391.3 kW	$848k	45.2 s	483.6 km	49.9 km/h
FCHEV optimized	60.2 kg	3.03 kWh	369.7 kW	$836k	40.5 s	483.9 km	47.8 km/h
FCREx initial	43.0 kg	770 kWh	340.0 kW	$1469k	30.6 s	482.8 km	51.5 km/h
FCREx optimized	65.3 kg	24.2 kWh	391.0 kW	$904k	25.2 s	486.5 km	48.0 km/h

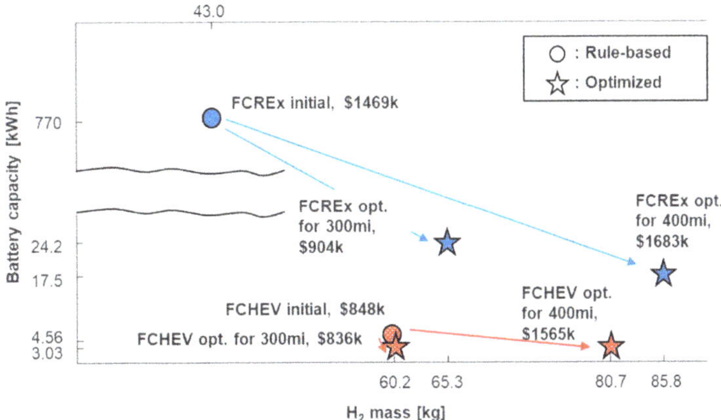

Figure 8. Summary of optimized hydrogen tank and battery pack for class 8 fuel cell vehicles.

The optimized FCREx has larger hydrogen tank than that of the optimized FCHEV. As the weight of the powertrain components increases, the optimization to satisfy the performances increases hydrogen tank as well as battery. The increased battery capacity increases the weight, and then the hydrogen tank also increases to cover longer distance.

The RCOs of class 8 linehaul trucks are compared in the rule-based technique and optimized as shown in Figure 9. A fuel cell-dominant hybrid is the most economical design choice for a class 8 linehaul fuel cell truck. The present value of costs is almost as high as the purchase price. This indicates that this design solution depends on VMT, energy cost, and duration of ownership. An FCHEV has the lower fuel cost in this case.

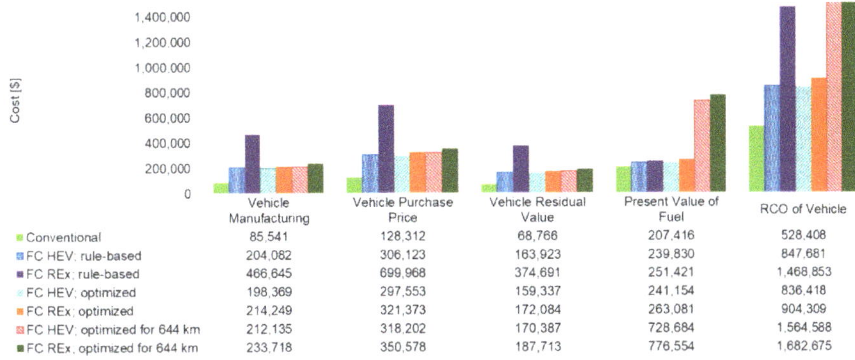

	Vehicle Manufacturing	Vehicle Purchase Price	Vehicle Residual Value	Present Value of Fuel	RCO of Vehicle
Conventional	85,541	128,312	68,766	207,416	528,408
FC HEV: rule-based	204,082	306,123	163,923	239,830	847,681
FC REx: rule-based	466,645	699,968	374,691	251,421	1,468,853
FC HEV: optimized	198,369	297,553	159,337	241,154	836,418
FC REx: optimized	214,249	321,373	172,084	263,081	904,309
FC HEV: optimized for 644 km	212,135	318,202	170,387	728,684	1,564,588
FC REx, optimized for 644 km	233,718	350,578	187,713	776,554	1,682,675

Figure 9. Comparison of relevant cost of ownership (RCO) for a class 8 linehaul truck.

For longer distance class 8 linehaul trucks, we changed the optimization objective to 644 km. We assume that the cost of hydrogen increases to $12 per gge. Onboard hydrogen storage increases by optimizing components. Table 8 summarizes the results. Hydrogen storage increases by 20 kg for 161 km. It needs more fuel cell power to sustain both grade speed and acceleration performance. FCREx sizing increases fuel cell power by about 50 kW and FCHEV sizing also increases fuel cell power by 10 kW to compensate for the loss in battery power. EPA 65 may not offer many opportunities for regenerative braking either. The impact of battery size on fuel economy is not very prominent.

Table 8. Optimization results for a class 8 linehaul truck driving 643.7 km.

Vehicle Type	Fuel Mass	Battery Capacity	Fuel Cell Power	RCO	Acc. Time	Range	Grade Speed
FCHEV optimized	80.7 kg	3.02 kWh	377.6 kW	$1565k	40.6 s	643.9 km	48.0 km/h
FCREx optimized	85.8 kg	17.5 kWh	440.0 kW	$1683k	27.6 s	643.6 km	53.6 km/h

We add the RCO obtained by optimizing a class 8 linehaul truck for 644 km to Figures 7 and 8. In a longer-range case, a fuel cell–dominant hybrid is also the most economical design for a class 8 linehaul truck. Higher hydrogen costs result that the present value of total fuel costs is more than twice the vehicle purchase price.

6. Conclusions

The optimum component size for an FCREx depends on the cost of hydrogen and the powertrain components, the length of the ownership period, and VMT. Optimum component sizing for an FCHEV is less sensitive to these factors. The optimum design for an FCREx relies primarily on onboard hydrogen storage for energy. The proposed sizing process finds economically optimum design solutions for fuel cell vehicles while ensuring that there is no tradeoff in performance. For class 4 delivery trucks, optimized FCRExs and FCHEVs have comparable component sizes and RCO estimates. The FCREx ownership costs is slightly less than that of FCHEV by 1.3%. For a class 8 linehaul truck, the RCO of an FCHEV is 7.5% lower than that of an FCREx design. The energy cost may be equal to the initial cost. Therefore, a higher hydrogen cost could affect the solution. In this study, we do not consider the cost associated with infrastructure or downtime for charging. They need to be factored in later when comparing the RCOs of different powertrains. We expect that the proposed sizing process will use representative real-world cycles for sizing and cost estimates and for linking fuel cell cost to the power and operating conditions of load levels, duration, and so on.

Author Contributions: Conceptualization, R.V.; methodology, R.V. and N.K.; software, K.S.; validation, R.V., N.K. and K.S.; formal analysis, R.V.; investigation, K.S.; resources, N.K.; data curation, K.S.; writing—original draft preparation, K.S.; writing—review and editing, K.S., N.K, and R.V.; visualization, K.S.; supervision, R.V.; project administration, A.R.

Acknowledgments: The authors would like to acknowledge the financial support of Jason Marcinkoski (Fuel Cell Technologies Office, U.S. Department of Energy) to conduct this work. The submitted manuscript has been created by the UChicago Argonne, LLC, Operator of Argonne National Laboratory (Argonne). Argonne, a U.S. Department of Energy Office of Science laboratory, is operated under Contract No. DE-AC02-06CH11357. The U.S. Government retains for itself, and others acting on its behalf, a paid-up nonexclusive, irrevocable worldwide license in said article to reproduce, prepare derivative works, distribute copies to the public, and perform publicly and display publicly, by or on behalf of the Government.

Conflicts of Interest: The authors declare no conflict of interest.

References

1. Kast, J.; Morrison, G.; Gangloff, J.J.; Vijayagopal, R.; Marcinkoski, J. Designing Hydrogen Fuel Cell Electric Trucks in a Diverse Medium and Heavy Duty Market. *Res. Transp. Econ.* **2017**, *70*, 1–10. [CrossRef]
2. California Fuel Cell Partnership. *Medium- & Heavy-Duty Fuel Cell Electric Truck Action Plan for California*; California Fuel Cell Partnership: West Sacramento, CA, USA, 2016.
3. Walters, M.; Kuhlmann, A.; Ogrzewalla, J. Fuel Cell Range Extender for Battery Electric Vehicles. In Proceedings of the International Conferences on Electrical Systems for Aircraft, Railway, Ship Propulsion and Road Vehicles (ESARS), Aachen, Germany, 3–5 March 2015; IEEE: Aachen, Germany, 2015.
4. He, H.; Zhang, Y.; Wan, F. Control Strategies Design for a Fuel Cell Hybrid Electric Vehicle. In Proceedings of the IEEE Vehicle Power and Propulsion Conference (VPPC), Harbin, China, 3–5 September 2008; IEEE: Harbin, China, 2008.
5. Hames, Y.; Kaya, K.; Baltacioglu, E.; Turksoy, A. Analysis of the Control Strategies for Fuel Saving in the Hydrogen Fuel Cell Vehicles. *Int. J. Hydrog. Energy* **2018**, *43*, 10810–10821. [CrossRef]

6. Fauvel, C.; Napal, V.; Rousseau, A. Medium and Heavy Duty Hybrid Electric Vehicle Sizing to Maximize Fuel Consumption Displacement on Real World Drive Cycles. In Proceedings of the 26th International Electric Vehicle Symposium and Exhibition (EVS), Los Angeles, CA, USA, 6–9 May 2012.

7. Kim, N.; Moawad, A.; Vijayagopal, R.; Rousseau, A. Validation of Sizing Algorithm for Several Vehicle Powertrains. In Proceedings of the FISITA World Automotive Congress, Busan, Korea, 26–30 September 2016; FISITA: Busan, Korea, 2016.

8. Marcinkoski, J.; Vijayagopal, R.; Kast, J.; Duran, A. Driving an Industry: Medium and Heavy Duty Fuel Cell Electric Truck Component Sizing. *World Electr. Veh. J.* **2016**, *8*, 78–89. [CrossRef]

9. Bendjedia, B.; Rizoug, N.; Boukhnifer, M.; Bouchafaa, F.; Benbouzid, M. Influence of Secondary Source Technologies and Energy Management Strategies on Energy Storage System Sizing for Fuel Cell Electric Vehicles. *Int. J. Hydrog. Energy* **2018**, *43*, 11614–11628. [CrossRef]

10. Lee, H.; Cha, S.; Kim, N.; Jeong, J.; Vijayagopal, R.; Rousseau, A. Development of Vehicle Component Sizing Process Using Optimization Algorithm. In Proceedings of the IEEE Vehicle Power and Propulsion Conference (VPPC), Belfort, France, 11–14 December 2017; IEEE: Belfort, France, 2017.

11. Wild, S.M. *Solving Derivative-Free Nonlinear Least Squares Problems with Pounders*; Argonne National Laboratory: Lemont, IL, USA, 2014.

12. Eren, Y.; Gorgun, H. An Applied Methodology for Multi-Objective Optimum Sizing of Hybrid Electric Vehicle Components. *Int. J. Hydrog. Energy* **2015**, *40*, 2312–2319. [CrossRef]

13. Al-Alawi, B.M.; Bradley, T.H. Review of Hybrid, Plug-In Hybrid, and Electric Vehicle Market Modeling Studies. *Renew. Sustain. Energy Rev* **2013**, *21*, 190–203. [CrossRef]

14. Mock, P. *Entwicklung eines Szenariomodells zur Simulation der zukünftigen Marktanteile und CO2-Emissionen von Kraftfahrzeugen (VECTOR21)*; Deutsches Zentrum für Luft- und Raumfahrt: Cologne, Germany, 2010.

15. Simeu, S.; Brokate, J.; Stephens, T.; Rousseau, A. Factors Influencing Energy Consumption and Cost-Competitiveness of Plug-in Electric Vehicles. *World Electr. Veh. J.* **2018**, *9*, 23. [CrossRef]

16. Rousseau, A.; Stephens, T.; Brokate, J.; Özdemir, E.D.; Klötzke, M.; Schmid, S.A.; Plötz, P.; Badin, F.; Ward, J.; Lim, O.T. Comparison of Energy Consumption and Costs of Different Plug-in Electric Vehicles in European and American Context. In Proceedings of the 28th International Electric Vehicle Symposium and Exhibition (EVS), Goyang, Korea, 3–6 May 2015.

17. Alternative Fuel Data Center. Maps and Data—Vehicle Weight Classes & Categories. Available online: https://afdc.energy.gov/data/ (accessed on 18 March 2019).

18. Autonomie. Available online: www.autonomie.net (accessed on 27 February 2018).

19. Federal Highway Administration. Highway Statistics 2013, Table VM-1. Available online: http://www.fhwa.dot.gov/policyinformation/statistics/2013/ (accessed on 4 June 2015).

20. Islam, E.; Ayman, M.; Kim, N.; Rousseau, A. *An Extensive Study on Sizing, Energy Consumption, and Cost of Advanced Vehicle Technologies*; Technical Report ANL/ESD-17/17; Argonne National Laboratory: Lemont, IL, USA, 2018.

21. Kim, N.; Moawad, A.; Vijayagopal, R.; Rousseau, A. Impact of Fuel Cell and Storage System Improvement on Fuel Consumption and Cost. *World Electr. Veh. J.* **2016**, *8*, 305–314. [CrossRef]

22. USDRIVE. Fuel Cell Technical Team Roadmap. June 2013. Available online: http://energy.gov/sites/prod/files/2014/02/f8/fctt_roadmap_june2013.pdf (accessed on 16 July 2018).

23. Eberle, U.; Muller, B.; Helmolt, R.V. Fuel Cell Electric Vehicles and Hydrogen Infrastructure: Status 2012. *Energy Environ. Sci.* **2012**, *5*, 8780–8798. [CrossRef]

24. Rogers, S. *EV Everywhere Grand Challenge: Electric Drive Status and Challenges, Vehicle Technologies Program—Advanced Power Electronics and Electric Motors*; U.S. Department of Energy: Washington, DC, USA, 24 July 2012.

25. SAE International. *Performance Requirements for Determining Tow-Vehicle Gross Combination Weight Rating and Trailer Weight Rating*; SAE International: Detroit, MI, USA, 2016.

Article

A Portable Direct Methanol Fuel Cell Power Station for Long-Term Internet of Things Applications

Chung-Jen Chou [1,*], Shyh-Biau Jiang [1], Tse-Liang Yeh [2], Li-Duan Tsai [3], Ku-Yen Kang [3] and Ching-Jung Liu [3]

[1] Institute of Opto-Mechatronics Engineering, National Central University, Taoyuan City 32001, Taiwan; sbjiang@viewmove.com
[2] Institute of Mechanical Engineering, National Central University, Taoyuan City 32001, Taiwan; tlyeh@cc.ncu.edu.tw
[3] Material and Chemical Research Laboratories (MCL), Industrial Technology Research Institute (ITRI), B77, 195, Sec. 4, Chung Hsing Rd. Chutung, Hsingchu City 31057, Taiwan; LiDuanTsai@itri.org.tw (L.-D.T.); KYK@itri.org.tw (K.-Y.K.); mo@itri.org.tw (C.-J.L.)
* Correspondence: petpoku@gmail.com; Tel.: +886-3-4227151 (ext. 34387)

Received: 22 April 2020; Accepted: 7 July 2020; Published: 9 July 2020

Abstract: With regard to the best electro-chemical efficiency of an active direct methanol fuel cell (DMFC), the stacks and their balance of plant (BOP) are complex to build and operate. The yield of making the large-scale stacks is difficult to improve. Therefore, a portable power station made of multiple simpler planar type stack modules with only appropriate semi-active BOPs was developed. A planar stack and its miniature BOP components are integrated into a semi-active DMFC stack module for easy production, assembly, and operation. An improved energy management system is designed to control multiple DMFC stack modules in parallel to enhance its power-generation capacity and stability so that the portability, environmental tolerance, and long-term durability become comparable to that of the active systems. A prototype of the power station was tested for 3600 h in an actual outdoor environment through winter and summer. Its performance and maintenance events are analyzed to validate its stability and durability. Throughout the test, it maintained the daily average of 3.3 W power generation with peak output driving capability of 12 W suitable for Internet of Things (IoT) applications.

Keywords: direct methanol fuel cell; multi stacks; portable power; energy management system; Internet of Things; long-term; in-field test

---

1. Introduction

Developing intelligent services to cover remote areas, Internet of Things (IoT) networks need to be deployed in places such as farms, forest, outlying islands, infrastructure piping, etc., where there is no electricity power grid. Acquiring power supply has always been a problem. Thus, this has led to demands for long-term reliable power supplies [1]. Secondary batteries are commonly used on IoT equipment. However, due to limited energy storage, connection to a power source for charging is still necessary for long operations. In addition to engine-driven generators, two types of power sources, namely, the energy-harvesting sources [2–5] and fuel cells (FCs) [6,7], can be used to charge. When installing energy-harvesting device, we need to consider the availability of the natural energy source and location suitability. Moreover, to meet steady demand, large enough capacity of the secondary batteries is needed to buffer the inherently uncontrollable variations in the natural energy resource [8–10]. Therefore, these power supplies are difficult to be portable. On the other hand, FCs demonstrate stable power generation needing relatively low battery buffering capacity. They can be installed quickly with the ability to work for a long time.

The FCs generate power by electrochemical reactions of different fuel types in different reaction modes. These are: polymer electrolyte fuel cell (PEMFC), direct methanol fuel cell (DMFC), direct formic acid fuel cell (DFAFC), direct ethanol fuel cell (DEFC), and solid oxide fuel cell (SOFC) [11–13]. Using liquid fuel, a DMFC has a higher fuel volume energy density so that the system is lightweight, easy to replenish, safe, and can be operated at room temperature. Therefore, DMFCs are suitable candidates portable power supplies for emergency deployments and IoT applications [14–16].

Facilitating an optimal environment for efficient electrochemical reactions, the balance of plant (BOP) of a DMFC can be an aggressive active type or a sub-optimal passive type. A passive DMFC maintains balance during electrochemical reactions without external forces. The architecture is simple. However, its electro-chemical efficiency is sensitive to the environment, the power generation is unstable, and the system durability is poor [17,18]. In an active DMFC, BOP components are installed to regulate the conditions for optimal reactions so that high power generation efficiency, stability, environmental tolerance, and long-term durability can be maximized for marketability. Therefore, the majority of DMFC products on the market are the active type [19]. Nevertheless, active DMFCs are disadvantaged by the complexity of the stack assembly process [20], the excessive number of BOP components, and complex in-system piping. Therefore, active DMFCs have low manufacturing yield, are difficult to miniaturize, and difficult to repair [21,22]. To ensure optimal operating conditions, the complexity of an active DMFC system is necessary. On the other hand, for simplicity, the passive DMFCs may be stable for only limited power output and operation time. However, for portability and long-term power generation reliability, a marketable semi-active DMFC design needs to be weighed between complexity and optimization. To address the complexity shortcomings of the active DMFCs, modularization would be one of the primary solutions. Additionally, several studies on the modularization of FCs, including stack modular design, combination of multiple stack modules, and control management of multiple stack modules have been proposed [23–27].

While a modularized stack can be individually installed, put in operation and serviced conveniently, there are great advantages to having multiple stack modules functioning collectively to increase power capacity, and to enhance the durability with added redundancies. Following this strategy, a novel architecture and control of the DMFC power station is developed with specifications suitable for IoT applications. Simplifying the system architecture, the functionalities of BOP components in the active DMFCs were reviewed. Only the necessary components were miniaturized and integrated onto the passive DMFC to form a compact semi-active DMFC stack module. The mechanism of a power station was reworked with much simplified wiring and piping to accept convenient multi-stack module plug-ins and to be operated collectively by one energy management system (EMS). Space for a larger fuel tank can, therefore, be made available for a longer period before refueling.

Although managing the fuel, heat and water of a DMFC for efficiency and durability is complicated and difficult compared with other types of FC [22,28], integrating controls of the electromechanical system with the EMS to accommodate the load demand can make improvements. Simultaneous controls of the limited BOP components in multiple DMFC stacks can maintain proper operating conditions of the semi-active DMFC power station effectively such that its environmental tolerance, endurance, and power generation stability of the system can approach that of an active DMFC. In verifying our new design, long-term evaluation of the system performance, the regulation of operating conditions by the BOP components, and the changes in stack characteristics is essential [29,30]. Therefore, a prototype of our power station was tested for 3600 h in actual outdoor environments throughout winter and summer under different weathers, temperatures, and humidity to verify its suitability for IoT applications in the field.

2. Design of a Modular Semi-Active Direct Methanol Fuel Cell (DMFC) Power Station

A DMFC power station contains two major systems. One is the DMFC stack system for electrochemical reactions, and the other is the EMS to control the reaction conditions and energy balance.

2.1. DMFC Stack System

The DMFC stack system consists of a fuel-cell stack, BOP components and piping for air or liquid delivery. The DMFC stack is the core for the electrochemical reaction, and the BOP components assist in maintaining appropriate reaction conditions, supplying appropriate amount of reactants, and discharging by-products.

Figure 1 is a comparison of the BOP components in an active and our semi-active DMFC stack system. The whole block diagram represents an active system architecture, while the semi-active system consists of only a fuel tank plus the members encircled by the red dashed line. All four members enclosed in the dashed line can be constructed using only miniaturized components such that the complex active DMFC stack system can be slimmed down to a compact stack module. In addition, the semi-active DMFC stack module is made of only two separate sets of planar stack and flow field plates without the need for precise channel alignment and pressurized spacing, thereby greatly reducing the production difficulties. With a sufficient air contact area, an air blower instead of an air pump is enough and one piezoelectric dosing pump is enough to supply fuel. Thus, miniature BOPs can be integrated into the stack module.

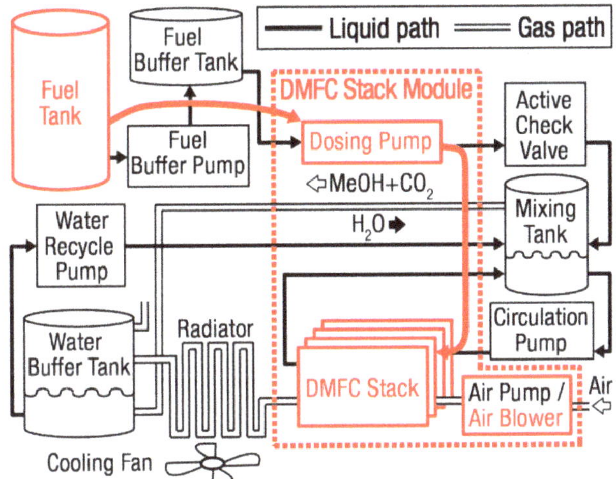

Figure 1. Comparing full balance of plant (BOP) components in an active direct methanol fuel cell (DMFC) stack system with that of a semi-active system marked in red, where a DMFC stack module is enclosed by the red dashed line.

Simplifying the active architecture, the semi-active stack module does not have active temperature, humidity control, fuel buffer tank, nor check valve, therefore, supporting measures are needed. The feedback control on the fuel-supplying dosing pump was implemented to compensate the influence of changing the fuel level height on the flow rate. Fuel tank was relocated below the DMFC stack modules to prevent leakage due to either gravity or pressure imbalances. Due to the passive water recovery mechanism, the air supply flow rate was reduced and the reaction temperature was also decreased from the electrochemical optimum of 75 °C to 45 °C to reduce excessive water evaporative loss.

Figure 2 shows the photo of the semi-active DMFC stack module proposed. The stack module has a sandwich structure, with planar DMFC stacks on both sides with the metal electrodes protruding on the top. The flow field plate is in the middle with two air blowers on the right edge blowing air supplying oxygen. The fuel dosing pump is mounted at the lower right corner with an inlet connected to the fuel tank through a hose. It has membrane electrode assembly (MEA) of 61.6 cm^2, 8 cells series, operating at 2.8 V to generate 1 W nominally. Its dimension is $120 \times 60 \times 15$ mm^3 and it weighs less than 200 g.

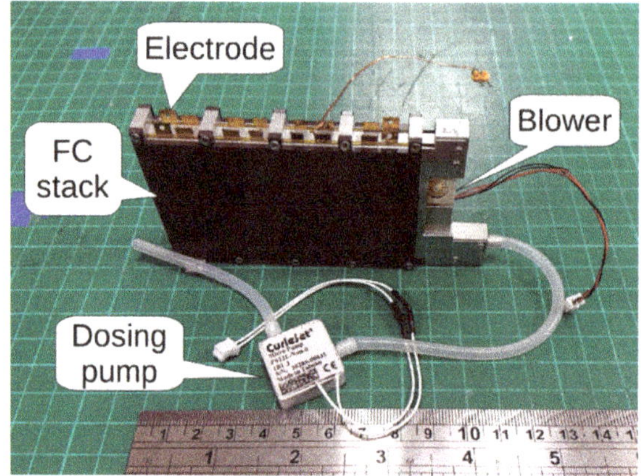

Figure 2. Semi-active DMFC stack module.

2.2. Energy Management System (EMS)

Figure 3 is the functional block diagram of the semi-active DMFC power station. The DMFC stack system is located on the top consisting of a fuel tank and a set of four DMFC stack modules. The lower half is the function block diagram of the EMS controlling the collective operating conditions and status report communication.

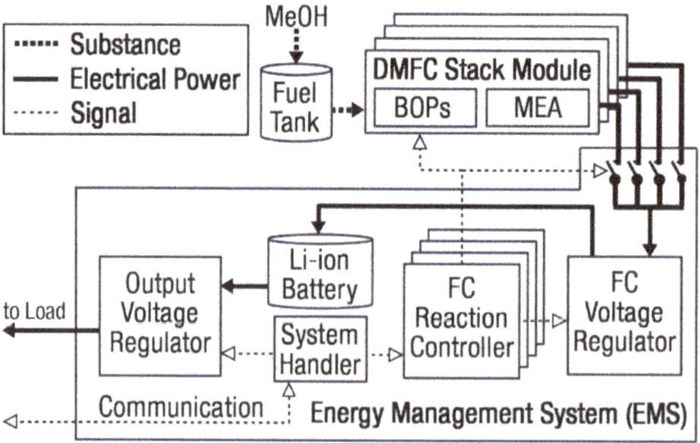

Figure 3. Semi-active DMFC power station functional block diagram.

The EMS has a system handler at the core of the multiple sets of FC reaction controller. FC reaction controllers coordinate the BOPs in the stack modules to work, while the FC voltage regulators keep the modules properly loaded to maintain the electrochemical reactions and manage the energy storage into the battery. The switches between the stack modules and the regulator determine the operating mode of the corresponding module, whether it is offline for maintenance or online in parallel power generation. Li-ion batteries act as energy buffer between generation and output demand. The output voltage regulator serves the output demand. There is a core supervisor monitoring the status of the batteries to control the inhibition of the power generation and the output power drive so that batteries are protected from over charge or discharge.

When DMFC stacks are activated, they can be switched online and offline independently, and their working voltages and currents can change rapidly as a variable energy source. In order to have their electricity output connected in parallel for stable power generation, the dynamic energy management control is designed to maintain the stability of their reaction voltages to the setpoint V_{sp} by the FC voltage regulator. Given the characteristics of the MEA, the generator current estimator estimates I_{es} as the target current to draw from stacks. It also adjusts the current estimate based on the error between the present FC voltage V_{FC} and the setpoint voltage V_{sp} so that V_{FC} converges to V_{sp}. In the inner loop, the current feedback controller measures the actual output current I_o and quickly adjusts the direct current to direct current (DC/DC) converter so that I_o tracks I_{es}. Meanwhile, all electricity generated, namely I_o drawn from the stack, gets pumped into the battery for storage.

The right side of Figure 4 shows the output voltage regulator taking energy from the Li-ion battery to drive the output demand. Wherein, the output converter is a boost DC/DC converter controlled by the system handler. When enabled, it draws from the battery voltage to drive the output voltage. The output would be turned off when disabled to prevent over drawing on the batteries. An eFuse module which prevents excessive output current and reversed current against load variation improves the reliability of the power station.

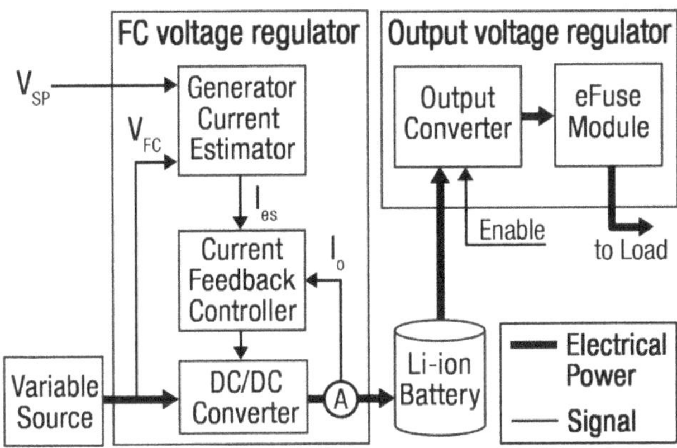

Figure 4. The fuel-cell (FC) voltage regulator and the output voltage regulator block diagram.

The electrochemical reaction conditions of the DMFC stack modules are maintained by their individual BOPs under the coordinated control of the FC reaction controller as shown in Figure 5a. The reaction voltage calculator determines the appropriate stack reaction voltage setpoint V_{sp}, based on the ambient conditions, the stack temperature and the characteristics of the DMFC stack module, for the FC voltage regulator to follow. The fuel supply calculator determines the fuel consumption to control the dosing pumps to replenish according to the sensor-less approach equation, Equation (1). Assuming only a portion of the fuel consumed by the DMFC stack module is effectively converted into electricity I_{FC} and the rest is the temperature dependent crossover, thus derived the fuel consumption estimator according to sensing measurements.

$$S = K_T \int_0^{\Delta t} I_{FC} \cdot dt + K_c \cdot \Delta t \qquad (1)$$

where S is the total fuel consumption over a given period Δt, K_T is the fuel consumption coefficient resulting in effective power generation represented by the DMFC output current I_{FC}, and K_c is the fuel consumption rate resulting in the crossover. Both K_T and K_c coefficients are determined from the MEA size, the stack module characteristics, and the ambient temperature. The backbone of the controls is to maintain the proper heat generation for temperature management. The fuel consumption is estimated by Equation (1) to determine the appropriate fuel supply to maintain, to increase, or to decrease current operating temperature.

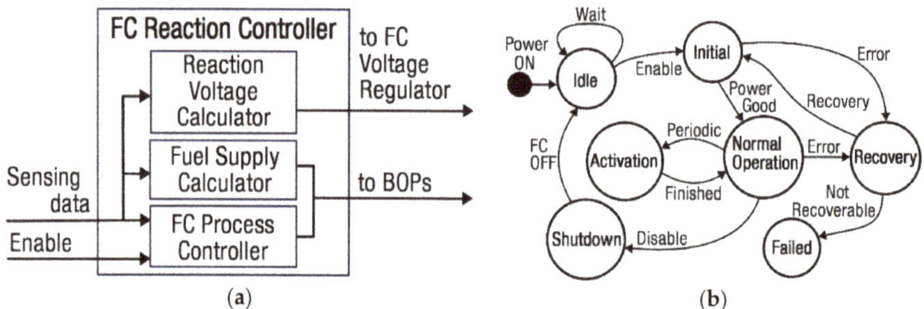

Figure 5. (**a**) Functional block diagram of an FC reaction controller; (**b**) the state machine of the FC process controller.

Figure 5b illustrates the state machine of the FC process controller explaining the logic behind the controls. When the power station is turned on, it first enters the idle state. When the enable signal is triggered, it enters the initial state that the controller controls the fuel supply and stack voltages to increase the stack temperature progressively. When the initial process ends, the state enters the normal operation state. If it yields an abnormal result, then the process enters the recovery state. In the normal operation state, the stack module continues to generate power, perform self-tests, and periodically switches to the activation state. If an error occurs, then the process switches to the recovery state. When the disable command is received, it switches to the shutdown state for safe turn off. While operating in the recovery state, the process attempts to resolve the stack problem. If it succeeds, the controller returns to the initial state to restart or the controller enters the fail state to stop any operations on this stack. In the activation state, the process blocks the stack current, stops supplying oxygen and waits 60 s. Afterwards, the controller returns to normal operation to generate electricity again. This process refreshes the MEAs, the reaction efficiency of which deteriorated after long usage.

The system handler of the EMS sets the system operating modes according to the commands received from the communication as well as the Li-ion battery voltage monitored. The power output is disabled when the battery voltage is less than 3.5 V, and it will be reactivated only until the voltage exceeds 3.7 V. The station power generating is enabled when the battery voltage falls below 3.8 V and will be disabled when the charged voltage rises above 4.1 V. The overall power balance of whole DMFC power station is described by Equation (2)

$$P_{BAT} = P_{FC} \cdot \eta_{Reg} - P_{Load} \cdot D_{Load} - P_{EMS} \tag{2}$$

where P_{BAT} denotes the power charging the batteries, P_{FC} is the total FC power generation, η_{Reg} is the efficiency of the FC voltage regulator, P_{Load} is the power station output power to the external load, D_{Load} is the working duty of the external load, and P_{EMS} is the power consumption of EMS. With the battery buffering and the output voltage regulator, P_{Load} is designed up to 12 W exceeding P_{FC} and to be suitable to drive IoT applications.

2.3. DMFC Power Station

Figure 6a shows the front view of the portable DMFC power station. Its dimension is $26 \times 22 \times 22 \text{ cm}^3$, and it weighs approximately 2 kg. A fuel injection port is located at the left of the front cover, a heat sink was attached on the right side cover, and vent holes were provided on the top and side for good air circulation.

Figure 6b shows the rear view of the inside. A 4 L fuel tank approximately half of the height of the power station was placed at the bottom providing 1.5 to 2 kWh energy generation capacity to last longer than 20 days of continuous operation. DMFC stack module sockets to accept four stack modules were placed on the upper left side. The number of modules to be plugged in can be determined by the power demand of the targeted application. The EMS control board is located on the upper right. The Li-ion batteries are placed at the right front of the fuel tank. The installation and replacement of the DMFC stack modules are simple and quick, connecting the dosing pump to the fuel tank with a hose, plugging electronic signal and power cables into the corresponding EMS sockets, and it is done. The power station is easy to assemble, to maintain, and portable. We conducted a long-term field test to demonstrate its feasibility for IoT applications in outdoor environments.

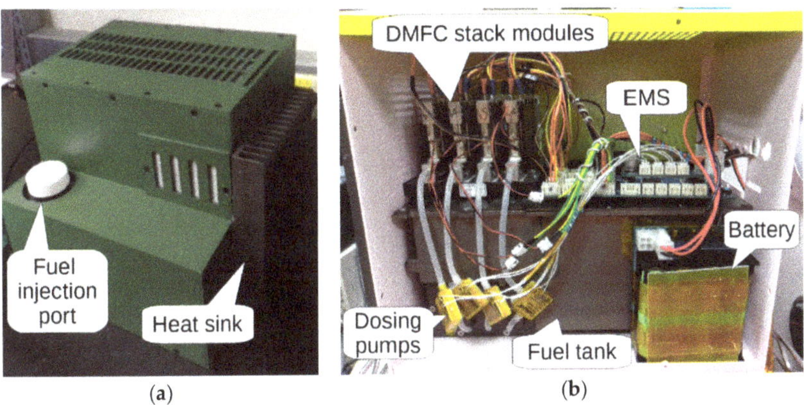

Figure 6. Portable DMFC power station. (**a**) Front view; (**b**) rear view internal configuration

3. Experimental Setup

To verify the viability for field applications, the ability of long-term stable electric power generation in a real outdoor environment is a key issue. The experimental test was performed over half a year, beginning in winter and ending in summer, thus covering a variety of weather conditions with wide temperature and humidity variations. From the test, the degradation of components was assessed, and the controls, the fueling, and the maintenance strategies was further refined.

Figure 7a depicts the experimental environment and instrument configuration. The power station was placed in a small outdoor rain-proof cabinet on the roof of a building to simulate an actual field environment. A supervising personal computer was placed indoors for remote status check and data logging. Figure 7b shows the inside of the cabinet where the power station was placed in the test and the dummy load acting as its power demand. Throughout the test, supervising the system for debugging as well as validation purposes, the data recorded are the voltage, current, temperature of the stack modules, the Li-ion battery, the outdoor ambient, and the inside of the cabinet. The performance data were accumulated to one sample per minute for analysis.

Figure 7. DMFC power station experiment setup. (**a**) Outside view; (**b**) inside view.

To test the power station to its full capacity, with 4 stack modules installed, the power generating capability is 3 to 4 W. To guarantee continuous electricity generation at full power for shortening test time, a 10 W thermoelectric cooler was selected as the dummy load. Therefore, P_{Load} was set to 10 W and D_{Load} was set to 1 in the power balance Equation (2), and the battery voltage varied between 3.5 V to 3.7 V due to the EMS over-drainage protection.

4. Results and Discussion

The full 3600 h of long-term outdoor power generation test data are shown in Figure 8. The figure shows the history of temperatures, the variations of stack power generations, and the maintenance events.

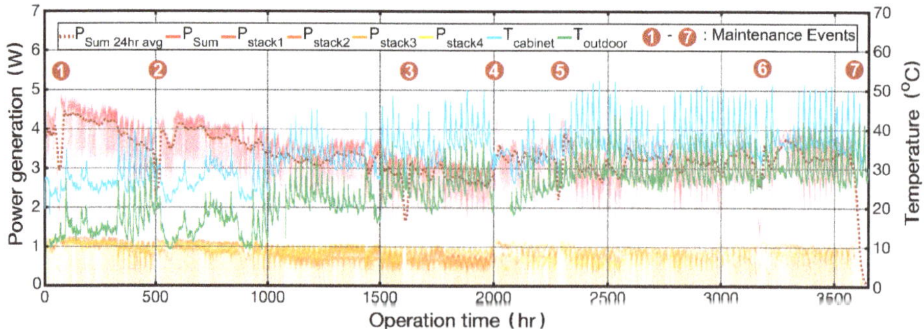

Figure 8. Long-term power-generation test data in an actual outdoor environment.

4.1. Temperature

The temperature outdoors and that inside of the cabinet is shown in green and blue respectively. The temperature inside can be higher by as much as 10–15 °C due to direct sunlight and self-heating by the power station. In the winter, the outdoor temperature fell below 10 °C, while it was always higher than 20 °C inside with the power station in operation except for a short time from a cold start. In the summer, the maximum outdoor temperature reached 40 °C while it rose to 50 °C inside the cabinet.

4.2. Power

The four thin lines in brown, red-orange, yellow-orange, and yellow, at the bottom of Figure 8 represent the power generation of the four stacks, respectively. The power fell to zero periodically as the power generation was suspended for the stack refreshing in the activation state. The theoretical optimal power generation of a single stack was approximately 1 W, and the actual power generation decreased when the reaction temperature deviated to either side from the optimal.

The thin line in red depicts the total power generated, while its 24-h daily moving average is shown by the dark red dotted line. The total power generated reached 4 W at the beginning then decreased gradually down to less than 3 W after hour 1600 when the weather warmed up. In order to compensate for the evaporation of water at high summer temperature, the methanol concentration was reduced to 80 *v/v*% after hour 2006 from the pure methanol for the winter. The power generation recovered to 3.3 W throughout the rest of the experiment.

4.3. Maintenance

In Figure 8, the sudden changes in the daily power average represented an abnormal event of the power station and they are marked by event numbers in chronological order. Event at hour 55 (❶), the power station suffered a shutdown due to initial wiring mistake. Normal operation resumed after correction. Events at hour 497 (❷), hour 1602 (❸), and hour 2280 (❺) were shutdowns caused by exhaustion of fuel. The power station resumed operation after replenishment. At hour 2006 (❹), the fuel concentration was changed for the summer. Event at hour 3165 (❻) was the first failure when the rusted buckle of dosing pump 2 broke and the deteriorated dosing pump 4 was also replaced. The power station resumed power generation after maintenance. After hour 3581 (❼) the characteristics of 4th stack became abnormal, and the rest of the stacks became flooded and failed subsequently. The power station stopped its operation and concluded the long-term test after hour 3628.

4.4. Failure Analysis

Investigating the episodes resulting in the final failure, Figure 9 zooms in on the records of the four stacks in the range of hour 3581 to 3631. The frames show the voltages, the power generation, and the temperature from top to bottom. The figure shows a normal operation initially in which the power outputs from the stacks were all approximately 1 W, the voltages were controlled at 2.8 V, and the stack temperatures were rising above 55 °C as the ambient temperature rose.

The failure event started to develop at 3582 h into the experiment when the power generation and, thus, also the temperature, of the 4th stack on one side of the station, as shown in yellow, became lower than other stacks. The 4th stack was cut off and shut down because an abnormality was determined by the EMS at hour 3589 (❶) and its voltage peaked briefly and dropped gradually to zero followed by its temperature reaching 7 °C lower than others. It started to sink heat from the others. After hour 3591, the power of the 2nd stack, shown in red-orange, began to oscillate and become frail even though the ambient temperature fell back. At hour 3600 (❷), the 2nd stack got cut off, and stopped generating power causing temperature drops. Since it is sandwiched by the still normal and heat generating stacks 1 and 3, its temperature did not decrease as much as the 4th stack. The power of the 1st stack also became frail, as shown in brown, until it stopped at hour 3606 (❸). At this time, the power generation of the 3rd stack, shown in orange, briefly increased to compensate for the over-cooling temperature, and then finally diminished. At hour 3621 (❹), the researcher came to check and restarted the power station. Although the 1st and 2nd stacks could still be restarted, their powers quickly failed again. The EMS determined that the system could not operate anymore at hour 3628 (❺) and stopped. Based on this final episode, from the first stack power failure to the total shutdown took about 32 h. With multiple stack modules in the system, the redundancies enhanced durability and provided lead time to call for maintenances.

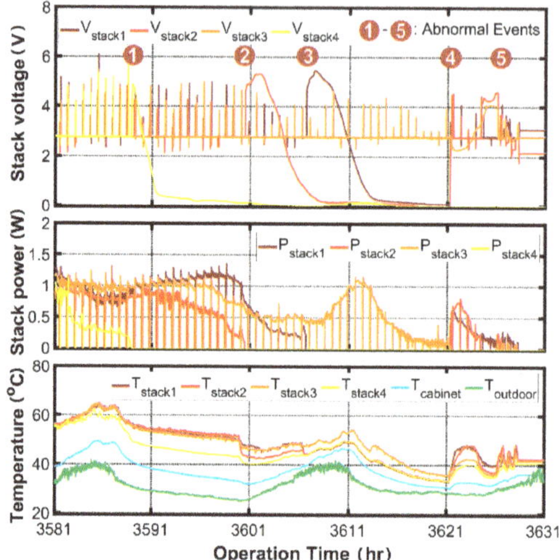

Figure 9. High-resolution experimental record from 3581 to 3631 h.

A forensic engineering study was carried out on the failed stack modules, and the final failure was caused by the aging of MEAs. The degradation of electrochemical efficiency leads to increasing fuel demand at the targeted operating condition beyond MEA capability and successively causing damage by flooding.

4.5. Fault Management

Analyzing the episode of the aforementioned failure events for causes and effects, the design and control logics can be refined for better durability. In the final failure, it commenced when a stack had an abnormality and was turned off, whereby the abnormal stack stopped generating power and started sinking the heat from other stacks causing the EMS to make wrongful judgments that led to other flooding failures successively. While the logic of the aggressive fault tolerant auto-recovery feature is still under development, a conservative fault-management strategy could be preferred during the verification to reduce the risk of chained faults. In the case of any stack failure, just shut down the station, and send a notification of abnormality for forensic investigation, maintenance, and improvements.

4.6. Discussion

The long-term performance of our system was determined from the test record, which is compared with other DMFC systems reported in the literature as shown in Table 1.

In designing the power driving capability, although our semi-active DMFC with 4 stack modules has only rated 4 W, we added Li-ion battery buffering as well as an output regulator to boost intermittent load-driving capability to 12 W. With 12 W, it can drive high-power sensors, actuators, or a wireless device for radio communications. Therefore, the design is suitable for IoT applications. Due to the simplified components and piping, our effective system volume-based energy density reached a comparable 152 Wh/L despite a low 10% electrochemical conversion efficiency and low fuel volume-based energy density of 500 Wh/L. Moreover, due to mechanical compactness in exchange for a larger fuel tank while maintaining portability, 480 h of operating time between refueling is more than 3 times that of other systems under persistent full power demand.

Table 1. Performance specifications of the DMFC generating systems reported in the literature [19,31].

	Generating Power	System Volume-Based Energy	Fuel Volume-Based Energy	Conversion Efficiency	Cartridge Volume and Generating Time
Motorola Lab [32]	2 W	155 Wh/L	956 Wh/L	20%	48 h with 200 cc methanol
LG Chem [33]	25 W	250 Wh/L	1250 Wh/L	26%	10 h with 200 cc methanol
Toshiba [34]	1 W	143 Wh/L	800 Wh/L	16%	20 h with 25 cc methanol
SFC EFOY COMFORT 80i [35]	40 W	163 Wh/L [1]	1100 Wh/L	22%	137.5 h with 5 L methanol
Our system	4 W (output 12 W)	152 Wh/L	500 Wh/L	10%	480 h with 4 L methanol

[1] Consider with M5 fuel cartridge.

This system demonstrated a stable average power generation above 3.3 W throughout the 3600 h of test. With semi-active controls, adjusting the methanol concentration from pure to 80 *v/v*% helped to improve the power-generation efficiency in hot summer. For the long run, operating characteristics could be updated to reflect the deterioration, or the geometric thermal conditions, so that premature flooding can be avoided and the life can be further extended. To increase power capacity, enlarging the stack for a larger reaction area, increasing the number of stack modules, and improving the reaction efficiency and stability by further optimizing the thermal balance controls are all possible. While increasing the volume energy density for portability, the space utilization of the system can be further polished.

5. Conclusions

A novel portable DMFC power station has been developed for long-term field IoT applications with modularized semi-active stacks and mechatronic controls. Its configuration and controls are introduced in detail. The architecture of the active DMFC stack system was simplified to solve the problems of low manufacturing yield and high maintenance difficulties of active systems. Meanwhile, the EMS was redesigned to achieve coordinated controls on multi semi-active DMFC stack modules to increase configuration flexibility and power capability while overcoming the instability of the passive systems.

The prototype of the power station was tested for 3600 h operation in an actual outdoor environment throughout winter and summer with the ambient temperature ranging from 10 °C to 40 °C and a variety of weather conditions to verify its in-field feasibility and reliability. The causes of the 3 faults were discovered and resolved fundamentally. Therefore, the power station is proven to have long-term outdoor weather endurance, stable power generation, high system volume-based energy density, and easy portability. It is also enhanced with Li-ion batteries buffering and an output regulator to provide an intermittent high-power output drive. Therefore, it is feasible and marketable for remote mobile IoT applications.

Author Contributions: The project was conceptualized by K.-Y.K. and C.-J.C. under the project leadership of L.-D.T. Subsequent methodology development and validation was carried out by C.-J.C., K.-Y.K., and C.-J.L. on the EMS with mechatronics controls and data curation and analysis by C.-J.C. The academic writing was initially drafted by C.-J.C. and S.-B.J. and was later refurbished by C.-J.C. and T.-L.Y. for publication. All authors have read and agreed to the published version of the manuscript.

Funding: This research received no external funding.

Acknowledgments: We thank the Industrial Technology Research Institute of Taiwan for providing the resources, the technology of fuel cells, as well as their supports so that the development and verification was possible.

Conflicts of Interest: The authors declare no conflict of interest.

References

1. Groom, T.B.; Gabl, J.R.; Pourpoint, T.L. Portable Power Generation for Remote Areas Using Hydrogen Generated via Maleic Acid-Promoted Hydrolysis of Ammonia Borane. *Molecules* **2019**, *24*, 4045. [CrossRef]
2. Chalasani, S.; Conrad, J.M. A survey of energy harvesting sources for embedded systems. In Proceedings of the IEEE SoutheastCon 2008, Huntsville, AL, USA, 3–6 April 2008; pp. 442–447.
3. Shaikh, F.K.; Zeadally, S. Energy harvesting in wireless sensor networks: A comprehensive review. *Renew. Sustain. Energy Rev.* **2016**, *55*, 1041–1054. [CrossRef]
4. Ko, H.; Lee, J.; Jang, S.; Kim, J.; Pack, S. Energy Efficient Cooperative Computation Algorithm in Energy Harvesting Internet of Things. *Energies* **2019**, *12*, 4050. [CrossRef]
5. Zareei, M.; Vargas-Rosales, C.; Anisi, M.H.; Musavian, L.; Villalpando-Hernandez, R.; Goudarzi, S.; Mohamed, E.M. Enhancing the Performance of Energy Harvesting Sensor Networks for Environmental Monitoring Applications. *Energies* **2019**, *12*, 2794. [CrossRef]
6. Wilberforce, T.; Alaswad, A.; Palumbo, A.; Dassisti, M.; Olabi, A.G. Advances in stationary and portable fuel cell applications. *Int. J. Hydrog. Energy* **2016**, *41*, 16509–16522. [CrossRef]
7. Chen, Y.-S.; Lin, S.-M.; Hong, B.-S. Experimental Study on a Passive Fuel Cell/Battery Hybrid Power System. *Energies* **2013**, *6*, 6413–6422. [CrossRef]
8. Bajpai, P.; Dash, V. Hybrid renewable energy systems for power generation in stand-alone applications: A review. *Renew. Sustain. Energy Rev.* **2012**, *16*, 2926–2939. [CrossRef]
9. Lagorse, J.; Paire, D.; Miraoui, A. Sizing optimization of a stand-alone street lighting system powered by a hybrid system using fuel cell, PV and battery. *Renew. Energy* **2009**, *34*, 683–691. [CrossRef]
10. Alsharif, M. A Solar Energy Solution for Sustainable Third Generation Mobile Networks. *Energies* **2017**, *10*, 429. [CrossRef]
11. Ellis, M.W.; Von Spakovsky, M.R.; Nelson, D.J. Fuel cell systems: Efficient, flexible energy conversion for the 21st century. *Proc. IEEE* **2001**, *89*, 1808–1818. [CrossRef]
12. Velisala, V.; Srinivasulu, G.N.; Reddy, B.S.; Rao, K.V.K. Review on challenges of direct liquid fuel cells for portable application. *World J. Eng.* **2015**, *12*, 591–606. [CrossRef]
13. Carrette, L.; Friedrich, K.; Stimming, U. Fuel cells–fundamentals and applications. *Fuel Cells* **2001**, *1*, 5–39. [CrossRef]
14. Zhang, T.; Wang, Q.-M. Performance of miniaturized direct methanol fuel cell (DMFC) devices using micropump for fuel delivery. *J. Power Sources* **2006**, *158*, 169–176. [CrossRef]
15. Das, V.; Padmanaban, S.; Venkitusamy, K.; Selvamuthukumaran, R.; Blaabjerg, F.; Siano, P. Recent advances and challenges of fuel cell based power system architectures and control—A review. *Renew. Sustain. Energy Rev.* **2017**, *73*, 10–18. [CrossRef]
16. Kuan, Y.-D.; Lee, S.-M.; Sung, M.-F. Development of a Direct Methanol Fuel Cell with Lightweight Disc Type Current Collectors. *Energies* **2014**, *7*, 3136–3147. [CrossRef]
17. Zhu, Y.; Liang, J.; Liu, C.; Ma, T.; Wang, L. Development of a passive direct methanol fuel cell (DMFC) twin-stack for long-term operation. *J. Power Sources* **2009**, *193*, 649–655. [CrossRef]
18. Achmad, F.; Kamarudin, S.K.; Daud, W.R.W.; Majlan, E.H. Passive direct methanol fuel cells for portable electronic devices. *Appl. Energy* **2011**, *88*, 1681–1689. [CrossRef]
19. Kamaruddin, M.Z.F.; Kamarudin, S.K.; Daud, W.R.W.; Masdar, M.S. An overview of fuel management in direct methanol fuel cells. *Renew. Sustain. Energy Rev.* **2013**, *24*, 557–565. [CrossRef]
20. Sgroi, M.; Zedde, F.; Barbera, O.; Stassi, A.; Sebastián, D.; Lufrano, F.; Baglio, V.; Aricò, A.; Bonde, J.; Schuster, M. Cost Analysis of Direct Methanol Fuel Cell Stacks for Mass Production. *Energies* **2016**, *9*, 1008. [CrossRef]
21. Kamarudin, S.K.; Achmad, F.; Daud, W.R.W. Overview on the application of direct methanol fuel cell (DMFC) for portable electronic devices. *Int. J. Hydrog. Energy* **2009**, *34*, 6902–6916. [CrossRef]
22. Li, X.; Faghri, A. Review and advances of direct methanol fuel cells (DMFCs) part I: Design, fabrication, and testing with high concentration methanol solutions. *J. Power Sources* **2013**, *226*, 223–240. [CrossRef]
23. Marx, N.; Boulon, L.; Gustin, F.; Hissel, D.; Agbossou, K. A review of multi-stack and modular fuel cell systems: Interests, application areas and on-going research activities. *Int. J. Hydrog. Energy* **2014**, *39*, 12101–12111. [CrossRef]

24. Vulturescu, B.; De Bernardinis, A.; Lallemand, R.; Coquery, G. Traction power converter for PEM fuel cell multi-stack generator used in urban transportation. In Proceedings of the 2007 European Conference on Power Electronics and Applications, Aalborg, Denmark, 2–5 September 2007; pp. 1–10.

25. Cardozo, J.; Marx, N.; Boulon, L.; Hissel, D. Comparison of Multi-Stack Fuel Cell System Architectures for Residential Power Generation Applications Including Electrical Vehicle Charging. In Proceedings of the 2015 IEEE Vehicle Power and Propulsion Conference (VPPC), Montreal, QC, Canada, 19–22 October 2015; pp. 1–6.

26. Wu, C.-C.; Chen, T.-L. Dynamic Modeling of a Parallel-Connected Solid Oxide Fuel Cell Stack System. *Energies* **2020**, *13*, 501. [CrossRef]

27. Yang, C.-H.; Chang, S.-C.; Chan, Y.-H.; Chang, W.-S. Dynamic Analysis of the Multi-Stack SOFC-CHP System for Power Modulation. *Energies* **2019**, *12*, 3686. [CrossRef]

28. Hu, X.; Wang, X.; Chen, J.; Yang, Q.; Jin, D.; Qiu, X. Numerical Investigations of the Combined Effects of Flow Rate and Methanol Concentration on DMFC Performance. *Energies* **2017**, *10*, 1094. [CrossRef]

29. Kim, Y.; Shin, D.; Seo, J.; Chang, N.; Cho, H.; Kim, Y.; Yoon, S. System integration of a portable direct methanol fuel cell and a battery hybrid. *Int. J. Hydrog. Energy* **2010**, *35*, 5621–5637. [CrossRef]

30. Leo, T.; Raso, M.; Navarro, E.; Mora, E. Long Term Performance Study of a Direct Methanol Fuel Cell Fed with Alcohol Blends. *Energies* **2013**, *6*, 282–293. [CrossRef]

31. Zhao, T.S.; Yang, W.W.; Chen, R.; Wu, Q.X. Towards operating direct methanol fuel cells with highly concentrated fuel. *J. Power Sources* **2010**, *195*, 3451–3462. [CrossRef]

32. Xie, C.; Bostaph, J.; Pavio, J. Development of a 2W direct methanol fuel cell power source. *J. Power Sources* **2004**, *136*, 55–65. [CrossRef]

33. LG Chem commercializes portable fuel cell. *Fuel Cells Bull.* **2005**, *2005*. [CrossRef]

34. Latest DMFC prototypes from Toshiba, Hitachi. *Fuel Cells Bull.* **2003**, *2003*. [CrossRef]

35. EFOY COMFORT Technical Data. Available online: https://www.efoy-comfort.com/technical-data (accessed on 1 March 2020).

Article

Comparative Analysis of On-Board Methane and Methanol Reforming Systems Combined with HT-PEM Fuel Cell and CO_2 Capture/Liquefaction System for Hydrogen Fueled Ship Application

Hyunyong Lee [1,2], Inchul Jung [1], Gilltae Roh [1], Youngseung Na [3] and Hokeun Kang [2,*]

[1] R&D Division, Korean Register, 36, Myeongji Ocean City 9-ro, Gangseo-gu, Busan 46762, Korea; leehy@krs.co.kr (H.L.); icjung@krs.co.kr (I.J.); gtroh@krs.co.kr (G.R.)

[2] Division of Marine System Engineering, Korea Maritime and Ocean University, 727 Taejong-ro, Yeongdo-Gu, Busan 49112, Korea

[3] Department of Mechanical and Information Engineering, University of Seoul, Seoulsiripdaero 163, Dongdaemun-gu, Seoul 02504, Korea; ysna@uos.ac.kr

* Correspondence: hkkang@kmou.ac.kr; Tel.: +82-51-410-4260; Fax: +82-404-3985

Received: 4 December 2019; Accepted: 25 December 2019; Published: 2 January 2020

Abstract: This study performs energetic and exergetic comparisons between the steam methane reforming and steam methanol reforming technologies combined with HT-PEMFC and a carbon capture/liquefaction system for use in hydrogen-fueled ships. The required space for the primary fuel and captured/liquefied CO_2 and the fuel cost have also been investigated to find the more advantageous system for ship application. For the comparison, the steam methane reforming-based system fed by LNG and the steam methanol reforming-based system fed by methanol have been modeled in an Aspen HYSYS environment. All the simulations have been conducted at a fixed $W_{net,\ electrical}$ (475 kW) to meet the average shaft power of the reference ship. Results show that at the base condition, the energy and exergy efficiencies of the methanol-based system are 7.99% and 1.89% higher than those of the methane-based system, respectively. The cogeneration efficiency of the methane-based system is 7.13% higher than that of the methanol-based system. The comparison of space for fuel and CO_2 storage reveals that the methanol-based system requires a space 1.1 times larger than that of the methane-based system for the total voyage time, although the methanol-based system has higher electrical efficiency. In addition, the methanol-based system has a fuel cost 2.2 times higher than that of the methane-based system to generate 475 kW net of electricity for the total voyage time.

Keywords: steam methane reforming; steam methanol reforming; electrical efficiency; exergy efficiency; LNG

---

1. Introduction

The International Maritime Organization assessed that international shipping accounted for about 2.2% of total carbon dioxide emissions in 2012, which is approximately 796 million tonnes of CO_2, and forecasted that this amount will increase between 50% and 250% in the period to 2050 under a business-as-usual scenario [1]. However, when the Paris Agreement on climate change mitigation was adopted in 2015 to deal with the global-warming concerns, shipping was not included [2]. Instead, in April 2018, the IMO established a strategy to reduce the total amount of annual greenhouse gas (GHG) emissions from shipping by at least 50% by 2050 compared to 2008 [3]. To achieve this, several technical and operational measures that improve energy efficiency and reduce CO_2 emissions in the shipping industry should be introduced, such as increasing energy efficiency of engines, adopting waste heat recovery systems, improving the hull form, implementing speed reduction and alternative sea routes [4]. In addition to the above measures, using different propulsion systems, such as fuel cells, are also considered possible alternatives [5]. Hydrogen fuel cells emit no direct GHGs, but the emissions generated during hydrogen production should be considered. The emissions from hydrogen production are highly dependent on feedstock and primary energy sources [6]. In the shipping industry, fuel cell power generation can eliminate NOx, Sox, and particulate material (PM) emissions, and reduce CO_2 emission compared to emissions from conventional diesel engines [2]. The advantage of fuel cells for maritime applications is the reduction of noise, vibrations, and infra-red signatures, along with their modular and flexible design, water generation, etc., although they may be application specific [5]. Seven fuel cell types were evaluated from the perspectives of relative cost, module power level, lifetime, tolerance for cycling, flexibility towards type of fuel, technological maturity, size, sensitivity to fuel impurities, emissions, safety, and efficiency. As a result of the evaluation, the low temperature proton exchange membrane (PEM) fuel cell and the high temperature PEM fuel cell (HT-PEMFC) received, respectively, the first and second highest score in the ranking, and this implied that those technologies are the most promising for marine use [7]. An LT-PEMFC has relatively higher power density than an HT-PEMFC; however, the HT-PEMFC has several other advantages [8]. The operating temperature of LT-PEMFCs is between 50 and 90 °C and that for HT-PEMFCs is between 120 and 200 °C. The lower operational temperature of LT-PEMFCs results in limited tolerance to fuel impurities such as sulfur and carbon monoxide (CO), which reduce its performance drastically. To be specific, LT-PEMFCs require fuels containing less than 30 ppm of CO and less than 1 ppm of sulfur, whereas HT-PEMFCs can work with concentrations of up to 3% of CO and 20 ppm of sulfur in the fuel without permanent degradation [7]. This higher tolerance to impurities of HT-PEMFC makes it possible to develop a simpler fuel reforming system [8]. Additionally, the water management of HT-PEMFCs is easier because the water produced in the fuel cell is in vapor, and the waste heat from HT-PEMFCs can be recovered and used for steam or hot water generation [9].

Since the gravimetric energy density of hydrogen is approximately 120 MJ/kg, 2.47 times higher than natural gas and 2.8 times higher than diesel, hydrogen provides higher gravimetric energy than other fossil fuels. However, volumetric energy density of liquid hydrogen is approximately 8.51 GJ/m^3, which corresponds to 40.8% of natural gas and 23.7% of diesel [10]. The lower volumetric energy density could be a drawback for some vessels in those applications that cannot support a large volume of storage or higher frequency of refueling [11]. To overcome this, several maritime fuel cell studies have considered on-board reforming of methane and methanol to hydrogen, although the applied fuel cell types are different [5,12–20]. This is because methane and methanol stored at liquid state have higher volumetric energy density than hydrogen [10], and the operation expenditure (OPEX) can be reduced owing to the lower price of both fuels compared to that of hydrogen. In addition, the bunkering infrastructure for methane and methanol is not an issue, unlike that for hydrogen. The bunkering infrastructure for methane (namely LNG) is rapidly expanding, and that for methanol requires minimal modification from the existing conventional infrastructure [5,12].

There are several technologies for reforming carbon-based fuels to hydrogen, which include steam reforming, partial oxidation, and auto-thermal steam reforming. Among these processes, steam reforming provides higher efficiency, higher production yield, and lower rate of side reactions [21].

Steam methane reforming is one of the most proven and commercially available technologies for hydrogen production [22], and at present, 80% to 85% of global hydrogen production is derived via this technology [23,24]. Steam methane reforming technology has been widely investigated from the energy and exergy efficiency and economic and environmental perspectives in the past decades.

Simpson et al. evaluated the performance of steam methane reforming system using an exergy analysis. The results revealed that irreversibility of chemical reactions results in the largest amount of exergy destruction, and exergy loss through the exhaust gas stream was significant. Results of a parametric study show that the highest exergy efficiencies of 62.73% was achieved at operating temperatures of 974 K. The highest exergy efficiencies of 62.85% was attained at an operating pressure of 6.8 atm. The effect of steam carbon ratio (S/C ratio) was also investigated, and the result revealed that the highest efficiencies were achieved at an S/C ratio of approximately 3.2 [23]. Welaya et al. evaluated the partial oxidation and steam reforming process to convert a carbon-based marine fuel, such as natural gas, gasoline, and diesel, into hydrogen-rich gases suitable for application to the PEMFCs on board ships. Among several options evaluated, the natural gas steam reforming system showed the highest fuel processing efficiency [14]. Authayanun et al. investigated the theoretical performance of an HT-PEMFC integrated with the steam reformer using various primary fuels, i.e., methane, methanol, ethanol, and glycerol. Results revealed that for the steam methane reforming, CO fraction lower than the acceptable limit for the HT-PEMFC can be attained with higher S/C ratios and lower temperature. For S/C ratios (3–6), operating temperature lower than 1000 K should be maintained in order to keep the CO fraction at an acceptable level for an HT-PEMFC or a water–gas shift (WGS) reactor should be included. Steam methanol reforming produces the lowest CO fraction among the studied fuels and can be directly fed to the HT-PEMFC for all of the studied cases (S/C ratio: 1–3, reformer temperature: 423–523 K). The steam methanol reforming system without a WGS reactor and steam methane reforming with a WGS reactor achieved the highest system efficiency, approximately 50%, among several options in the study [25]. Nerem et al. evaluated hydrogen, LNG, or methanol as PEMFC fuel on a cruise vessel, based on the space required on board, environmental impact, and life cycle cost (LCC) aspects. An external reformer other than hydrogen fuel was considered. Results show that the LNG system requires the smallest dimensions, whereas hydrogen and methanol require equal dimensions. From the perspective of environmental impact, LNG is a better solution than methanol for use in fuel cells. Further, LNG achieved the lowest LCC, 1.10 times higher than heavy fuel oil (HFO), while hydrogen and methanol are 1.14 and 1.15 times more expensive than HFO [12]. Arsalis et al. evaluated a micro combined heat and power system integrated with HT-PEMFC and steam methane reformer. They reported that the cogeneration and electrical efficiencies of the system are 55.46% and 27.62%, respectively [26].

Methanol is an advantageous fuel for mobile fuel cell applications since it has low boiling temperature (65 °C). Therefore, it can be stored in a liquid state at atmospheric pressure and normal environment temperature, unlike liquefied methane (−163 °C) [27,28]. In addition, as no carbon–carbon bond exists, methanol can be converted to hydrogen at lower temperature (150–350 °C) than other carbon-based fuels, and it can be activated at lower temperature than methane [28]. With these advantages, methanol steam reforming has been widely developed. Faungnawakij et al. has investigated the effect of varying S/C ratio (0–10), reforming temperatures (25–1000 °C), and pressures (0.5–3 atm) on the steam methanol reforming process. Results show that the optimized operating condition based on efficiency was the temperature range of 100–225 °C, S/C range of 1.5–3, and pressure at 1 atm. In addition, an operating temperature higher than approximately 150 °C and operating pressure varying from 0.5 to 3 atm did not affect the methanol conversion and hydrogen yield [21]. Herdem et al. modeled the methanol steam reforming system to produce power using a HT-PEMFC for portable power generation and examined performance variation of the HT-PEMFC with varying composition of reformate gas. The result reveals that lower S/C ratio and higher reforming temperature increase CO mole fraction in the reformed gas. However, higher fuel cell temperatures decrease the effect of CO mole fraction on the HT-PEMFC performance [27]. Mousavi Ehteshami and Chan analyzed the steam reforming of methanol, ethanol, and diesel in a technical and economical point of view. It was found that steam methanol reforming showed the easy conversion and the highest energy efficiency. Therefore, methanol is considered to be one of the promising fuels for hydrogen production by using steam reforming. However, the model used in the study did not take into account heat recovery and heat integration in the system. Therefore, it is possible that the efficiencies of fuels with higher reforming temperature than methanol can be increased when heat recovery and integration are applied [29]. Romero-Pascual and Soler investigated an HT-PEMFC-based CHP system integrated with a methanol steam reformer. The result reveals that 24% of system power efficiency and a CHP efficiency over 87% were achieved [9]. Table 1 summarizes the simulation parameters about steam methane and steam methanol reforming system from other works.

Table 1. Simulation parameters about steam methane and steam methanol reforming from the open literature.

Ref.	Primary Fuels	S/C Ratio	P_{ref} >(Bar)	T_{ref} >(°C)	Notes
[23]		3.2	10	700	Purpose of paper was to evaluate performance of hydrogen production via steam methane reforming.
[30]	Methane	4	25	900	Steam methane reforming system was modeled and reformer in simulation was developed using a Gibbs equilibrium model in Aspen Plus. For CO_2 capture, MEA scrubbing process was applied as black box model.
[31]		3	1	700	Steam methane reforming system integrated with HT-PEMFC was simulated and performance was evaluated by exergy analysis.
[32]		1.2	-	350	Steam methanol reforming system integrated with PEMFC was simulated and performance was evaluated by exergy analysis.
[33]	Methanol	1.5	3.8	260	Steam methanol reforming system was experimented and the obtained results were used for simulation of power train integrated system.
[34]		1–2	-	240–300	Steam methanol reforming system integrated with HT-PEMFC was simulated. Parametric study with varying S/C ratio, T_{ref}, reformate composition, etc. were implemented.

Although steam reforming of methane and methanol on board ships offers several advantages as aforementioned, CO_2 emission will be unavoidable because these are carbon-based fuels [15]. One of the future options for carbon-based fuels could be to install carbon capture and storage (CCS) systems on board ships [2]. If hydrogen production by reforming a carbon-based fuel is applied on board ships, the resulting CO_2 can be captured by using a CCS, and then the produced hydrogen can be considered as a zero-emission fuel at the ship level [2]. The use of CCSs on board ships was investigated in an Eurostar project. On-board chemical capture, CO_2 liquefaction, and a temporary storage system for ships were developed. The result shows that the concept is technically feasible and capable and can reduce CO_2 emissions by 65% [35].

There are several technologies for CO_2 capture in hydrogen production, which include adsorption, absorption, use of membranes, and cryogenic/low temperature processes [36,37]. Among them, the absorption process using monoethanolamine (MEA) is the most mature and promising process for CO_2 capture [38,39]. Many authors have presented hydrogen production processes associated with the MEA absorption process [30,40,41]. The MEA absorption process comprises two major stages: Absorption of CO_2 in the absorber and desorption of CO_2 in the stripper to regenerate the amine solvent. One of the drawbacks of this process is the large amount of heat required to regenerate the amine solvent [42,43]. Typically, regeneration occurs at an elevated temperature (100–140 °C) and pressure not much higher than the atmospheric [38,44,45]. The heat required for regeneration is supplied to the re-boiler by a separate steam cycle [43].

The challenge in the marine environment is the handling and storage of any captured CO_2. Storage of CO_2 in gaseous form requires a huge space, whereas the storage of CO_2 in liquid form requires a large amount of power consumption for its liquefaction. Considering the limited space of the vessel, the captured CO_2 should be stored in a liquid form even if the energy consumption for liquefaction is high. The pressure and temperature of the liquid CO_2 for storage should be higher than the triple-point pressure (5.2 bar) and lower than the critical-point temperature (31 °C). In the published reports, the pressure of liquid CO_2 for shipping is mainly in the range from 6 to 20 bar, and in this range of pressure the temperature varies form −52 to −20 °C [46]. Feenstra et al. conducted a technical and economic evaluation of ship-based carbon capture on diesel or LNG fueled ships. For the LNG fueled ships, the cooling energy from evaporation of the LNG was used for liquefying the captured CO_2. For diesel fueled ships, the ammonia-based refrigeration cycle was used for liquefaction of CO_2 [47]. Berstad et al. evaluated the energy consumption and CO_2 liquefaction ratio for CO_2 phase separation from the flue gas. The result reveals that the specific power consumption and CO_2 liquefaction ratio are significantly affected by CO_2 concentration in flue gas. To maximize the CO_2 capture ratio by liquefaction, the concentration of gases other than CO_2 in the flue gas should be minimum and pressure should be as high as practically possible [48].

The authors of this study consider that to achieve zero emission from shipping alternative fuels such as green hydrogen from renewable energy should be applied. However, the storage volume for hydrogen, lack of infrastructure for hydrogen bunkering, and high cost of hydrogen, among other factors, are challenging at the current level of technology. Therefore, on-board methane and methanol steam reforming with CO_2 capture could be one of the transition solutions.

Therefore, in this study, we present on-board methane and methanol steam reforming systems integrated with an HT-PEMFC system for power generation, along with a CO_2 capture/liquefaction system for storage of the captured CO_2 on board. The performance of the integrated systems was evaluated through an exergy and energy analysis. In addition, the spaces required for the captured CO_2 and primary fuel storage were also compared. Those evaluations were carried out for a reference ship. The following features distinguish this study from previous works:

(1)   Steam reforming, HT-PEMFC, and CO_2 capture/liquefaction systems are simultaneously considered. Heat integration and recovery were implemented for practical comparison. Excess heat from the HT-PEMFC and steam reforming system were used in the CO_2 capture system.
(2)   For the steam methane reforming-based system, liquefied natural gas (LNG) was used as primary fuel because it is the most cost effective for ship storage of natural gas and has been well proved in LNG fueled-ship applications.
(3)   For CO_2 liquefaction, the steam methane reforming-based system used the cold energy of LNG, whereas the steam methanol reforming-based system has a separate refrigeration cycle.

The main objectives of this study are as follows:

(1)   To develop methane and methanol steam reforming systems combined with HT-PEMFC and CO_2 capture/liquefaction systems suitable for the reference ship.
(2)   To carry out exergy and energy analyses for the developed integrated systems to assess the energy efficiency, exergy efficiency, and exergy destruction of components within each system.
(3)   To evaluate the overall fuel cost and overall space required for storage of the liquefied CO_2 and primary fuels.
(4)   To carry out parametric studies with varying operating conditions, such as the S/C ratio, operating temperature of the reforming process, and CO_2 capture ratio.

2. System Description

2.1. Reference Ship Description

A general cargo ship with main engine power of 3800 kW is chosen as the reference ship for integrating the steam reforming, HT-PEMFC and CO_2 capture/liquefaction systems. Although the operational engine load depends on the ship design, in general, it can be much smaller than the engine total capacity [48]. Therefore, in this study, the systems are designed based on the average shaft power. This approach allows a more precise evaluation of the amount of CO_2 emissions and energy consumption for CO_2 capture/liquefaction on board. To calculate the average shaft power, a load factor that indicates the fraction of power needed by the engine to navigate at the average speed was calculated by using the average and maximum speeds. The product of load factor and total installed engine power (MCR) provides the average shaft power. The formulas for load factor and average shaft power are presented in the equations below. The detailed specifications of the reference ship, including the calculated load factor and average shaft power, are presented in Table 2 [49].

$$\text{Load factor} = \left(\frac{\text{Average Speed}}{\text{Max Speed}} \right)^3 \tag{1}$$

$$\text{Average shaft power [kW]} = \text{Load factor} \cdot \text{MCR [kW]} \tag{2}$$

In the present study, for simplicity of system design, we only considered the power required for propulsion excluding other hotel powers.

Table 2. Specification of the reference ship.

Specifications	Values
Type	General Cargo
Overall length	120 m
Beam	13 m
Deadweight	3000 tonnage
Main engine power	3800 kW
Maximum speed	14 knots
Average speed	7 knots
Total voyage time	209 h
Load factor	0.125
Average shaft power	475 kW

2.2. Description of Steam Methane Reforming-Based System

Figure 1 shows the block diagram of the steam methane reforming system combined with HT-PEMFC and CCS on board ships. The integrated system consists of six main unit/systems: Reformer for producing reformate gas, combustor for providing heat to reformer, WGS reactor for conversion of CO to CO_2, HT-PEMFC for power generation, CO_2 capture unit, and CO_2 liquefaction system for storage.

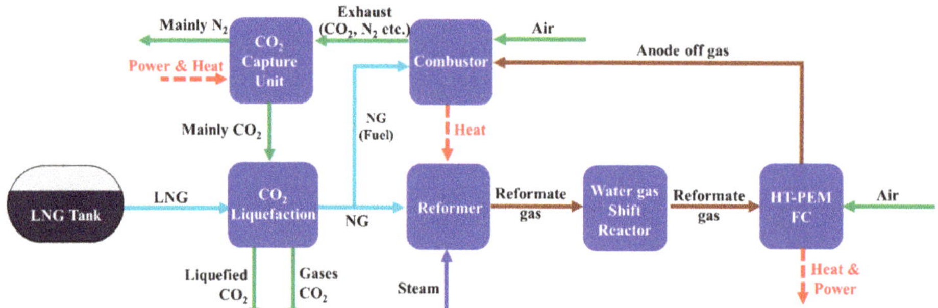

Figure 1. Block diagram of steam methane reforming-based system.

Natural gas is normally used as feedstock for steam methane reforming and it normally contains small amounts of sulfur compounds, which must be removed to avoid contamination of the catalyst in the reformer and low temperature shift reactor [50]. However, in the proposed system, LNG was considered as feedstock for reforming. As an industry practice, prior to liquefaction, natural gas is further treated to remove sulfur compounds along with water and any residual CO_2 to avoid freezing [51]. Therefore, a desulfurization unit was not included in this study. As shown in Figure 2, liquefied from an on-board storage tank is pumped and vaporized by heat exchange with the captured CO_2 in HEX-1. Then, the vaporized CH_4 is divided into two streams: One as feedstock for reforming (stream F4), the other as fuel for the combustor (stream F7). The CH_4 used as feedstock is mixed with high temperature steam (stream W3) and further preheated by heat exchange with the reformate gas stream from the reformer at HEX-2. Then, the steam methane mixture (stream F6) is supplied to the reformer, where the reforming reaction occurs as expressed in Equations (4) and (5), and converted to H_2, CO, and CO_2. The steam methane reforming reaction is highly endothermic; therefore, a large amount of heat must be supplied by the combustor by burning supplemental methane as fuel and by burning off-gas (mostly unreacted CH_4 and unused H_2) from the fuel cell. The operating temperature and pressure for the steam methane reformer in this model were set as 700 °C and 3 bar, respectively. The steam to carbon molar ratio (S/C) of 3:1 was applied to avoid coke formation.

$$CH_4 + H_2O = CO + 3H_2 \quad \Delta H_{298} = 206.3 \text{ kJmol}^{-1} \tag{3}$$

$$CH_4 + 2H_2O = CO_2 + 4H_2 \quad \Delta H_{298} = 165 \text{ kJmol}^{-1} \tag{4}$$

The reformate gas exiting HEX-2 (stream R2) is further used to preheat air entering to the combustor, and then enters the WGS reactor. The WGS reaction is moderately exothermic and it converts undesired CO in reformate gas to CO_2 and H_2, as shown in Equation (6). The WGS reactor is modeled as a single stage, and the reaction occurs at 250 °C. The heat produced during the WGS reaction is used for steam generation.

$$CO + H_2O = CO_2 + H_2 \quad \Delta H_{298} = -41 \text{ kJmol}^{-1} \tag{5}$$

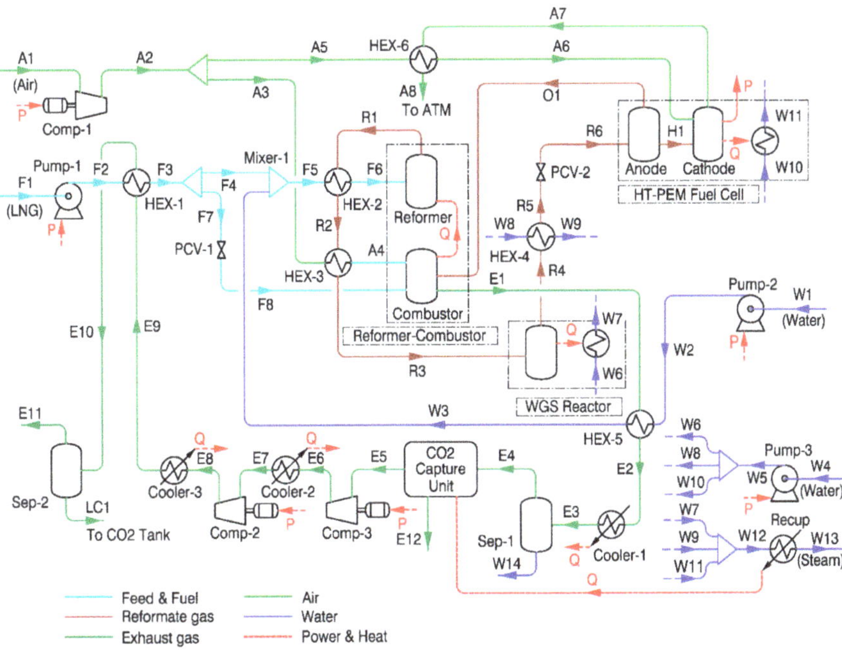

Figure 2. Process flow diagram for steam methane reforming-based system.

The reformate gas leaving the WGS rector (stream R4), which has an acceptable level of CO content (<1%) for the HT-PEMFC, is firstly cooled down to 160 °C and then supplied to the anode side of the HT-PEMFC. Dry air is supplied to the cathode side and used for the fuel cell reaction, converting the chemical energy of hydrogen to electricity. Unreacted hydrogen is supplied to the combustor for heat generation. The exhaust gas from the combustor preheats the water supplied to the reformer and then is cooled down to 40 °C when exiting Cooler-1. After removing the condensed water at Sep-1, the exhaust gas stream (E4) enters the post combustion CO_2 capture unit using aqueous MEA as solvent. As large quantities of heat are required to regenerate the solvent within the MEA CO_2 capture process, the fractions of heat produced from the WGS reactor, HT-PEMFC, and HEX-4 are supplied to the CO_2 capture unit. The MEA CO_2 capture unit is modeled as a "black box" in the present study. The detailed explanation for the black box model of the MEA process is provided in the next section. Then, the captured CO_2 is compressed to 7 bar by two stage compressors with cooling in between and then partially liquefied by heat exchange with the LNG feedstock at HEX-1. The liquefied CO_2 is separated in Sep-2 and stored at 7 bar, −48.5 °C in a temporary tank on board the ship. Noteworthy, although CO_2 capture by using low temperature and vapor–liquid phase separation could be advantageous from the perspective of CO_2 liquefaction, the capture ratio is significantly affected by concentration of other gases. Therefore, in this study, the MEA CO_2 capture unit and CO_2 liquefaction system are modeled separately.

2.3. Description of Steam Methanol Reforming-Based System

Figure 3 shows the block diagram of the steam methanol reforming system combined with HT-PEMFC and CCS on board a ship. The integrated system consists of five main unit/sub-systems: Reformer for producing reformate gas, combustor for providing heat to the reformer, HT-PEMFC for power generation, CO_2 capture unit, and CO_2 liquefaction system for storage. Unlike the steam methane reforming system, the WGS reactor is not added because the hydrogen rich gas produced in steam methanol reforming includes CO contents tolerable to the HT-PEMFC in this study.

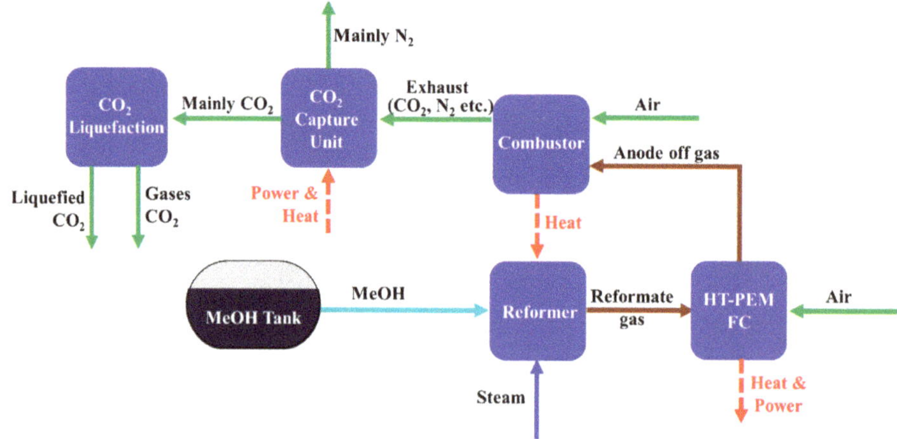

Figure 3. Block diagram of steam methanol reforming-based system.

As shown in Figure 4, methanol and water are mixed at 25 °C. The water and methanol mixture is preheated by the steam generated at the HT-PEMFC at HEX-1 and further vaporized by the reformate gas stream (stream R1) from the reformer at HEX-2. Then, H_2 rich reformate gas is produced in the reformer. The main reactions that take place in the reformer are as follows [52]:

$$CH_3OH + H_2O = CO_2 + 3H_2 \quad \Delta H_{298} = +49.7 \text{ kJmol}^{-1} \tag{6}$$

$$CO + H_2O = CO_2 + H_2 \quad \Delta H_{298} = -41 \text{ kJmol}^{-1} \tag{7}$$

$$CH_3OH = CO + 2H_2 \quad \Delta H_{298} = +90.7 \text{ kJmol}^{-1} \tag{8}$$

Equation (6) represents the steam methanol reforming reaction, Equation (7) represents the water gas shift reaction, and Equation (8) represents the methanol decomposition reaction. Only the WGS reaction is exothermic and the other two are endothermic. The operating temperature and pressure for the steam methanol reformer in this model are 200 °C and 3 bar, respectively. A S/C ratio of 1.5:1 was selected based on literature reviews [27,33]. Reformate gases containing H_2, CO_2, CO, and CH_3OH exiting HEX-2 are further adjusted to 160 °C at HEX-3 and 1.1 bar by PCV-1. Then, the reformate gas is fed to the anode side of the HT-PEMFC for power generation. As off-gas (stream O1) from the fuel cell contains H_2 unreacted in the fuel cell and CH_3OH unconverted in the reformer, the fraction of these is supplied to the combustor to produce heat. The remaining is recycled to the anode inlet stream. The exhaust gas stream (E1) from the combustor preheat air is supplied to the combustor, and the remaining heat in the exhaust gas is recovered in HEX-5. The exhaust gas stream (E4), from which most water is removed in Sep-1, is fed to the CO_2 capture unit, which is modeled as a "black box," in the same way as the steam methane reforming system.

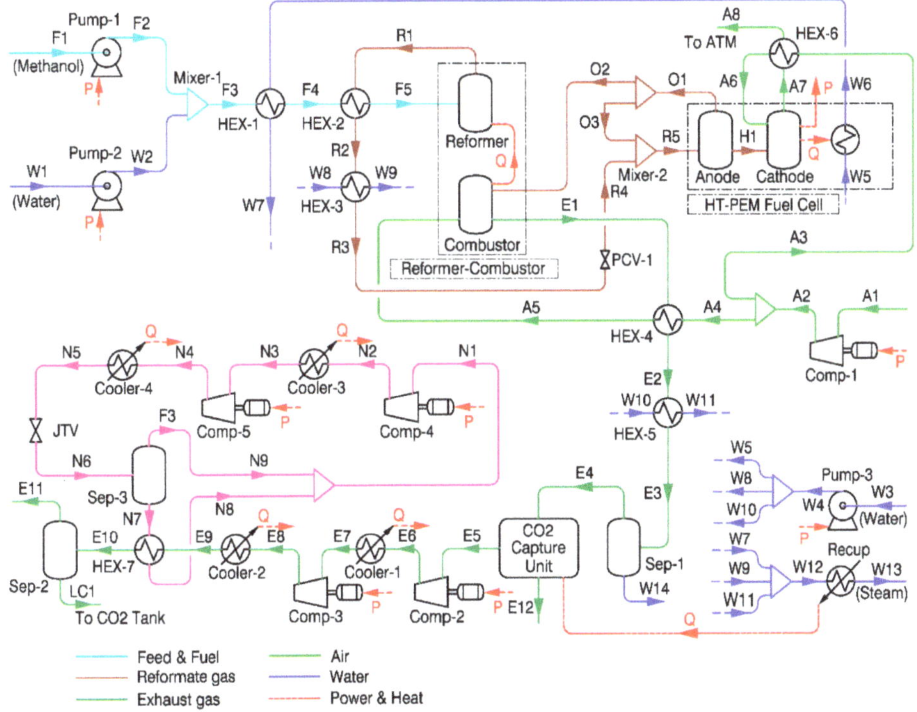

Figure 4. Process flow diagram for steam methanol reforming-based system.

The captured CO_2 stream (E5) is compressed to 14 bar by Comp-1, -2, and liquefied passing through HEX-7. Unlike the steam methane reforming system, a separate liquefaction cycle using ammonia as refrigerant was modeled. The liquefied CO_2 is separated in Sep-2 and stored in a temporary tank on board the ship. The operating and design parameters of the steam methane and methanol reforming system combined with HT-PEMFC, CO_2 capture, and liquefaction systems are presented in Table 3.

Table 3. Base condition of simulations.

Unit Name	Parameter	Values	
		Steam Methane Reforming-Based System	Steam Methanol Reforming-Based System
Steam reformer	Operating temperature	700 °C [23,31]	200 °C [34]
	Operating pressure	3 bar	3 bar
	S/C ratio	3 [31,53]	1.5 [27]
WGS reactor	Operating temperature	250 °C [54]	–
	Operating pressure	1.1 bar	–
Combustor	Operating temperature	800 °C	300 °C
	Operating pressure	1.1 bar	1.1 bar
	Air-fuel ratio	1.05 [30,41]	1.05 [27]
HT-PEMFC	Fuel utilization factor	0.83 [31,55]	
	Cathode stoichiometric ratio	2 [31,55]	
	Operating temperature	160 °C [31]	
	Operating pressure	1.1 bar	
	Output voltage per cell	0.637 V [56]	
	Current density	0.2 A cm^{-2} [56]	
CO$_2$ capture unit [38]	Solvent	MEA	
	Feed gas temperature	40 °C	
	Steam temperature	130 °C	
	Specific reboiler duty	5112 kJ/kg CO$_2$	
	CO$_2$ Capture ratio	90%	
CO$_2$ Liquefaction system	Refrigerant	LNG	NH$_3$
	Liquefaction condition	−8.5 °C at 7 bar	−29.75 °C at 14 bar
	Total emitted CO$_2$ to ATM (Stream E11 and E12)	55.83 kg/h	
Compressors	Polytropic efficiency	75%	
Pump	Adiabatic efficiency	85%	
Converter	Efficiency	98% [57]	
Heat exchangers	Min. temperature approach	Heat exchangers for CO$_2$ liquefaction: 3 °C [46] Other heat exchangers: 10 °C [41]	
Net electrical power (AC)		475 (±0.2) kW	

Note: For comparison purpose, the mass flow rate of total emitted CO$_2$ of the steam methanol reforming-based system is adjusted to have the same value as that of the steam methane reforming-based system.

3. System Simulation and Assumptions

The process simulation and heat and mass balance calculations were carried out using the ASPEN HYSYS process simulator. The heat and mass balances obtained from the converged simulation are utilized for exergy and energy calculations in a separate spreadsheet. The Peng–Robinson equation of state is used for different units/sub-systems including steam reforming, HT-PEMFC, and CO$_2$ liquefaction system [29,58]. A "Gibbs reactor" model, which considers the condition of the Gibbs free energy of the reacting system being at a minimum at equilibrium to calculate the product mixture composition [59], is used to simulate the reformer and WGS reactor, whereas the "conversion reactor" model is used to simulate the combustor [50]. There is no separate module to simulate a fuel cell in ASPEN HYSYS. Therefore, in the present study, the HT-PEMFC is modeled as a "conversion reactor" and a "splitter," which attain a conversion ratio equal to the hydrogen utilization factor [8,9]. For simplicity, the carbon capture unit using MEA is modeled as a "splitter" in ASPEN HYSYS, which can be considered as a black box assumption. In a black box assumption, the complex process for capturing CO$_2$ is modeled by one single box, which has two outlet streams of differing composition from a single inlet stream. Although the black box model enables estimating the minimum work or heat required for CO$_2$ capture, the calculated value for capture is generally a little different from the actual work and heat reported in the literature [30,41]. The concept for the black box model of the CO$_2$ capture has been used in several studies [30,41,60]. To obtain information about the energy and exergy consumption per kg of captured CO$_2$, the values reported in reference [37] are used in the present work.

The concentration of H$_2$, CO$_2$, and CO in the anode inlet stream can affect performance of HT-PEMFC. Andreasen et al. investigated the variation of HT-PEMFC performance by the feeding mixture of H$_2$, CO (0%, 0.15%, 0.25%, 0.5%, 1%), CO$_2$ (0%, 25%) to emulate methanol or methane reformate gas. Results reveal that increasing both CO and CO$_2$ concentration decreases output voltage. In total, eight cases of experiments, 1% CO and 25% CO$_2$ case and 0.5% CO and 25% CO$_2$ case show the first and second lowest output voltage, approximately 0.645 and 0.652 V, respectively, at 0.2 A cm^{-2}

and operating temperature of 160 °C [61]. Other experimental studies show similar results. Devrim et al. evaluated the combined effect of CO and CO_2 in anode inlet stream and results show that no significant performance degrade due to the addition of only CO_2 into H_2, however addition of CO in H_2 and CO_2 mixture increase degrade of performance. H_2, CO_2, CO (75%, 24%, 3% and 75%, 24%, 1%) mixtures show output voltage of approximately 0.629 and 0.634 V at 0.2 A cm^{-2} and operating temperature of 160 °C [62]. It was reported that the impact of the CO presence (up to 2.0%) at higher operating temperature (160 °C and above) and lower current densities (below 0.3 A cm^{-2}) is very low [63,64]. Therefore, in this study, the performance degrade due to CO contents is neglected and output voltage of HT-PEMFC is fixed as 0.637 V [56].

Since the amount of electricity consumption vary with several operation modes of ship, it is assumed that the target ship operates with constant average shaft power, namely 475 kW for simplicity. Further, a current density of 0.2 A cm^{-2} is assumed for constant power generation. Noteworthy, lower current densities lead to higher electrical efficiency, and require a larger cell area. However, considering the feasibility check and comparison for ship application of the methane-, methanol-based system is main purpose of this study, the above assumptions are deemed reasonable.

The general assumptions used in the modeling of the integrated energy system are as follows:

- The simulations are implemented in a steady state and are not suitable for start-up operations.
- The composition of air is considered to be 79% N_2 and 21% O_2 on a mole basis.
- For simplicity, LNG is represented by pure, liquefied CH_4.
- The reaction time is considered long enough to achieve phase and chemical equilibrium.
- The reformate gases exiting the reformer are at the reformer temperature.
- Heat and pressure losses are assumed to be negligible in all operational units.
- Complete fuel oxidation is assumed in the combustor.
- In the CO_2 capture unit, only heat consumption is considered because the power consumption at cooling pumps, solvent pumps, and other devices is relatively small.
- The heat ejected from coolers is not recovered.

4. Performance Evaluation

To conduct a comparative analysis between the steam methane reforming-based system and the steam methanol reforming-based system, energy, exergy efficiency, and exergy destruction are used along with the carbon capture storage ratio and the required volumes of both fuels and captured CO_2 during navigation.

4.1. Energy Analysis of the Integrated Systems

The objective functions that are used in the energy analysis for system performance evaluation are electrical efficiency ($\eta_{en,sys,electrical}$) and cogeneration efficiency (η_{cogen}). The electrical efficiency of the systems is defined as the ratio of the net electrical power output of the system to the lower heating value of the feed and fuel entering the system, as expressed in Equation (9) for the steam methane reforming-based system and in Equation (10) for the steam methanol reforming-based system [27,54].

$$\eta_{en,CH4,electrical} = \frac{P_{net,\ electrical}}{\left(\dot{n}_{feed,CH_4} + \dot{n}_{fuel,CH_4}\right) \cdot LHV_{CH_4}} \tag{9}$$

where LHV_{CH4} is the lower heating value of CH_4. The net electrical power ($P_{net,electrical}$) is calculated by subtracting the power consumed in the system ($P_{pump-1} + P_{pump-2} + P_{pump-3} + P_{comp-1} + P_{comp-2} + P_{comp-3}$) from the power generated in the HT-PEMFC ($P_{HT-PEMFC,AC}$).

$$\eta_{en,CH3OH,electrical} = \frac{P_{net,\ electrical}}{\dot{n}_{feed,CH_3OH,} \cdot LHV_{CH_3OH}} \tag{10}$$

where LHV_{CH3OH} is the lower heating value of CH_3OH. The net electrical power ($P_{net,electrical}$) is calculated by subtracting the power consumed in the system ($P_{pump-1} + P_{pump-2} + P_{pump-3} + P_{comp-1} + P_{comp-2} + P_{comp-3} + P_{comp-4} + P_{comp-5}$) from the power generated in the HT-PEMFC ($P_{HT-PEMFC, AC}$).

The electrical power generated by HT-PEMFC ($P_{HT-PEMFC, AC}$) can be calculated as follows [65]:

$$P_{HT-PEMFC,AC} = \eta_{HT-PEMFC} \cdot \dot{n}_{H_2} \cdot LHV_{H_2} \cdot \eta_{Converter} \tag{11}$$

where $\dot{n}_{H2}$ is the molar flow rate of hydrogen that reacts in the HT-PEMFC, LHV_{H2} is the lower heating value of hydrogen, $\eta_{HT-PEMFC}$ is the electrical efficiency of the HT-PEMFC, and $\eta_{Converter}$ is the efficiency of the converter.

The efficiency of HT-PEMFC ($\eta_{HT-PEMFC}$) can be found from [65]

$$\eta_{HT-PEMFC} = \mu_f \cdot \frac{V_c}{EMF_{max}} \cdot 100 \tag{12}$$

where μ_f is the fuel utilization factor, V_c is the produced voltage of the cell, and EMF_{max} is the electromotive force when all the energy from the hydrogen fuel cell, the heating value or enthalpy of formation, was converted to electrical energy. Fuel utilization factor, μ_f and EMF_{max} can be determined as follows:

$$\mu_f = \frac{H_{2, \, consumed}}{H_{2,supplied}} \tag{13}$$

$$EMF_{max} = -\frac{\Delta h_f}{2F} \tag{14}$$

The heat produced in the fuel cell stack can be determined from [66]

$$\dot{Q}_{heat, \, HT-PEMFC} = \Sigma(h_{in, \, c} \cdot \dot{n}_{in, \, c}) - \Sigma(h_{out, \, c} \cdot \dot{n}_{out, \, c}) + \Sigma(h_{in, \, a} \cdot \dot{n}_{in, \, a}) - \Sigma(h_{out, \, a} \cdot \dot{n}_{out, \, a}) - P_{HT-PEMFC,DC} \tag{15}$$

The cogeneration efficiency of the system ($\eta_{en,sys,cogen}$) is defined as the ratio between the summation of the rate of available heat output and the net electrical power to the lower heating value of the fuel and feed entering the system, as expressed in Equation (16) for the steam methane reforming-based system and in Equation (17) for the steam methanol reforming-based system [54].

$$\eta_{en,CH_4,cogen} = \frac{P_{net, \, electrical} + Q_{net}}{\left(\dot{n}_{CH_4, \, feed} + \dot{n}_{CH_4,fuel}\right) \cdot LHV_{CH_4}} \tag{16}$$

$$\eta_{en, \, CH_3OH,cogen} = \frac{P_{net, \, electrical} + Q_{net}}{\left(\dot{n}_{CH_3OH, \, feed}\right) \cdot LHV_{CH_3OH}} \tag{17}$$

where

$$Q_{net} = Q_{hot \, water(W13)} - Q_{water(W3)} \tag{18}$$

4.2. Exergy Analysis of the Integrated Systems

Exergy is defined as the maximum amount of useful energy that can be obtained from a stream when it reaches an equilibrium condition with the reference environment while interacting only with this environment [67]. The exergy analysis is applied to measure the exergy destruction and exergy efficiency for each component of the system proposed in this study. The equation of exergy destruction for each component can be derived from the equation of exergy balance. Considering a control volume at steady state, the general form of exergy balance can be written as follows:

$$\sum \dot{Ex}_Q - \sum \dot{Ex}_w + \sum \dot{Ex}_{flow, \, in} - \sum \dot{Ex}_{flow, \, out} - \sum \dot{Ex}_{dest} = 0 \tag{19}$$

where Ex_Q represents the rate of exergy transfer due to heat exchange with the environment, Ex_w is the rate of exergy transfer related to work, Ex_{dest} represents exergy destruction, and Ex_{flow} corresponds to the exergy transfer rate associated with the flow of the stream. $\dot{Ex}$ represents the stream exergy, which can be expressed as follows [68]:

$$\dot{Ex} = F \cdot \left(Ex_{chem} + Ex_{phys} + \Delta_{mix}Ex \right) \tag{20}$$

F, Ex_{chem}, Ex_{phys}, and $\Delta_{mix}Ex$ denote the molar flow rate and the physical, chemical, and mixing exergy of the stream.

The chemical exergy of the stream, accounting for phases and their composition at the reference condition, is given by Equation (21), as follows [68]:

$$Ex_{chem} = L_0 \cdot \sum x_{0,i} \cdot Ex_{chem,i}^{0l} + V_0 \cdot \sum y_{0,i} \cdot Ex_{chem,i}^{0v} \tag{21}$$

L_0 and V_0 are the liquid and vapor mole fraction of the stream at the reference condition, respectively. x_i and y_i are the mole fraction of species i in the liquid and vapor phases, respectively. $Ex_{chem,i}^{0l}$ and $Ex_{chem,i}^{0v}$ denote the standard chemical exergy of component i in the liquid and vapor phases, respectively. The superscripts l and v denote liquid and vapor phases, respectively, and subscript 0 refers to the reference conditions. The reference conditions, T_0 and P_0, are set to 25 °C and 101.325 kPa in this work.

The physical exergy of the stream can be calculated from the enthalpies and entropies of the pure components, the amount of each phase, and their respective compositions, as follows [68]:

$$Ex_{phys} = \left[L \cdot \sum \left(x_i \cdot H_i^l - T_0 \cdot \sum x_i \cdot S_i^l \right) + V \cdot \sum \left(y_i \cdot H_i^v - T_0 \cdot \sum y_i \cdot S_i^v \right) \right]_{actual\ T,P} - \left[L_0 \cdot \sum \left(x_i \cdot H_i^l - T_0 \cdot \sum x_i \cdot S_i^l \right) + V_0 \cdot \sum \left(y_i \cdot H_i^v - T_0 \cdot \sum y_i \cdot S_i^v \right) \right]_{T_0,P_0} \tag{22}$$

Hi and Si are the molar enthalpy and molar entropy of pure component i, respectively.

The mixing exergy, which has always a negative value, can be found from [68]:

$$\Delta_{mix}Ex = \Delta_{mix}H - T_0 \cdot \Delta_{mix}S \tag{23}$$

where

$$\Delta_{mix}M = L \cdot \left(M^l - \sum x_i \cdot M^l \right) + V \cdot \left(M^v - \sum y_i \cdot M^v \right) \tag{24}$$

in which M is any thermodynamic property of the mixture and pure component i, respectively. The detailed methodology of exergy calculation of the stream can be referred from reference [68]. The equations for the exergy destruction of each component are summarized in Table 4.

To define exergy efficiency, the input and output of the system should be defined. In these integrated systems, output is the summation of the net electrical power generated and stream exergy change of the available hot water, whereas input is the summation of chemical exergy of the feed and fuel entering the system. The exergy efficiencies of the systems are expressed as follows [54]:

$$\eta_{ex,CH_4,sys} = \frac{P_{net,\ electrical} + \dot{Ex}_{net,\ hotwater}}{\dot{Ex}_{feed,CH_4} + \dot{Ex}_{fuel,CH_4}} \tag{25}$$

$$\eta_{ex,CH_4,sys} = \frac{P_{net,\ electrical} + \dot{Ex}_{net,\ hotwater}}{\dot{Ex}_{feed,CH_3OH}} \tag{26}$$

$$\dot{Ex}_{net,\ steam} = \dot{Ex}_{hot\ water\ generated} - \dot{Ex}_{water\ supplied} \tag{27}$$

Table 4. Equation of exergy destruction and efficiency of the components.

Components	Exergy Destruction
Compressors [53]	$\dot{Ex}_{dest} = \dot{Ex}_{in} + P_{in,\,comp} - \dot{Ex}_{out}$
Pumps [53]	$\dot{Ex}_{dest} = \dot{Ex}_{in} + P_{in,\,pump} - \dot{Ex}_{out}$
Heat exchangers [53]	$\dot{Ex}_{dest} = \sum \dot{Ex}_{in} - \sum \dot{Ex}_{out} = \sum (\dot{Ex}_{in} - \dot{Ex}_{out})_{hot} + \sum (\dot{Ex}_{in} - \dot{Ex}_{out})_{cold}$
Reformer-Combustor [53]	$\dot{Ex}_{dest} = \sum \dot{Ex}_{in} - \sum \dot{Ex}_{out} =$ $\dot{Ex}_{in,\,feed} + \dot{Ex}_{in,\,fuel} + \dot{Ex}_{in,\,off\,gas\,from\,HT-PEMFC} -$ $\dot{Ex}_{out,\,reformate\,gas\,from\,reformer} - \dot{Ex}_{out,\,exhaust\,gas}$
WGS [53]	$\dot{Ex}_{dest} = \sum \dot{Ex}_{in} - \sum \dot{Ex}_{out} = \dot{Ex}_{in,\,reformate\,gas\,from\,reformer} +$ $\dot{Ex}_{in,\,cooling\,water} - \dot{Ex}_{out,\,reformate\,gas\,to\,HT-PEMFC} - \dot{Ex}_{out,\,cooling\,water}$
Separators	$\dot{Ex}_{dest} = \dot{Ex}_{in} - \dot{Ex}_{out}$
HT-PEM fuel cell [69,70]	$\dot{Ex}_{dest} = \sum \dot{Ex}_{in} - \sum \dot{Ex}_{out}$ $= \dot{Ex}_{in,\,reformate\,gas\,from\,WGS} + \dot{Ex}_{in,\,air}$ $+ \dot{Ex}_{in,\,cooling\,water} - \dot{Ex}_{out,\,air} - \dot{Ex}_{out,\,cooling\,water}$ $- \dot{Ex}_{out,\,off\,gas} - P_{out,\,electrical}$
Coolers [53]	$\dot{Ex}_{dest} = \dot{Ex}_{in} - \dot{Ex}_{out}$
Valves	$\dot{Ex}_{dest} = \dot{Ex}_{in} - \dot{Ex}_{out}$

5. Results and Discussion

5.1. Energy and Exergy Analyses

The steam reforming systems presented in this study are evaluated in terms of energy, exergy efficiency of the overall systems, and exergy destruction rate of each component. For the energy efficiency, electrical and cogeneration efficiencies are utilized. Electrical efficiency, cogeneration efficiency, and exergy efficiency at the base condition presented in Table 3 for the methane-based and methanol-based systems are compared in Figure 5. Electrical efficiencies of 39.53% for the methane-based system and 47.53% for the methanol-based system are obtained. It can be interpreted that the methanol-based system uses the feed energy input (LHV basis) more efficiently than the methane-based system for 475 kW of net electricity generation. Noteworthy, the overall electricity consumption of the methane-based and methanol-based systems are 18.51 and 30.97 kW, respectively, as indicated in Figure 6. The higher electricity consumption in the methanol-based system is mainly attributed to electricity consumption of the compressors (Comp-4 and -5) in the ammonia refrigerant cycle, whereas the methane-based system, which uses the cold energy of LNG for CO_2 liquefaction, does not consume additional electricity. Regarding the cogeneration efficiency, the methane-based system has higher value, 63%, than that of the methanol-based system, 55.86%, and this can be explained by Figure 7. As can be observed in Figure 7, the methane based system consumes a higher amount of overall energy (LHV basis) than the methanol-based system and generates more available heat for 475 kW of net electricity generation. The higher value of heat generated can be explained by the fact that the steam methane reforming reaction is more endothermic than the steam methanol reforming reaction.

To evaluate the performance of the components, the exergy destructions of each component in the methane-based and methanol-based systems at the base condition provided in Table 3 are calculated. It is observed that there is a total exergy destruction of 690.68 kW for the methane-based system and 591.16 kW for the methanol-based system. The exergy destructions in each system are broken down to the component level and the results are presented in Figures 8 and 9. The HT-PEMFC, reformer-combustor, and CCU have the largest percentages of total exergy destruction. For the methane-based system, the HT-PEMFC is the component having the highest exergy destruction with 365.5 kW (51.5%), followed by the reformer-combustor with 145.76 kW (20.6%) and CCU with 86.03 kW (14.0%). The exergy destructions in the methanol-based system follow a similar trend, having the highest exergy destruction in the HT-PEMFC with 342.34 kW (57.9%), reformer-combustor with 145.76 kW (20.6%), and CCU with 85.74 kW (14.5%). Exergy destruction of the HT-PEMFC and reformer-combustor mainly results from high irreversibility of the chemical reaction [54]. Exergy destructions of CCUs are calculated by assuming 1529 kW/kg CO_2 [37], which are caused by the unavoidable heat exergy required for MEA regeneration, and therefore, are dependent on the amount of generated CO_2 in the overall process. The larger total exergy destruction in the methane-based system is mainly attributed to the larger exergy destruction in the reformer-combustor. This relies on the fact that the steam methane reforming reaction is operated at a higher temperature condition of 700 °C (combustion temperature of 800 °C) instead of 200 °C (combustion temperature of 300 °C) in the steam methanol reforming reaction.

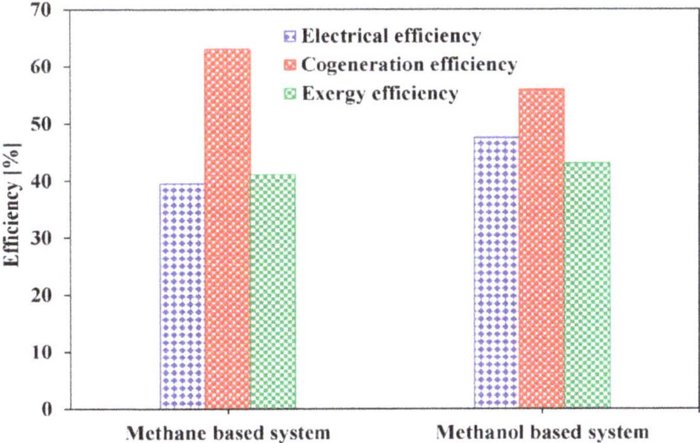

Figure 5. Electrical, cogeneration, and exergy efficiencies of the methane-based and methanol-based systems for 475 kW of net electricity generation.

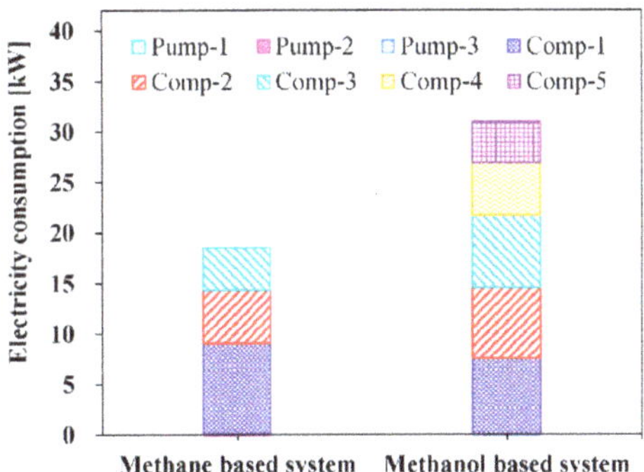

Figure 6. Break-down of electricity consumption of the methane-based and methanol-based systems for 475 kW of net electricity generation.

The other reasons explaining why the total exergy destruction is higher in methane-based systems than in methanol-based systems is that higher exergy destruction occurs in the heat exchangers (mainly HEX-1, HEX-3, and HEX-5), Mixer-1. For the heat exchangers, the larger exergy destruction derives from the larger temperature difference between cold and hot streams. This is because fuel in cryogenic temperature (LNG) is supplied in the methane-based system and streams from the reformer and combustor have inherent temperature. For Mix-1, more exergy destruction in the methane-based system than that in the methanol-based system is generated from mixing of streams with larger temperature difference. Therefore, reducing the temperature differences in these heat exchangers and Mix-1 by optimization can effectively reduce the system exergy destruction [71]. The exergy destruction generated in the WGS reactor, which is additionally equipped for the methane-based system to lower the CO fraction, is another reason.

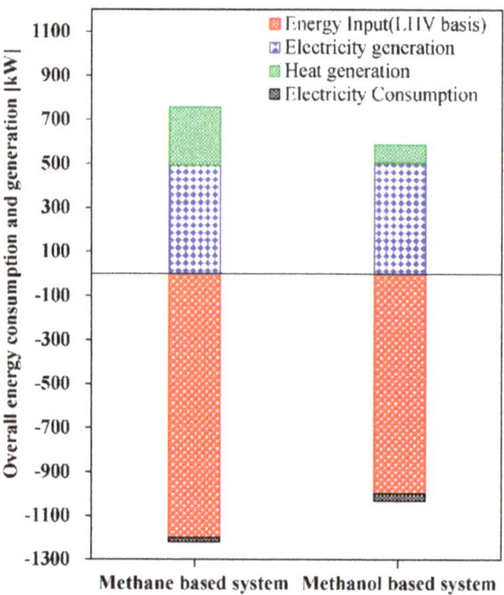

Figure 7. Overall energy consumption and generation of the methane-based and methanol-based systems for 475 kW of net electricity generation.

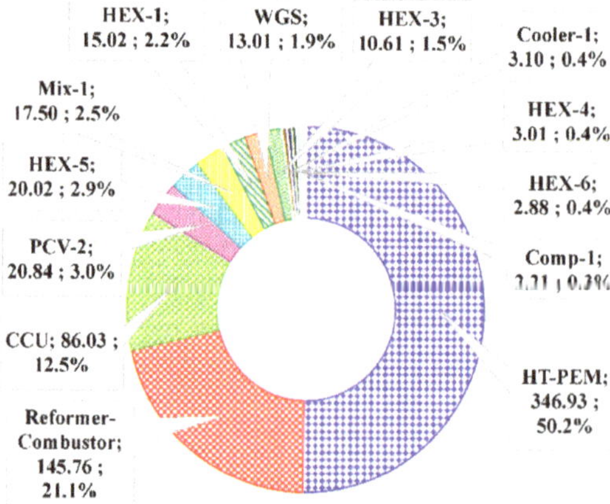

Figure 8. Break-down of exergy destruction for the methane based-system. Total exergy destruction 690.68 kW. (Unit names: Exergy destruction, kW; exergy destruction ratio, %).

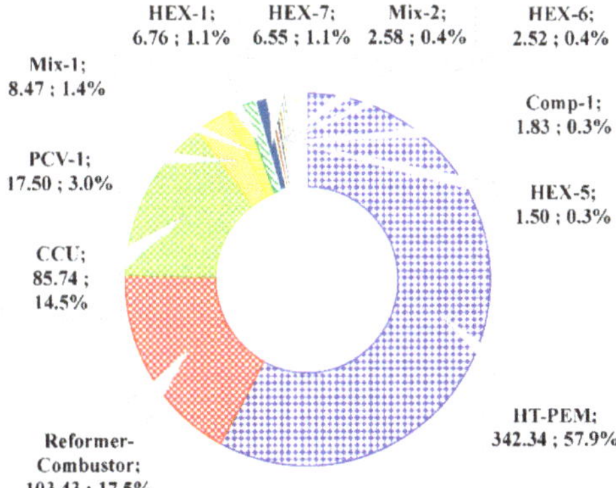

Figure 9. Break-down of exergy destruction for the methanol based-system. Total exergy destruction 591.16 kW. (Unit names: Exergy destruction, kW; exergy destruction ratio, %).

5.1.1. Effect of Varying Reforming Temperature

The variation of H_2 molar flow rate at the outlet of the steam methane and steam methanol reformers in thermodynamic equilibrium with varying reforming temperature and S/C ratio was derived to understand the system behavior. As can be observed in Figure 10a, the methane steam reformer shows the tendency that at S/C ratios below 3.5, the hydrogen flow rate increases when the reforming temperature increases up to 750 °C. After then, the H_2 molar flow rate decreases as the reforming temperature increases. At S/C ratios above 3.5, the H_2 molar flow rate continuously decreases as the reforming temperature increases. The steam methanol reformer shows a trend that as the reforming temperature increases from 180 to 260 °C, the molar flow rate of H_2 continuously decreases, although the decrease rates are different, as shown in Figure 10b. Figure 10a,b represent the behavior of each reformer and can be used for interpretation of each system in the next section. The effects of varying the reforming temperature on the efficiencies and molar flow rate of CO_2 in the methane-based and methanol-based systems were studied and illustrated in Figure 11.

As can be observed in Figure 11a, for the methane-based system, as the reforming temperature increases from 700 to 950 °C, the electrical, cogeneration, and exergy efficiency continuously decrease, from 39.53%, 63%, and 41.14% to 37.92%, 58.03%, and 38.47%, respectively. This trend occurs because as the reforming temperature increases, the amount of additional fuel needed for the reformer and combustor increases by 5.8% and leads to a decrease in efficiencies of the system. It can be noticed that the slope of the efficiency curves between 700 and 750 °C is less steep than that at other temperature ranges. The reason for this is that the amount of hydrogen produced increases when the temperature is increased from 700 to 750 °C; after then, the amount of hydrogen produced starts to decrease, as shown in Figure 10a. In addition, the slope of the cogeneration efficiency curve is a little steeper than that of the electrical and exergy efficiencies as the amount of the produced CO_2 increases; accordingly, the heat required to capture CO_2 in the CCU increases. The methanol-based system shows a similar behavior to that of the methane-based system, as shown in Figure 11b. As the reforming temperature increases from 180 to 260 °C, the electrical and exergy efficiencies decrease from 47.85% and 43.25% to 46.27% and 42.23%, respectively, whereas the cogeneration efficiency stays almost constant. Noteworthy, unlike the methane-based system, as the reforming temperature increases above 200 °C, the amount of produced CO_2 decreases. This happens because an increase in the reforming temperature above 200 °C will favor the CO formation and the increase in CO content will decrease the production of H_2 and CO_2. Therefore, the heat required for CO_2 capture increases and this results in a slight increase in cogeneration efficiency as the reforming temperature increases.

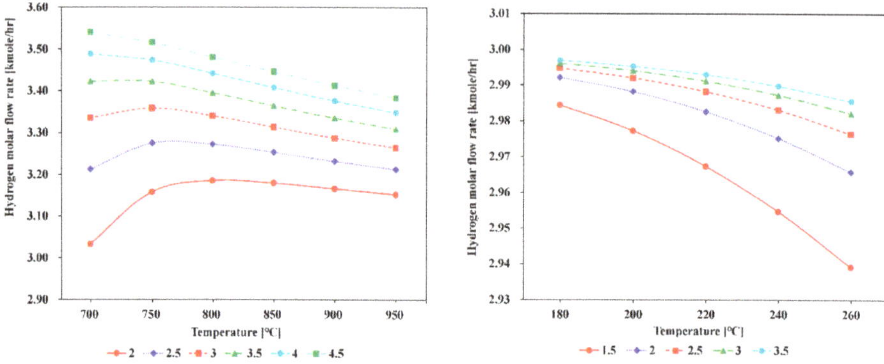

(**a**) Steam methane reformer: Methane supply of 1 kmole/h.

(**b**) Steam methanol reformer: Methanol supply of 1 kmole/h.

Figure 10. Variation of H_2 molar flow rates of the exit gases from the reformer as a function of the S/C ratio and reforming temperature.

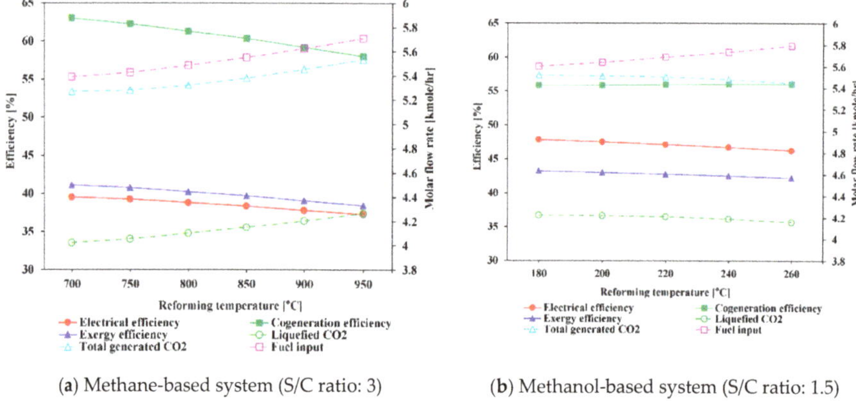

(**a**) Methane-based system (S/C ratio: 3)

(**b**) Methanol-based system (S/C ratio: 1.5)

Figure 11. Influence of reforming temperature on system efficiencies and variation of molar flow rates of CO_2 and fuel.

5.1.2. Effect of Varying Steam to Carbon Ratio

Figure 12 shows the change in the electrical, cogeneration, and exergy efficiencies with varying S/C ratio. For the steam methane-based system, when the S/C ratio is increased from 2 to 4.5, the electrical efficiency, exergy efficiency, and cogeneration efficiency decrease from 46.64%, 63.77%, and 42.06% to 30.30%, 62.07%, and 40.14%, respectively. This tendency occurs because a higher S/C ratio requires a considerable amount of heat to produce steam, resulting in more fuel consumption in the combustor, as can be observed in Figure 13a. The increase in parasitic power consumption for the pump and CO_2 compressor is another reason for the decrease in electrical and exergy efficiencies. Regarding the cogeneration efficiency, the increased heat consumption in the CCU due to increased CO_2 generation is one of the reasons. Noteworthy, when the S/C ratio increases, H_2 contents in the product gas is increased, although this increase rate becomes slower, as depicted in Figure 10a. This effect compensates slightly the required fuel consumption. Nevertheless, the efficiencies decrease with S/C ratio because the increase rate of additional fuel is greater than the increase rate of hydrogen production. For the methanol-based system, the efficiencies profiles are relatively flat with increasing S/C ratio from 1.5 to 3.5, as shown in Figure 12b. This trend is attributable to the lower reforming temperature and combustion temperature in the methanol-based system, which can be confirmed by the slight increase in fuel consumption. This trend can also be due to the trade-off between the increase in hydrogen production and the increase in parasitic power consumption with increasing S/C ratio.

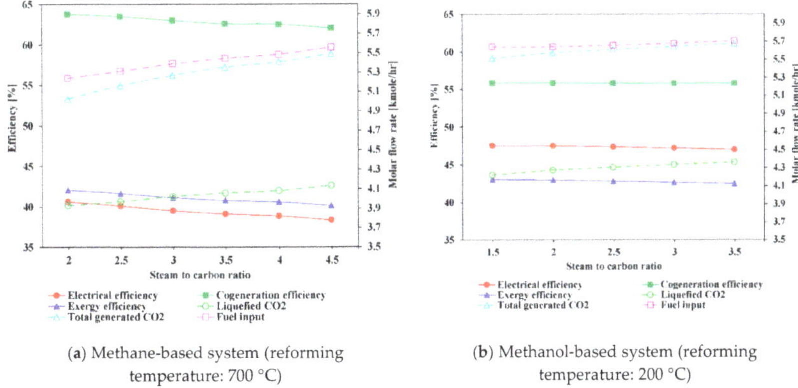

(**a**) Methane-based system (reforming temperature: 700 °C)

(**b**) Methanol-based system (reforming temperature: 200 °C)

Figure 12. Influence of S/C ratio on system efficiencies and variation of molar flow rates of CO_2 and fuel.

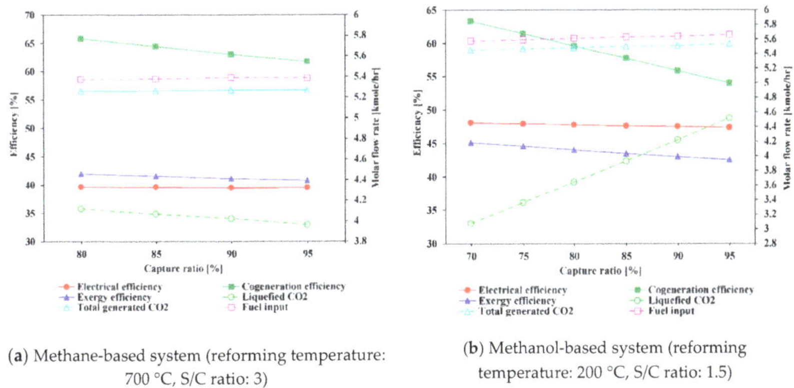

(**a**) Methane-based system (reforming temperature: 700 °C, S/C ratio: 3)

(**b**) Methanol-based system (reforming temperature: 200 °C, S/C ratio: 1.5)

Figure 13. Influence of CO_2 capture ratio on system efficiencies and variation of molar flow rates of CO_2 and fuel.

5.1.3. Effect of Varying CO_2 Capture Ratio

Figure 13 shows the change of the electrical, cogeneration, and exergy efficiencies with carbon capture ratio. For the steam methane-based system, the profile of electrical efficiency is almost flat, whereas the exergy and cogeneration efficiency continuously decrease with increasing capture ratio from 80% to 95%. The exergy efficiency is reduced by 1.1% and the cogeneration efficiency by 4.06%. This trend is attributed to the fact that as the capture ratio decreased, heat consumption in the CCU decreased, leading to exergy and cogeneration efficiency. Note that a CO_2 capture ratio below 80% in the methane-based system causes the formation of dry ice due to the lower temperature of LNG. Although the formation of dry ice can be managed by temperature control, it is not considered in this study. In the case of the methanol-based system, the electrical efficiency, cogeneration efficiency, and exergy efficiency decrease from 48.06%, 63.32%, and 45.12% to 47.35%, 54.03%, and 42.56%, respectively, with increasing carbon capture ratio from 70% to 95%. The higher decrease rate of electrical efficiency in the methanol-based system than that in the methane-based system is mainly attributed to the increase in power consumption of the CO_2 compressor and NH_3 compressors for CO_2 liquefaction. In addition, as the capture ratio increases, the liquefied CO_2 ratio proportionally increases because the total emitted CO_2 is fixed, as presented in Table 3. In methanol-based systems, the amount of available heat is relatively small compared to that in the methane-based systems, and thus, fuel consumption increases faster than in the methane-based systems as capture rates increase.

5.2. Space and Operational Cost

Figure 14 illustrates the volume for storage of the fuel and liquefied CO_2 along with the cost of fuels for the methane-based and methanol-based systems, which are required for 475 kW of net electricity generation during the total navigation time. A specific fuel cost of 9.76 USD/mmBtu for LNG [72] and 26.08 USD/mmBtu [73] for methanol are used, and both are the average cost in 2018 in the references. The result shows that that methane-based system requires 43.69 m^3 for LNG storage and 32.30 m^3 for the liquefied CO_2 storage, whereas the methanol-based system requires 48.03 m^3 for methanol and 36.17 m^3 for the liquefied CO_2. Accordingly, the methanol-based system needs approximately 1.1 times the volume (equivalent to 8 m^3 more) for fuel and liquefied CO_2 storage. In other words, the methanol-based system consumes a higher amount of methanol as fuel and generate a higher amount of CO_2. Regarding the fuel cost, the methanol-based system has a 2.2 times higher fuel cost than the methane-based system for 475 kW of net electricity generation during the total navigation time. Therefore, the methane-based system is more competitive than the methanol-based system from the economic point of view, when only the fuel cost and volume are taken into account. However, note that the overall investment cost for both systems, which is beyond the scope of the current study, may also be an important factor in the selection of system.

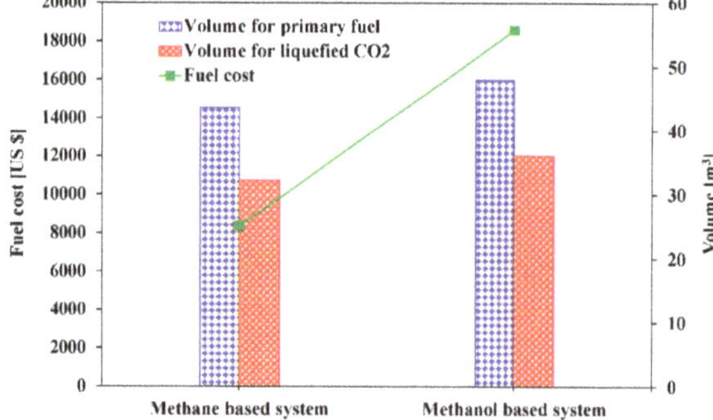

Figure 14. Fuel volumes and cost of methane-based and methanol-based systems for 475 kW of net electricity generation during the total navigation time.

6. Conclusions

In this work, the authors have performed a comparison between the steam methane reforming and steam methanol reforming technologies combined with HT-PEMFC and carbon capture systems for hydrogen-fueled ship applications. To find the most suitable technologies, an energy/exergy analysis, along with a space and fuel cost investigation, have been conducted. All the simulations have been conducted at a fixed $W_{net, electrical}$ (475 kW).

It is shown that, at the base condition, the energy and exergy efficiencies of the methanol-based system are 7.99% and 1.89% higher than those of the methane-based system, respectively. The different efficiencies between systems mainly arises from the reforming temperature difference. For fuel and CO_2 storage, the methanol-based system requires a space 1.1 times larger than that of the methane-based system for the total navigation time, although the methanol-based system has higher electrical efficiency. Accordingly, the methanol-based system has 2.2 times higher fuel cost than the methane-based system for 475 kW of net electricity generation during the total navigation time. In the parametric study, both systems show a similar trend, in which with increasing reforming temperature and S/C ratio, the electrical, exergy, and cogeneration efficiencies gradually decreased.

The comparative analysis reveals that the methanol-based system has many technological advantages directly related to its low reforming temperature, which leads to better integration to the HT-PEMFC. However, the methane-based system showed economic advantages from the perspective of fuel cost and better availability in the maritime sector.

Furthermore, this work shows the feasibility of combining reforming, HT-PEMFC, CO_2 capture, and liquefaction systems for both methanol and methane fuels in heat and power integration point of view. Excessive heat from HT-PEMFC and reformer in both methane-, methanol-based system are enough for CO_2 capture unit, which require a large amount of heat to regenerate the amine solvent. In addition, for the methane based-system, cold energy of LNG, which should be vaporized, can be utilized for CO_2 liquefaction, therefore power consumption for compressors can be reduced. Although separate ammonia refrigerant cycle is required for CO_2 liquefaction for methanol based-system and power consumptions for compressors are slightly high, methanol based-system has still higher efficiency.

Several limitations were identified for consideration in the future study. In the present study, constant current density of 0.2 A cm^{-2} was assumed and resulted in a little higher electrical efficiency. More simulations in several current density within the operating window of HT-PEMFC are required in the future study. In addition, future study should use output voltage with real reformate gas for the detailed assessment. Present study compared two systems in the process simulation level, however, future study should include sizing and on-board arrangement of systems for reforming, CO_2 capture, and liquefaction systems since those systems may take large spaces and lead to different results. Furthermore, other fuels such as ethanol and liquefied petroleum gas (LPG) which is getting attention together with methanol and LNG in maritime industry should be assessed in the future study. Although the present study has some limitations, the concepts suggested in this study can give other perspectives on applying hydrogen fuel cell on board and can be a good reference for the further development of hydrogen fuel cell ship.

Author Contributions: Formal analysis, H.L.; Investigation, Y.N.; Methodology, I.J. and G.R.; Supervision, H.K.; Writing—original draft, H.L.; Writing—review and editing, H.K. All authors have read and agreed to the published version of the manuscript.

Funding: This material is based upon work supported by the Technology Innovation Program (20004659, the development of LNG Fuel Gas Supply System for Coastal Ships) which was funded by the Ministry of Trade, Industry & Energy (MOTIE, Korea) and Core Technology Development and System Construction for LNG Bunkering program (20180048, Test evaluation for LNG bunkering equipment and development of test technology) which was funded by the Ministry of Oceans and Fisheries (Korea).

Conflicts of Interest: The authors declare no conflict of interest.

Nomenclature

Symbols

e	Specific exergy, kJ/kg
$\dot{E}x$	Exergy flow rate, kW
s	Specific entropy, kJ/kg·°C
h	Specific enthalpy, kJ/kg
P	Pressure, bar
T	Temperature, °C
h	Time, h
$\dot{m}$	Mass flow rate, kg/h
$\dot{n}$	Molar flow rate, kmole/h
Vc	Output voltage
μ_f	Fuel utilization factor
Q	Heat rate, kW
ρ	Density, kg/m^3

Abbreviations

LHV	Lower heating value
LNG	Liquefied natural gas
CCU	Carbon capture unit
MEA	Monoethanolamine
HT-PEMFC	High temperature proton-exchange membrane fuel cell
LT-PEMFC	Low temperature proton-exchange membrane fuel cell
OPEX	Operation expenditure
WGS	Water gas shift
HFO	Heavy fuel oil
CCS	Carbon capture and storage

References

1. International Maritime Organization. Third IMO Greenhouse Gas Study. Available online: http://www.imo.org/en/OurWork/Environment/PollutionPrevention/AirPollution/Pages/Greenhouse-Gas-Studies-2014.aspx (accessed on 30 December 2019).
2. DNV GL. Maritime Forecast to 2050, Energy Transition Outlook 2018. Available online: https://eto.dnvgl.com/2018/#Energy-Transition-Outlook-2018 (accessed on 30 December 2019).
3. Initial IMO Strategy on the Reduction of GHG Emissions From Ships. Resolution MEPC. Available online: http://www.imo.org/en/MediaCentre/HotTopics/GHG/Pages/default.aspx (accessed on 30 December 2019).
4. Zhu, M.; Fai, K.; Wei, J.; Li, K.X. Impact of maritime emissions trading system on fleet deployment and mitigation of CO_2 emission. *Transp. Res. Part D* **2018**, *62*, 474–488. [CrossRef]
5. Van Biert, L.; Godjevac, M.; Visser, K.; Aravind, P.V. A review of fuel cell systems for maritime applications. *J. Power Sources* **2016**, *327*, 345–364. [CrossRef]
6. Balcombe, P.; Brierley, J.; Lewis, C.; Skatvedt, L.; Speirs, J.; Hawkes, A.; Staffell, I. How to decarbonise international shipping: Options for fuels, technologies and policies. *Energy Convers. Manag.* **2019**, *182*, 72–88. [CrossRef]
7. Tronstad, T.; Astrand, H.H.; Haugom, G.P.; Langfeldt, L. Study on the Use of Fuel Cells in Shipping. Available online: http://www.emsa.europa.eu/component/flexicontent/download/4545/2921/23.html (accessed on 30 December 2019).
8. Jaggi, V.; Jayanti, S. A conceptual model of a high-efficiency, stand-alone power unit based on a fuel cell stack with an integrated auto-thermal ethanol reformer. *Appl. Energy* **2013**, *110*, 295–303. [CrossRef]
9. Romero-Pascual, E.; Soler, J. Modelling of an HTPEM-based micro-combined heat and power fuel cell system with methanol. *Int. J. Hydrogen Energy* **2014**, *39*, 4053–4059. [CrossRef]
10. Hydrogen Analysis Resource Center. Lower and Higher Heating Values of Fuels. Available online: https://h2tools.org/hyarc/hydrogen-data/lower-and-higher-heating-values-hydrogen-and-other-fuels (accessed on 30 December 2019).
11. Vessels, M. Fuel Cell Applications for Marine Vessels Why Fuel Cells Make Sense. Available online: https://info.ballard.com/fuel-cell-applications-for-marine-vessels?hsCtaTracking=8b99d2e2-4e14-44a2-b88d-9cda7551cd89%7Cdef386da-b1b5-49ff-983b-9cdcf065b567 (accessed on 30 December 2019).
12. Nerem, T. Assessment of Marine Fuels in a Fuel Cell on a Cruise Vessel. Available online: https://ntnuopen.ntnu.no/ntnu-xmlui/bitstream/handle/11250/2564492/19196_FULLTEXT.pdf?sequence=1&isAllowed=y (accessed on 30 December 2019).
13. Sattler, G. Fuel cells going on-board. *J. Power Sources* **2000**, *86*, 61–67. [CrossRef]
14. Welaya, Y.M.A.; El Gohary, M.M.; Ammar, N.R. Steam and partial oxidation reforming options for hydrogen production from fossil fuels for PEM fuel cells. *Alex. Eng. J.* **2012**, *51*, 69–75. [CrossRef]
15. e4ships FUEL CELLS IN MARINE APPLICATIONS. Available online: https://www.e4ships.de/app/download/13416971890/e4ships_Brochure_engl_2016.pdf?t=1568291372 (accessed on 30 December 2019).
16. Rajasekhar, D.; Narendrakumar, D. Fuel Cell Technology for Propulsion and Power Generation of Ships: Feasibility Study on Ocean Research Vessel Sagarnidhi. *J. Ship. Ocean Eng.* **2020**, *5*, 219–228.

17. Saito, N. The Maritime Commons: Digital Repository of the World. The Economic Analysis of Commercial Ships with Hydrogen Fuel Cell through Case Studies. Available online: https://commons.wmu.se/cgi/viewcontent.cgi?article=1617&context=all_dissertations (accessed on 30 December 2019).

18. Han, J.; Charpentier, J.F.; Tang, T. State of the Art of Fuel Cells for Ship Applications. In Proceedings of the 2012 IEEE International Symposium on Industrial Electronics, Hangzhou, China, 28–31 May 2012.

19. Alvarez, C.; Fern, C.; Carral, L. Analysing the possibilities of using fuel cells in ships. *Int. J. Hydrogen Energy* **2015**, *41*, 2853–2866. [CrossRef]

20. Leo, T.J.; Durango, J.A.; Navarro, E. Exergy analysis of PEM fuel cells for marine applications. *Energy* **2010**, *35*, 1164–1171. [CrossRef]

21. Faungnawakij, K.; Kikuchi, R.; Eguchi, K. Thermodynamic evaluation of methanol steam reforming for hydrogen production. *J. Power Sources* **2006**, *161*, 87–94. [CrossRef]

22. Chen, B.; Liao, Z.; Wang, J.; Yu, H.; Yang, Y. Exergy analysis and CO_2 emission evaluation for steam methane reforming. *Int. J. Hydrogen Energy* **2012**, *37*, 3191–3200. [CrossRef]

23. Simpson, A.P.; Lutz, A.E. Exergy analysis of hydrogen production via steam methane reforming. *Int. J. Hydrogen Energy* **2007**, *32*, 4811–4820. [CrossRef]

24. Alhamdani, Y.A.; Hassim, M.H.; Ng, R.T.L.; Hurme, M. The estimation of fugitive gas emissions from hydrogen production by natural gas steam reforming. *Int. J. Hydrogen Energy* **2017**, *42*, 9342–9351. [CrossRef]

25. Authayanun, S.; Saebea, D.; Patcharavorachot, Y.; Arpornwichanop, A. Effect of different fuel options on performance of high-temperature PEMFC (proton exchange membrane fuel cell) systems. *Energy* **2014**, *68*, 989–997. [CrossRef]

26. Arsalis, A.; Nielsen, M.P.; Kær, S.K. Modeling and parametric study of a 1 kWe HT-PEMFC-based residential micro-CHP system. *Int. J. Hydrogen Energy* **2011**, *36*, 5010–5020. [CrossRef]

27. Herdem, M.S.; Farhad, S.; Hamdullahpur, F. Modeling and parametric study of a methanol reformate gas-fueled HT-PEMFC system for portable power generation applications. *Energy Convers. Manag.* **2015**, *101*, 19–29. [CrossRef]

28. Palo, D.R.; Dagle, R.A.; Holladay, J.D. Methanol steam reforming for hydrogen production. *Chem. Rev.* **2007**, *107*, 3992–4021. [CrossRef]

29. Mousavi Ehteshami, S.M.; Chan, S.H. Techno-Economic Study of Hydrogen Production via Steam Reforming of Methanol, Ethanol, and Diesel. *Energy Technol. Policy* **2014**, *1*, 15–22. [CrossRef]

30. Zhu, L.; Li, L.; Fan, J. A modified process for overcoming the drawbacks of conventional steam methane reforming for hydrogen production: Thermodynamic investigation. *Chem. Eng. Res. Des.* **2015**, *104*, 792–806. [CrossRef]

31. Ye, L.; Jiao, K.; Du, Q.; Yin, Y. Exergy analysis of high-temperature proton exchange membrane fuel cell systems. *Int. J. Green Energy* **2015**, *12*, 917–929. [CrossRef]

32. Ishihara, A.; Mitsushima, S.; Kamiya, N.; Ota, K.I. Exergy analysis of polymer electrolyte fuel cell systems using methanol. *J. Power Sources* **2004**, *126*, 34–40. [CrossRef]

33. Wiese, W.; Emonts, B.; Peters, R. Methanol steam reforming in a fuel cell drive system. *J. Power Sources* **1999**, *84*, 187–193. [CrossRef]

34. Lotrič, A.; Sekavčnik, M.; Pohar, A.; Likozar, B.; Hočevar, S. Conceptual design of an integrated thermally self-sustained methanol steam reformer—High-temperature PEM fuel cell stack manportable power generator. *Int. J. Hydrogen Energy* **2017**, *42*, 16700–16713. [CrossRef]

35. TNO; Maritime Knowledge Centre. TU Delft Framework CO_2 Reduction in Shipping. Available online: https://www.mkc-net.nl/library/documents/893/download/ (accessed on 30 December 2019).

36. Voldsund, M.; Jordal, K.; Anantharaman, R. Hydrogen production with CO_2 capture. *Int. J. Hydrogen Energy* **2016**, *41*, 4969–4992. [CrossRef]

37. Ferrara, G.; Lanzini, A.; Leone, P.; Ho, M.T.; Wiley, D.E. Exergetic and exergoeconomic analysis of post-combustion CO_2 capture using MEA-solvent chemical absorption. *Energy* **2017**, *130*, 113–128. [CrossRef]

38. Bhown, A.S.; Freeman, B.C. Analysis and status of post-combustion carbon dioxide capture technologies. *Environ. Sci. Technol.* **2011**, *45*, 8624–8632. [CrossRef]

39. Aaron, D.; Tsouris, C. Separation of CO_2 from flue gas: A review. *Sep. Sci. Technol.* **2005**, *40*, 321–348. [CrossRef]

40. Fan, J.; Zhu, L. Performance analysis of a feasible technology for power and high-purity hydrogen production driven by methane fuel. *Appl. Therm. Eng.* **2015**, *75*, 103–114. [CrossRef]

41. Fan, J.; Zhu, L.; Jiang, P.; Li, L.; Liu, H. Comparative exergy analysis of chemical looping combustion thermally coupled and conventional steam methane reforming for hydrogen production. *J. Clean. Prod.* **2016**, *131*, 247–258. [CrossRef]

42. Oh, S.Y.; Binns, M.; Cho, H.; Kim, J.K. Energy minimization of MEA-based CO_2 capture process. *Appl. Energy* **2016**, *169*, 353–362. [CrossRef]

43. Mangalapally, H.P.; Notz, R.; Hoch, S.; Asprion, N.; Sieder, G.; Garcia, H.; Hasse, H. Pilot plant experimental studies of post combustion CO_2 capture by reactive absorption with MEA and new solvents. *Energy Procedia* **2009**, *1*, 963–970. [CrossRef]

44. Abu-Zahra, M.R.M.; Schneiders, L.H.J.; Niederer, J.P.M.; Feron, P.H.M.; Versteeg, G.F. CO_2 capture from power plants. Part I. A parametric study of the technical performance based on monoethanolamine. *Int. J. Greenh. Gas Control* **2007**, *1*, 37–46. [CrossRef]

45. Romeo, L.M.; Bolea, I.; Escosa, J.M. Integration of power plant and amine scrubbing to reduce CO_2 capture costs. *Appl. Therm. Eng.* **2008**, *28*, 1039–1046. [CrossRef]

46. Goo, S.; Bong, G.; Jun, C.; Min, J. Chemical Engineering Research and Design Optimal design and operating condition of boil-off CO_2 re-liquefaction process, considering seawater temperature variation and compressor discharge. *Chem. Eng. Res. Des.* **2017**, *124*, 29–45.

47. Feenstra, M.; Monteiro, J.; van den Akker, J.T.; Abu-Zahra, M.R.M.; Gilling, E.; Goetheer, E. Ship-based carbon capture onboard of diesel or LNG-fuelled ships. *Int. J. Greenh. Gas Control* **2019**, *85*, 1–10. [CrossRef]

48. Berstad, D.; Anantharaman, R.; Nekså, P. Low-temperature CO_2 capture technologies—Applications and potential. *Int. J. Refrig.* **2013**, *36*, 1403–1416. [CrossRef]

49. Minnehan, J.J.; Pratt, J.W. Practical Application Limits of Fuel Cells and Batteries for Zero Emission Vessels. Available online: https://energy.sandia.gov/wp-content/uploads/2017/12/SAND2017-12665.pdf (accessed on 30 December 2019).

50. Hajjaji, N.; Pons, M.N.; Houas, A.; Renaudin, V. Exergy analysis: An efficient tool for understanding and improving hydrogen production via the steam methane reforming process. *Energy Policy* **2012**, *42*, 392–399. [CrossRef]

51. Liquefied Natural Gas (LNG). Operations Consistent Methodology for Estimating Greenhouse Gas Emissions. Available online: https://www.api.org/~{}/media/Files/EHS/climate-change/api-lng-ghg-emissions-guidelines-05-2015.pdf (accessed on 30 December 2019).

52. Iulianelli, A.; Ribeirinha, P.; Mendes, A.; Basile, A. Methanol steam reforming for hydrogen generation via conventional and membrane reactors: A review. *Renew. Sustain. Energy Rev.* **2014**, *29*, 355–368. [CrossRef]

53. Tzanetis, K.F.; Martavaltzi, C.S.; Lemonidou, A.A. Comparative exergy analysis of sorption enhanced and conventional methane steam reforming. *Int. J. Hydrogen Energy* **2012**, *37*, 16308–16320. [CrossRef]

54. Nalbant, Y.; Colpan, C.O.; Devrim, Y. Energy and exergy performance assessments of a high temperature-proton exchange membrane fuel cell based integrated cogeneration system. *Int. J. Hydrogen Energy* **2019**, 1–11. [CrossRef]

55. Ribeirinha, P.; Abdollahzadeh, M.; Sousa, J.M.; Boaventura, M.; Mendes, A. Modelling of a high-temperature polymer electrolyte membrane fuel cell integrated with a methanol steam reformer cell. *Appl. Energy* **2017**, *202*, 6–19. [CrossRef]

56. Boaventura, M.; Sander, H.; Friedrich, K.A.; Mendes, A. The influence of CO on the current density distribution of high temperature polymer electrolyte membrane fuel cells. *Electrochim. Acta* **2011**, *56*, 9467–9475. [CrossRef]

57. Ersoz, A.; Olgun, H.; Ozdogan, S. Simulation study of a proton exchange membrane (PEM) fuel cell system with autothermal reforming. *Energy* **2006**, *31*, 1490–1500. [CrossRef]

58. Zohrabian, A.; Mansouri, M.; Soltanieh, M. Techno-economic evaluation of an integrated hydrogen and power co-generation system with CO_2 capture. *Int. J. Greenh. Gas Control* **2016**, *44*, 94–103. [CrossRef]

59. Haydary, J. *Chemical Process Design and Simulation: Aspen Plus and Aspen Hysys Applications*; John Wiley & Sons: Hoboken, NJ, USA, 2019; ISBN 1119089115.

60. Simpson, A.P.; Simon, A.J. Second law comparison of oxy-fuel combustion and post-combustion carbon dioxide separation. *Energy Convers. Manag.* **2007**, *48*, 3034–3045. [CrossRef]

61. Andreasen, S.J.; Vang, J.R.; Kær, S.K. High temperature PEM fuel cell performance characterisation with CO and CO_2 using electrochemical impedance spectroscopy. *Int. J. Hydrogen Energy* **2011**, *36*, 9815–9830. [CrossRef]

62. Devrim, Y.; Albostan, A.; Devrim, H. Experimental investigation of CO tolerance in high temperature PEM fuel cells. *Int. J. Hydrogen Energy* **2018**, *43*, 18672–18681. [CrossRef]

63. Oh, K.; Ju, H. Temperature dependence of CO poisoning in high-temperature proton exchange membrane fuel cells with phosphoric acid-doped polybenzimidazole membranes. *Int. J. Hydrogen Energy* **2015**, *40*, 7743–7753. [CrossRef]

64. Jo, A.; Oh, K.; Lee, J.; Han, D.; Kim, D.; Kim, J.; Kim, B.; Kim, J.; Park, D.; Kim, M.; et al. Modeling and analysis of a 5 kWe HT-PEMFC system for residential heat and power generation. *Int. J. Hydrogen Energy* **2017**, *42*, 1698–1714. [CrossRef]

65. Larminie, J.; Dicks, A.; McDonald, M.S. *Fuel Cell Systems Explained*; J. Wiley: Chichester, UK, 2003; Volume 2.

66. Authayanun, S.; Hacker, V. Energy and exergy analyses of a stand-alone HT-PEMFC based trigeneration system for residential applications. *Energy Convers. Manag.* **2018**, *160*, 230–242. [CrossRef]

67. Tsatsaronis, G. Definitions and nomenclature in exergy analysis and exergoeconomics. *Energy* **2007**, *32*, 249–253. [CrossRef]

68. Hinderink, A.P.; Kerkhof, F.P.J.M.; Lie, A.B.K.; De Swaan Arons, J.; Van Der Kooi, H.J. Exergy analysis with a flowsheeting simulator—I. Theory; calculating exergies of material streams. *Chem. Eng. Sci.* **1996**, *51*, 4693–4700. [CrossRef]

69. Mert, S.O.; Ozcelik, Z.; Dincer, I. Comparative assessment and optimization of fuel cells. *Int. J. Hydrogen Energy* **2015**, *40*, 7835–7845. [CrossRef]

70. Obara, S.; Tanno, I. Exergy analysis of a regional-distributed PEM fuel cell system. *Int. J. Hydrogen Energy* **2008**, *33*, 2300–2310. [CrossRef]

71. Lee, H.; Shao, Y.; Lee, S.; Roh, G.; Chun, K.; Kang, H. Analysis and assessment of partial re-liquefaction system for liquefied hydrogen tankers using liquefied natural gas (LNG)and H_2 hybrid propulsion. *Int. J. Hydrogen Energy* **2019**, *44*, 15056–15071. [CrossRef]

72. BP Statistical Review of World Energy Statistical Review of World. Available online: https://www.bp.com/content/dam/bp/business-sites/en/global/corporate/pdfs/energy-economics/statistical-review/bp-stats-review-2019-natural-gas.pdf (accessed on 30 December 2019).

73. Methanex Monthly Average Regional Posted Contract Price History. Available online: https://www.methanex.com (accessed on 30 December 2019).

Article

The ElectricalVehicle Simulator for Charging Station in Mode 3 of IEC 61851-1 Standard

Mihai Rata [1,*], Gabriela Rata [1], Constantin Filote [1], Maria Simona Raboaca [1,2,*], Adrian Graur [1], Ciprian Afanasov [1] and Andreea-Raluca Felseghi [1]

[1] Faculty of Electrical Engineering and Computer Science, Stefan cel Mare University of Suceava, 720229 Suceava, Romania; gabrielar@eed.usv.ro (G.R.); filote@usm.ro (C.F.); Adrian.Graur@usv.ro (A.G.); ciprafanasov@yahoo.com (C.A.); Raluca.FELSEGHI@insta.utcluj.ro (A.-R.F.)

[2] National Research and Development Institute for Cryogenic and Isotopic Technologies-ICSI, 240050 Rm. Valcea, Romania

* Correspondence: mihair@eed.usv.ro (M.R.); simona.raboaca@icsi.ro (M.S.R.)

Received: 9 December 2019; Accepted: 23 December 2019; Published: 31 December 2019

Abstract: As fuel consumption in the transport sector has increased at a faster pace than in other sectors, the use of electromobility represents the main strategy adopted by the automotive industry. In this context, as the number of electrical vehicles (EVs) will increase, it will also be necessary to increase the number of charging stations. The present paper presents a complete solution for charging stations that can be located in the office or mall parking area. This solution includes a mode 3 AC charging stations of International Electrotechnical Commission (IEC) 61851-1 Standard, an EV simulator for testing the good functionality of the charging stations (i.e., communications, residual-current device (RCD) protection) and a software application used for controlling the charging process by the programmable logic controller (PLC).

Keywords: electrical vehicle; AC charging station in mode 3; PLC

---

1. Introduction

Romania, as a signatory country of the Paris Agreement [1], has committed to reducing greenhouse gas emissions by 43% by 2030 compared to 2005 and to participate in the European Union's efforts to reduce greenhouse gas emissions by 30% by 2030.

In Reference [2], the authors depicted that the charging time is one of the main challenges that the electrical vehicle (EV) industry is facing. Generally, the EV charging levels are classified according to their power charging rates [3]. Overnight charging takes place in level I, as the EVs are plugged to a convenient power outlet (120 V) for slow charging (1.5–2.5 kW) over long hours. The main concern of level-I is the long charging time, which renders this charging level unsuitable for long driving cycles when more than one charging operation is needed. Moreover, from the electrical grid operation point of view, the long charging hours at night overloads the distribution transformers as they are not allowed to rest in a grid system with a high number of connected EVs [4]. Level-II charging requires a 240 V outlet; thus, it is characteristically used as the prime charging means for public and private facilities. This charging level is capable of supplying power in the range of 4–6.6 kW over a period of 3–6 h in order to restock the depleted EV batteries. The time required is still the main drawback in this charging level. Additionally, voltage sags and high-power losses in an electrical grid system with a high penetration of level II charging are some of the facing challenges for its widespread. Control and coordination in level II would reduce the negative impacts of level-II charging [5]; however, this requires an extensive communication system to be adopted. In general, both levels I and II require single-phase power sources with onboard vehicle chargers. On the contrary, three-phase power systems are used with off-board chargers for level III fast charging rates (50–75 kW). The use of fast charging stations significantly reduces the EV charging time for a complete charging cycle. Additionally, widespread deployment of fast EV charging stations across the urban and the residential areas would eliminate the EV range anxiety concern [6,7]. However, the high-power charging rates are essential over a short interval of time for level-III charging which imposes a very high demand on the utility grid [8,9]. The current grid infrastructure is not capable of supporting the desired high charging rates of level-III. Thus, accomplishing fast charging rates while solely depending on the electrical grid does require not only the improvement of the charging system but also the improvement of the electrical grid capacity. Additionally, drawing large amounts of current from the electrical grid will increase the utility charges, especially at the peak hours and, consequently, will increase the system cost. The impact of an EV charging station load on the electric grid systems is thoroughly discussed in Reference [10].

Romania, according to European Commission (EC) Directives 2016/30 November 2016 [11–13], has as strategic objective the clean energy [14,15]. Recently, the interest in producing electrical energy by using renewable sources is growing up. As conventional power generation sources, the wind energy is now a real competitive alternative [16]. If the renewable energy is used for charging electrical vehicles (EVs), then we can consider these vehicles with zero emissions [17]. In this context, as the number of electrical vehicles (EVs) will increase, it will also be necessary to increase the number of the charging stations which will have a negative impact on the power quality of the power network (Overvoltage, Voltage Sag) [18–22]. But, with the increase in the number of charging stations installed, the problem will be to carry out their maintenance. In this paper, a proposed EV simulator and a complete solution for charging stations (also called EVSE—electric vehicle supply equipment), that can be located in the office or mall parking area, are presented. The EV simulator can be a useful tool for those who do maintenance at the charging stations, as it can simulate all the cases in which an EV can be found, thus verifying the good functioning of the EVSE. The solution proposed in this paper can be implemented in the office or mall parking area, where the customers have the benefit of being able to charge their EVs for free. It is necessary to implement an efficient solution by which the customers are offered the certainty that their EV will not be disconnected from the charging station by an unauthorized person.

The originality of this paper is based on the specialized literature according to State of The Art, the originality of the research presented in the article consists in the development of an EV simulator for charging stations in mode 3, which helps the companies dedicated to the maintenance of the charging stations [22–24]. The innovation is given by the fact that the programmable logic controller (PLC) and human-machine interface (HMI) charging station control solution allows users to load EVs without the need for an radio-frequency identification (RFID) card, using a unique user-selected code that provides security during charging.

The unidirectional communication is achieved using the PWM (Pulse Width Modulation) signal with 1 KHz values [25,26]. The duty cycle of the PWM signal is closely related to the consumption of predefined current, which should not be exceeded. By modelling the PWM signal, the maximum current consumed by the EV can become restricted [17–19]. Knowing the number of phases used, the maximum permissible power can be calculated. Thus, using the PWM signal the load power can be controlled.

In the European Union and other countries that accept the standard International Electrotechnical Commission (IEC) 61851-1, it is used for charging stations. The standard is intended for defining general requirements for charging stations [20–22], used together with other standards (Ex. IEC 61851-22). The purpose of IEC 61851-1 is to cover EV charging equipment by providing AC power. This standard defines the input voltage limit at 1000 V.

This research tries to present a series of optimal solutions regarding the use and maintenance of charging stations. The present research represents an intermediate phase that is part of a complex research project [13]. The main objectives are the implementation of advanced theoretical and technological solutions to ensure some charging stations, fixed and mobile, for electric vehicles (EV) and hybrid electric vehicles plug-in (PHEV). In this context, this work represents a starting point for the development of fixed charging station in mode 3, according IEC 61851-1 Standard [24–26].

2. IEC 61851-1 Standard

The IEC 61851-1 is an international standard where the general requirements for EVs conductive charging systems are presented.

2.1. EV Charging Modes Defined in IEC 61851-1 Standard

According to IEC 61851-1 standard, the charging of electrical vehicles can be done in four ways (illustrated in Figure 1), as following:

- Mode 1 is the simplest solution for charging EV. In this case the EV is connected to the residences standard socket outlets but must have a circuit breaker for overload and earth leakage protections. In this mode the charging is realized without communication and it is rated up to 16 A.
- Mode 2 where the EV is connected to the domestic power grid via a particular cable with in-cable or in-plug control pilot and a protection device. The current must not exceed 32 A.
- Mode 3 where the EV is connected via specific socket on a dedicated charging station that has permanently installed the control and the protection functions. The rated charging current is up to 3 × 63 A.
- Mode 4, where the EV is fast charging in direct current (DC).

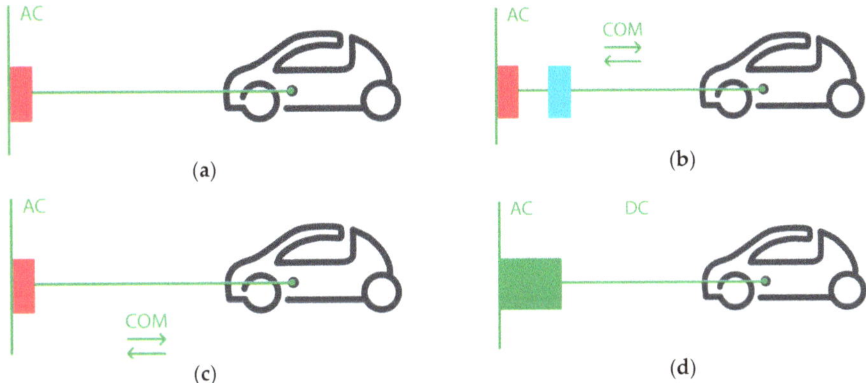

Figure 1. Different charging methods defined in IEC 61851-1: (**a**) Mode 1; (**b**) Mode 2; (**c**) Mode 3; (**d**) Mode 4.

2.2. Charging Station in Mode 3 According to IEC 61851-1 Standard

For charging EV in AC, the most used solution is mode 3 because it fully guarantees the safety of people. The station has a charging controller that, via a control pilot, can communicate with vehicle and checks (before starting charging process) the following:

- If the EV is correctly connected to the station;
- The maximum current capability of the cable assembly coded by a resistor. If the current is higher than this value, then the electric vehicle supply equipment (EVSE) interrupts the current supply. Most EVs are equipped with two cables one for slow charge (Mode 1) and other for fast charge (Mode 3);
- If the earthing system of the vehicle is connected correctly to EVSE.

The communication signal sent by the charging station to the EV on the control pilot and earth has a square wave shape of 1 kHz frequency and ±12.0–±0.4 V voltage range. The typically control pilot circuit according to IEC61851-1 is illustrated in Figure 2. The stage of charging cycle is identified by the station charger controller through positive voltage level of communication signal level, as follow:

- Vehicle unconnected, when the positive level of communication signal is 12 V;
- The cable assembly is plugged into both the EV and the EVSE, where the positive voltage level is 9 V;
- Vehicle is ready to receive energy, when S2 (Figure 2) is closed by the vehicle and the positive voltage level is 6 V.

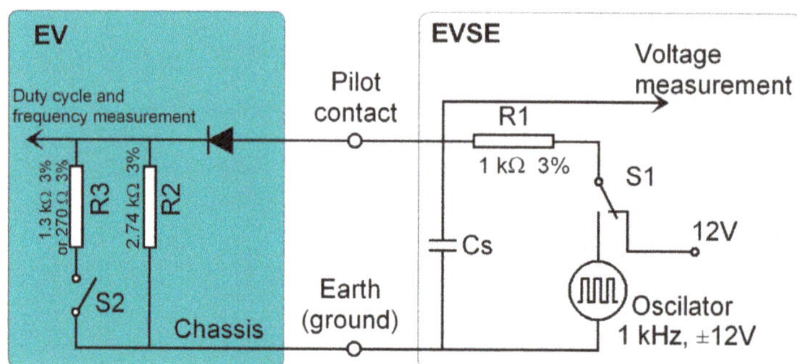

Figure 2. Typical control pilot circuit according to IEC61851-1.

The EVSE can send information through the duty cycle of the communication signal to the EV about the maximum current supported by the cable used. Consequently, the EV battery management system interprets the Control Pilot signal to limit charge rate.

3. Experimental Arrangement

The proposed solution for the EV simulator, which permits simulation of a real electrical vehicle and a prototype of charging station in mode 3 according to IEC 61851-1 standard are presented in Figure 3. The experimental arrangement for these tests is illustrated in Figure 3a and the circuit diagram for charging station controlled by PLC is illustrated in Figure 3b. The dimensions of our Charging Station are 58 cm/28 cm/21 cm (L × W × H), the mass is 8 Kg and the power is 24 kW.

Because most public charging stations are equipped with a type 2 socket, the EV simulator is equipped with a specific charging EVs cable with type 2 connector. This makes it possible for the simulator to be connected to any station which is equipped with socket type 2. All stages of EV charging process from a station can be created using the proposed EV simulator. In addition, for employee-owned electric vehicles customers' safety, the good working of station RCD protection has been implemented. To implement a complete solution for a charging station that can be placed in a mall or office parking, the authors choose the HMI-PLC system for station control. The EV Simulator is illustrated in Figure 3c.

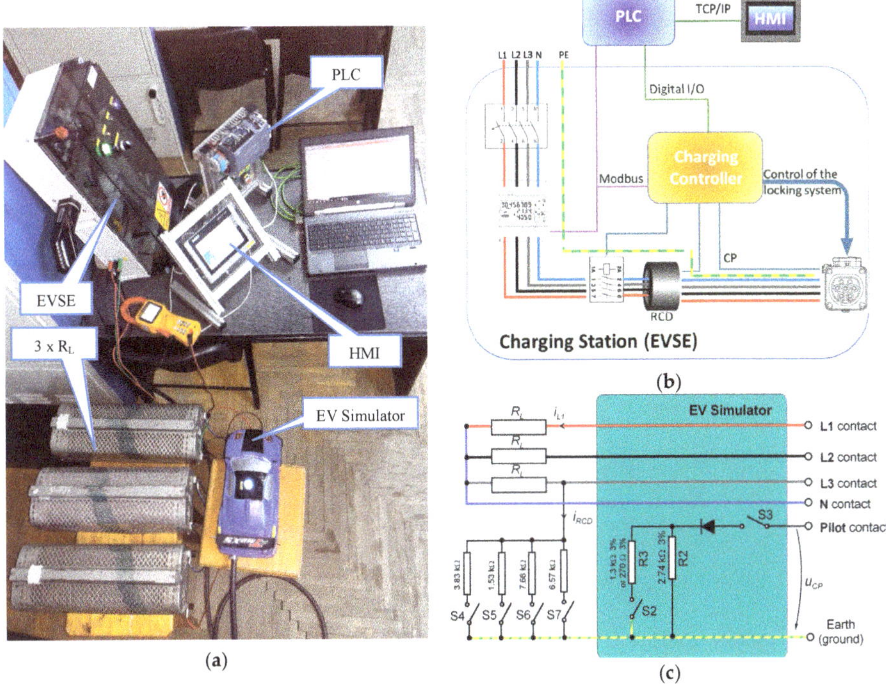

Figure 3. Experimental arrangement: (**a**) Experimental setup; (**b**) electric vehicle supply equipment (EVSE) Circuit block diagram; (**c**) electrical vehicle (EV) Simulator.

4. Experimental Results

On the control pilot, the EVSE generates a 1 kHz square wave at ±12 volts in order to detect whether the EV is correctly plugged, to communicate the maximum power allowed by the cable and for charging begin/end process control initiated by the EV (see Figure 4a) or initiated by EVSE (see Figure 4b).

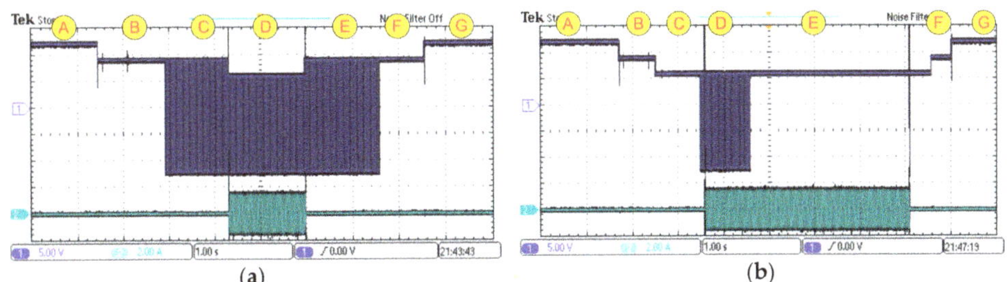

Figure 4. Communication signals between EVSE and EV. Ch 1 – u_{CP} and Ch 2 – i_{L1} (**a**) Control from EV; (**b**) Control from EVSE.

All charging stages can be generated by EVSE through the status of switch S1 (presented in Figure 2) and by EV simulator through the status of switches S2, S3 (presented in Figure 3c). These charging stages are detailed in Table 1.

Table 1. EV charging stages.

Stage	Switches			Description of the Stage
	S1 (Figure 2)	S2 (Figures 2 and 3)	S3 (Figure 3)	
Figure 4a. Charging begin/end regime controlled by the electrical vehicle				
A	OFF	OFF	OFF	**The vehicle is unconnected and** the voltage measured by EVSE on pilot contact is +12 V DC
B	OFF	OFF	ON	**EVSE is not ready.** In this case the cable assemble is connected to EV and to the EVSE and the voltage measured by EVSE is +9 V DC.
C	ON	OFF	ON	**EVSE is ready** and at pilot contact is generated PWM signal (+9 V− −12 V).
D	ON	ON	ON	**The vehicle is ready** and charging process is active. In this case the positive voltage level of PWM signal is depending of R3 resistor value. If ventilation in charging area is not required, the $R3 = 1.3$ kΩ and the positive voltage value of PWM signal is 6 V (case illustrated in Figure 4a). If in the charging area the ventilation is required, the $R3 = 270$ Ω the positive voltage value of PWM signal is 3 V.
E	OFF	ON	ON	**The vehicle is not ready** and charging process is aborted.
F	OFF	OFF	ON	**EVSE is not ready.**
G	OFF	OFF	OFF	**The vehicle is unconnected.**
Figure 4b. Charging begin/end regime controlled by the station (EVSE)				
A	OFF	OFF	OFF	**The vehicle is unconnected** and the voltage measured by EVSE on pilot contact is +12 V DC
B	OFF	OFF	ON	**EVSE is not ready.** In this case the cable assemble is connected to EV and to the EVSE and the voltage measured by EVSE is +9 V DC.
C	OFF	ON	ON	**The vehicle is ready** and at pilot contact is measured 6 V DC.
D	ON	ON	ON	**The EVSE is ready** and charging process is active. In this case at pilot contact is generated PWM signal.
E	OFF	ON	ON	**The EVSE is not ready** and charging process is aborted. In this case when S1 switches OFF the charging process continues for about 3 seconds
F	OFF	OFF	ON	**EV is not ready.**
G	OFF	OFF	OFF	**The vehicle is unconnected.**

Because the people safety in stations is very important, an RCD (residual current device) has been mounted. Of course, it must be periodically tested for the proper functionality of protection by the maintenance employer. The RCD protection of the EVSE is very easy to be tested using an EV simulator and the switches S4–S8 (see Figure 3c). The experimental results at different values of residual current are presented in Figure 5 and it is shown how the RCD tripping time is reduced as the residual current increases.

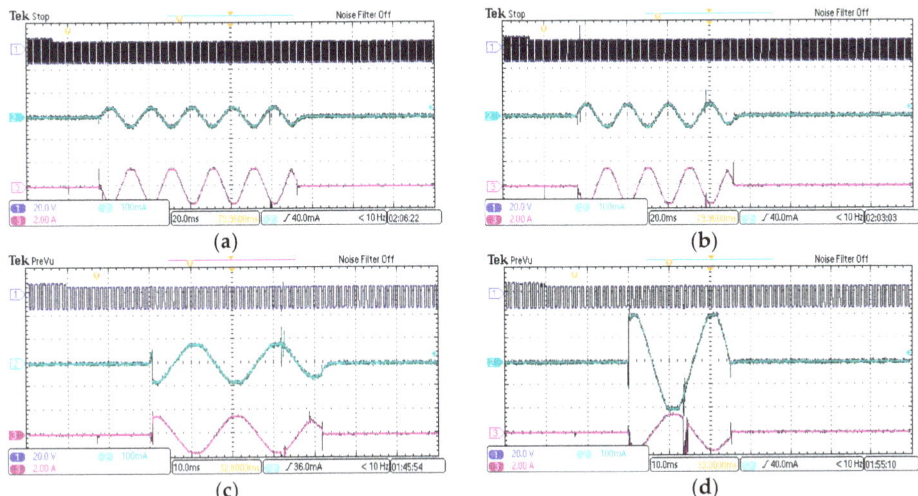

Figure 5. Residual current device (RCD) experimental tests; Ch 1 – u_{CP}, Ch 2 – i_{RCD}, Ch 3 – i_{L1}: (**a**) i_{RCD} = 30 mA; (**b**) i_{RCD} = 35 mA; (**c**) i_{RCD} = 60 mA; (**d**) i_{RCD} = 150 mA.

The charging stations can be installed in the office or mall parking area, where the customers have the benefit of being able to charge their EVs for free. The solution must offer customers the certainty that their EV will not be disconnected from the charging station by an unauthorized person. Usually, the customers must have an RFID card that will be used to control the beginning and the end of the charging processes. The proposed solution offers the same facility to the customers without any RFID card. To test the solution proposed by the authors for the charging station, a PLC has been used to control the implemented prototype. A friendly HMI connected with PLC has been implemented in order to be easily used by the customers. For example, when a customer plugs his EV in the charging station, he should wait the confirmation on HMI and, after that, he must introduce an identification code and press the START button. In this moment the locking system of the station socket will block the charging cable connector in the station socket, preventing the unauthorized disconnection of the EV from the charging station. This type of application can be simple extended for more than one station, all controlled by the same PLC and HMI. These stations can be in the office or mall parking area. Before starting the charging command, the user must choose in HMI the number of station and a unique identification code. When the user wants to command the end of charging, he must choose the number of station and to input the identification code. After the PLC recognizes the code, the charging process is interrupted and the cable connector is unlocked. So, the features of the PLC software are as follows:

- The user can control the charging station via an easy HDMI interface;
- All parameters from a three-phase energy meter device (model EEM-350-D-MCB from Phoenix Contact) through MODBUS protocol using RS485 interfaces can be read;
- An identification code is used to recognize the authorized person which commands the starting and finishing of the charging process.

The programmable logic controller (PLC) type S7-1200 and a human-machine interface (HMI) type TP700 Comfort, both from Siemens, have been used but other types of PLC or HMI can be used in order to implement the proposed solution as well. A sequence of program (realized in the dedicated software Totally Integrated Automation Portal—TIA Portal, from Siemens) and the HMI screen are presented in Figure 6a,b: The sequence from the main program, presented in Figure 6a, confirms the communication between PLC and three phase energy meter device.

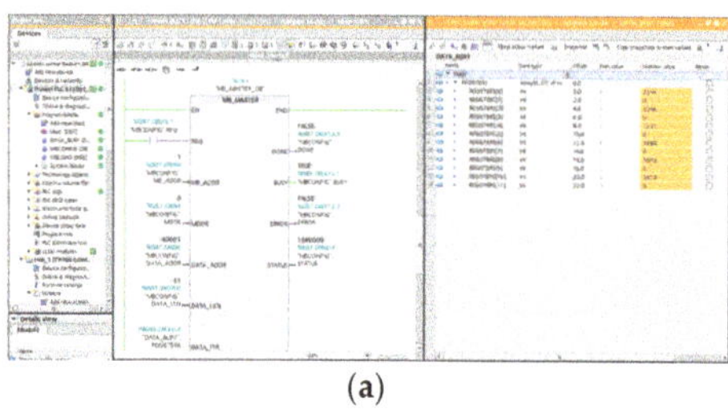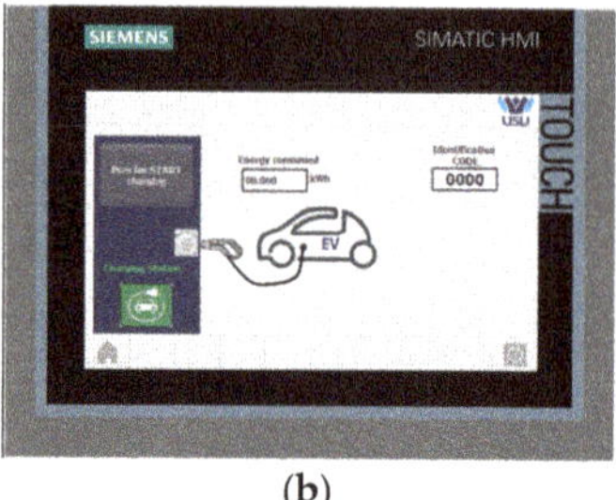

(a)   (b)

Figure 6. Software application: (**a**) programmable logic controller (PLC) software; (**b**) human-machine interface (HMI) screen.

This research helps the developers of Charging Station as [23–25]:

1. This simulation is used to realize the periodic maintenance of Charging Station, from exploitation of the RCD relay;
2. Verifying a correct functioning of the station controller;
3. The solution for controlling the Charging Stations with PLC and HMI offers users the possibility to charge the EV without an RFID card, using a unique identification code chosen by driver [26].

5. Conclusions

An EV can be charged from an EVSE only if communication between these is established and both send a confirmation that they are ready for the charging process. Periodically all charging stations must verify the good functioning by the maintenance persons. In this paper, an EV simulator has been developed that is a useful tool for those who do maintenance at the charging stations. This equipment is capable as follows:

- To create all charging stages for an EV to a charging station in mode 3 (EVSE) according to IEC 61851-1 standard and checks if the charging station works correctly;
- Permits measuring the charging station tripping time by generating different values of residual current.

The authors also propose a solution that can be implemented in the office or mall parking area, where the customers have the benefit of being able to charge their EVs for free. The charging stations are controlled by a PLC-HMI system and the customers are offered the certainty that their EV will not be disconnected from the charging station by an unauthorized person. When a customer plugged his EV to the charging station before START charging, it is necessary to introduce an identification code in HMI. For STOP charging, unlock the cable connector and disconnect the EV from the charging station; the PLC checks whether it introduced the same unique identification code in order to validate these actions.

Author Contributions: Conceptualization, M.R.; methodology, M.R. and G.R.; validation, C.F., M.S.R. and C.A.; formal analysis, M.R.; investigation, M.S.R. and G.R.; data curation, G.R.; writing—original draft preparation, A.-R.F., M.S.R. and M.R.; writing—review and editing, M.R. and C.F.; visualization, A.-R.F.; supervision, C.A.; project administration, A.-R.F. and C.F. All authors have read and agreed to the published version of the manuscript.

Acknowledgments: This work was supported by a grant from the Romanian Ministry of Research and Innovation, CCCDI-UEFISCDI, project number PN-III-P1-1.2-PCCDI-2017-0776/No. 36 PCCDI/15.03.2018, within PNCDI III.

Conflicts of Interest: The authors declare no conflict of interest.

References

1. United Nations Framework Convention on Climate Change (UNFCCC), Paris Agreement. Available online: https://unfccc.int/process-and-meetings/the-paris-agreement/the-paris-agreement (accessed on 28 August 2019).
2. Hidrue, M.K.; Parsons, G.R.; Kempton, W.; Gardner, M.P. Willingness to pay for electric vehicles and their attributes. *Resour. Energy Econ.* **2011**, *33*, 686–705. [CrossRef]
3. Yilmaz, M.; Krein, P.T. Review of battery charger topologies, charging power levels and infrastructure for plug-in electric and hybrid vehicles. *IEEE Trans. Power Electron.* **2013**, *28*, 2151–2169. [CrossRef]
4. Fairley, P. Speed bumps ahead for electric-vehicle charging. *IEEE Spectr.* **2010**, *47*, 13–14. [CrossRef]
5. Clement-Nyns, K.; Haesen, E.; Driesen, J. The impact of Charging plug-in hybrid electric vehicles on a residential distribution grid. *IEEE Trans. Power Syst.* **2010**, *25*, 371–380. [CrossRef]
6. Yunus, K.; la Parra, H.Z.D.; Reza, M. Distribution Grid Impact of Plug-In Electric Vehicles Charging at Fast Charging Stations Using Stochastic Charging Model. In Proceedings of the 2011 14th European Conference on Power Electronics and Applications, Birmingham, UK, 30 August–1 September 2011; pp. 1–11.
7. Botsford, C.; Szczepanek, A. Fast Charging vs. Slow Charging: Pros and cons for the New Age of Electric Vehicles. In Proceedings of the EVS24 International Battery, Hybrid and Fuel Cell Electric Vehicle Symposium, Stavanger, Norway, 13–16 May 2009.
8. Gnann, T.; Funke, S.; Jakobsson, N.; Plotz, P.; Sprei, F.; Bennehag, A. Fast charging infrastructure for electric vehicles: Today's situation and future needs. *Transp. Res. Part D Transp. Environ.* **2018**, *62*, 314–329. [CrossRef]
9. Morrow, K.; Karner, D.; Francfort, J. *Plug-in Hybrid Electric Vehicle Charging Infrastructure Review (U. S. Department of Energy Vehicle Technologies Program—Advanced Vehicle Testing Activity)*; Idaho National Laboratory (INL): Idaho Falls, ID, USA, 2008.
10. Deb, S.; Tammi, K.; Kalita, K.; Mahanta, P. Review of recent trends in charging infrastructure planning for electric vehicles. *Wiley Interdiscip. Rev. Energy Environ.* **2018**, *7*, e306. [CrossRef]
11. Răboacă, M.S.; Badea, G.; Enache, A.; Filote, C.; Răsoi, G.; Rata, M.; Lavric, A.; Felseghi, R.-A. Concentrating Solar Power Technologies. *Energies* **2019**, *12*, 1048. [CrossRef]
12. Badea, G.; Felseghi, R.-A.; Varlam, M.; Filote, C.; Culcer, M.; Iliescu, M.; Răboacă, M.S. Design and Simulation of Romanian Solar Energy Charging Station for Electric Vehicles. *Energies* **2019**, *12*, 74. [CrossRef]
13. Şişman, G.R.; Oproescu, M. Predictive maintenance of power industrial electronic equipment. *J. Electr. Eng. Electron. Control Comput. Sci.* **2015**, *1*, 21–30. Available online: https://jeeeccs.net/index.php/journal/article/view/18/10 (accessed on 1 December 2019).
14. Ahn, N.; Jo, S.Y.; Kang, S.-J. Constraint-Aware Electricity Consumption Estimation for Prevention of Overload by Electric Vehicle Charging Station. *Energies* **2019**, *12*, 1000. [CrossRef]
15. Chiper, A.M. The conception and execution of a simulation workbench for an automotive fuel gauge system. *J. Electr. Eng. Electron. Control Comput. Sci.* **2015**, *1*, 15–20. Available online: https://jeeeccs.net/index.php/journal/article/view/10/3 (accessed on 1 December 2019).
16. Bizon, N.; Thounthong, P. Real-time strategies to optimize the fueling of the fuel cell hybrid power source: A review of issues, challenges and a new approach. *Renew. Sust. Energy Rev.* **2018**, *91*, 1089–1102. [CrossRef]
17. Beleiu, H.G.; Beleiu, I.N.; Pavel, S.G.; Darab, C.P. Management of Power Quality Issues from an Economic Point of View. *Sustainability* **2018**, *10*, 2326. [CrossRef]
18. Maier, V.; Pavel, S.G.; Beleiu, H.G.; Farcas, V. Additional Indicators for Short Time Overvoltages and Voltage Impulses. In Proceedings of the 2016 International Conference and Exposition on Electrical and Power Engineering (EPE), Iasi, Romania, 20–22 October 2016; pp. 827–831. [CrossRef]
19. Rai, K.; Jandhyala, A.; Rai, S.C. Power Intelligence and Asset Monitoring for System. *J. Electr. Eng. Electron. Control Comput. Sci.* **2015**, *1*, 31–36. Available online: https://jeeeccs.net/index.php/journal/article/view/129/106 (accessed on 1 December 2019).
20. Maier, V.; Pavel, S.G.; Beleiu, H.G.; Buda, C. Survey Regarding Additional Indicators for Voltage Sags. In Proceedings of the 2016 International Conference on Development and Application Systems (DAS), Suceava, Romania, 19–21 May 2016; pp. 103–107. [CrossRef]
21. Liu, G.; Kang, L.; Luan, Z.; Qiu, J.; Zheng, F. Charging Station and Power Network Planning for Integrated Electric Vehicles (EVs). *Energies* **2019**, *12*, 2595. [CrossRef]

22. Bizon, N.; Thounthong, P. Fuel economy using the global optimization of the Fuel Cell Hybrid Power Systems. *Energy Convers. Manag.* **2018**, *173*, 665–678. [CrossRef]
23. Birleanu, F.G.; Bizon, N. Principles, Architectures and Challenges for Ensuring the Integrity, Internal Control and Security of Embedded Systems. *J. Electr. Eng. Electron. Control Comput. Sci.* **2015**, *1*, 37–45. Available online: https://jeeeccs.net/index.php/journal/article/view/68/60 (accessed on 1 December 2019).
24. Deb, S.; Tammi, K.; Kalita, K.; Mahanta, P. Impact of Electric Vehicle Charging Station Load on Distribution Network. *Energies* **2018**, *11*, 178. [CrossRef]
25. UEFISCDI. Project Number PN-III-P1-1.2-PCCDI-2017-0776/No. 36 PCCDI/15.03.2018, within PNCDI III, Grant from the Romanian Ministry of Research and Innovation, CCCDI. Available online: www.smile-ev.usv.ro (accessed on 28 August 2019).
26. Shareef, H.; Islam, M.M.; Mohamed, A. A review of the stage-of-the-art charging technologies, placement methodologies and impacts of electric vehicles. *Renew. Sustain. Energy Rev.* **2016**, *64*, 403–420. [CrossRef]

Article

Open-Source Dynamic Matlab/Simulink 1D Proton Exchange Membrane Fuel Cell Model

Arne L. Lazar, Swantje C. Konradt * and Hermann Rottengruber

Institute of Mobile Systems (IMS), Energy Conversion Systems for Mobile Applications, (EMA), Otto-von-Guericke University Magdeburg, Universitätsplatz 2, 39106 Magdeburg, Germany
* Correspondence: swantje.konradt@ovgu.de

Received: 6 August 2019; Accepted: 5 September 2019; Published: 9 September 2019

Abstract: This work presents an open-source, dynamic, 1D, proton exchange membrane fuel cell model suitable for real-time applications. It estimates the cell voltage based on activation, ohmic and concentration overpotentials and considers water transport through the membrane by means of osmosis, diffusion and hydraulic permeation. Simplified equations reduce the computational load to make it viable for real-time analysis, quick parameter studies and usage in complex systems like complete vehicle models. Two modes of operation for use with or without reference polarization curves allow for a flexible application even without information about cell parameters. The program code is written in MATLAB and provided under the terms and conditions of the Creative Commons Attribution License (CC BY). It is designed to be used inside of a Simulink model, which allows this fuel cell model to be used in a wide variety of 1D simulation platforms by exporting the code as C/C++.

Keywords: proton exchange membrane fuel cell; matlab; simulink; real-time capability; dynamic fuel cell model

---

1. Introduction

Proton exchange membrane fuel cells (PEMFC) can contribute to achieving the goal of a sustainable energy supply and production. High efficiency and power density as well as zero emissions are beneficial for both stationary and mobile applications. However, designing the water management, cooling and media supply of a PEMFC-system is challenging. A model-based approach for the simulation of such a system can be a valuable tool in this matter.

Many PEMFC-models have been developed in the past, with varying objectives. Some models intend to deliver highly accurate results through means of computational fluid dynamics (CFD) [1–4] whilst others target a faster simulation by reducing the complexity [5,6]. However, in terms of computational time, these models still operate in the range of min or hours. This makes them unsuitable for large system simulations like complete vehicles or even real-time evaluation. To account for this, simplified fuel cell models have been developed in recent history to reduce the required CPU-time at the expense of accuracy [7–9]. Some of the aforementioned models have been made publicly and freely available, others remain closed-source. Additionally, depending on the chosen programming language as well as structure of model inputs and outputs, the compatibility with various simulation environments might be restricted.

After considering the previous research on fuel cell modeling, the motivation for creating the presented model was to combine the following three aspects. First, the compatibility with commonly used 1D simulations' environments. Since each software has its strengths, weaknesses and price, individual programs might not be available for everyone who would benefit from a fuel cell model. In particular, educational facilities often lack the funds to provide expensive licenses. Offering a model that can be used in multiple environments, and therefore can be utilized by a large audience, was a primary motivation. Second, there is a low computational demand. Since large systems (e.g., cars or trains completing a drive cycle) consist of many components, each individual model is required to be of reduced complexity in order to keep the overall simulation time low. However, the operating conditions inside of such a dynamic system are constantly shifting. This is why a compromise between speed and accuracy, but with a focus on speed, was another objective. Third, there is an open-source software license. Since open-source models can be further expanded upon and offer value for research, education and development, this represented the third requirement.

The model at hand was designed to be used either as a cell model or as part of a fuel cell stack. In its core, it represents a cell model with internal (optional) scaling for stack parameters like cell area and quantity. In order to ensure compatibility, the MATLAB–Simulink (version used: R2016b by The MathWorks, Inc.) environment was chosen. Furthermore, to allow for fast calculations and even real-time applications, simplified equations are used, even though the underlying Nernst-equation has recently been found to show considerable inaccuracies [10]. Additionally, to further reduce the computational demand, no discrete model was used for the membrane electrode assembly (MEA). A compromise was made between simulation quality regarding water and gas transport and the speed of the calculations.

Possible use cases for the presented model are the creation of polarization curves or the cell performance estimation under varying conditions like in a moving vehicle. Figure 1 depicts the basic structure of the program code, which was designed to be run inside of a MATLAB-function-block as part of a Simulink model. Based on various inputs and physical parameters, the model estimates cell voltage as well as several other outputs in every time step. A complete list of the model inputs, outputs and parameters is provided as Supplementary Material. The calculations inside the code consist of simplified equations, presented in the following section.

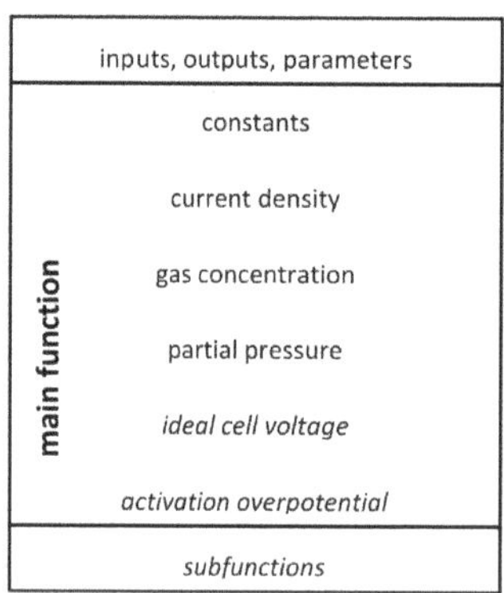

Figure 1. Model structure and calculation order. Italic functions are implemented as a subfunction.

2. Mathematical Model

In this fuel cell model, the real cell voltage E_{cell} is estimated by subtracting the voltage losses inside the cell from the ideal cell voltage $E_{T,p}$. These are summarized as activation E_{act_i}, ohmic E_{ohm} and concentration E_{con_i} overpotentials [11]:

$$E_{\text{cell}} = E_{T,p} - E_{\text{act}_i} - E_{\text{ohm}} - E_{\text{con}_i}. \tag{1}$$

2.1. Ideal Voltage

The theoretical maximum voltage of a PEMFC under reference conditions $E_{25C,1atm}$ can be calculated with the Gibbs free energy ΔG as well as the Faraday constant F and the number of electrons involved n [11]:

$$E_{25C,1atm} = -\frac{\Delta G}{nF} = \frac{237340}{2 \times 96485} = 1.23 \text{ V.} \tag{2}$$

Variations in temperature T and partial pressure of reactants p_i can be accounted for by using the Nernst-Equation [11]:

$$E_{T,P} = -\left(\frac{\Delta H}{nF} - \frac{T\Delta S}{nF}\right) + \frac{RT}{nF} \ln\left[\frac{p_{H_2} p_{O_2}^{0.5}}{p_{H_2O}}\right]. \tag{3}$$

Neglecting influences of changing enthalpy ΔH and entropy ΔS, as well as assuming the product water to be in the liquid phase, the ideal cell voltage $E_{T,P}$ can be expressed as follows [11]:

$$E_{T,p} = 1.482 - 0.000845\,T + 0.0000431\,T\,\ln\left(p_{H_2} p_{O_2}^{0.5}\right). \tag{4}$$

2.2. Activation Overvoltage

Based on the Butler–Volmer equation, the activation overpotential $E_{act,i}$ can be estimated as a function of current density i, exchange current density i_0 as well as temperature T and charge transfer coefficient α_i [12,13]:

$$E_{act,i} = \frac{RT_i}{\alpha_i F}\text{arcsinh}\left(\frac{i}{2i_{0,i}}\right), \tag{5}$$

where R denotes the universal gas constant and F the Faraday constant.

The exchange current density i_0 of a platinum electrode as a function of partial pressure p_i, temperature T, catalyst loading L_i and specific area a_i can be calculated on the basis of a reference value $i_{0,ref}$ [11]:

$$i_{0,i} = i_{0,ref,i} a_i L_i\left(\frac{p_i}{p_{ref}}\right)^{\gamma} \exp\left[-\frac{\Delta G_i}{RT_i}\left(1 - \frac{T_i}{T_{ref}}\right)\right], \tag{6}$$

$$T_{ref} = 298.15 \text{ K;} \quad p_{ref} = 1.0125 \text{ bar.}$$

For simplicity, a constant value was used for the activation energy ΔG_i, even though it can vary under real operating conditions [14].

2.3. Membrane Water Content

For estimating the water content λ inside of a Nafion-membrane, a function of water activity a is used [15]:

$$\lambda = \begin{cases} 0.043 + 17.81\,a - 39.85\,a^2 + 36\,a^3 \\ \text{for } a \leq 1 \\ 14 + 1.4(a - 1) \\ \text{for } 1 < a \leq 3 \end{cases}. \tag{7}$$

As a simplification, ab- and desorption of water were neglected. Furthermore, the distribution of water along the membrane geometry was assumed to be uniform.

The water activity a can be expressed as [16]

$$a = RH + 2\,s, \tag{8}$$

where RH denotes the relative humidity (for ideal gas properties) and s the liquid water volume fraction. For this model, it is assumed that liquid water is only present in the catalyst layer. For RH and s, a logarithmic average is used to account for nonhomogeneous distribution inside the channels (see Equation (17)). This can be disabled by suppling the model with equal values for input and output.

In case of a cold start at subzero temperatures, the behavior of frozen water inside of a Nafion-membrane is approximated using the following terms [16]:

$$\lambda_{\text{sat}} \begin{cases} = 4.837 \\ \quad \text{for } T < 223.15 \text{ K} \\ = \left[-1.304 + 0.01479 \, T - 3.594 \cdot 10^{-5} \, T^2 \right]^{-1} \\ \quad \text{for } 223.15 \text{ K} \le T < T_{\text{frost}} \\ > \lambda \\ \quad \text{for } T \ge T_{\text{frost}} \end{cases} . \tag{9}$$

To estimate the water concentration C_w, a simple proportional correlation between the membrane density ρ_{mem}, equivalent weight and water content λ is used [11]:

$$C_w = \frac{\rho_{\text{mem}}}{EW} \lambda. \tag{10}$$

2.4. Ohmic Overvoltage

Regarding ohmic resistances inside the cell, only the membrane resistance as the most influential factor is considered. Therefore, the overpotential can be calculated with the current density i, membrane conductivity σ_{mem} and (wet) thickness δ_{mem}. Since membrane thickness—at typical PEMFC operating conditions—is only marginally affected by swelling, the thickness is assumed to be constant [17,18]:

$$E_{\text{ohm}} = \frac{\delta_{\text{mem}} \, i}{\sigma_{\text{mem}}}. \tag{11}$$

To estimate the membrane conductivity σ_{mem}, the following correlation is used [9,12,19]:

$$\sigma_{\text{mem}} = 1.16 \, \max\{0, f - 0.06\}^{1.5} \exp\left[\frac{15000}{R} \left(\frac{1}{T_{\text{ref}}} - \frac{1}{T} \right) \right], \tag{12}$$

$$T_{\text{ref}} = 353.15 \text{ K}; \quad f = \frac{\lambda V_W}{\lambda V_W + V_m},$$
$$V_m = \frac{EW}{\rho_{\text{mem}}}; \quad V_W = \frac{18.01528}{\rho_W(T)}.$$

It is mostly dependent on membrane water content λ and temperature T_{mem}, whilst also being affected by the material properties of equivalent weight EW and membrane density ρ_{mem} as well as the density of water $\rho_W(T)$. A simple arithmetic average of anode and cathode site values is used for further calculations.

The density of water $\rho_w(T)$ at 1 [bar] as a function of temperature T can be approximated by [20]:

$$\rho_w(T) = 999.972 - 7 \cdot 10^{-3} \, (T - 4)^2 \cdot 10^{-3}, \tag{13}$$

$$T \text{ in } °C.$$

2.5. Concentration Overvoltage

Using the Nernst-equation, concentration overvoltage E_{con_i} is described as a function of temperature T, current density i and limiting current density i_L [11]

$$E_{\text{con}_i} = \frac{RT_i}{nF} \ln\left(\frac{i_{L_i}}{i_{L_i} - i} \right), \tag{14}$$

where R is the universal gas constant, n the number of electrons involved and F the Faraday constant. Since this ideal equation often underestimates the real overvoltage, e.g., because of uneven gas concentration, a correction factor (see Supplementary Materials) is used to adjust the results [11,21].

For the limiting current density i_L, a simplified expression including the Faraday F constant, number of electrons n, diffusion coefficient D_i, gas concentration C_i and diffusion distance (here: electrode thickness) δ_e is used [11]:

$$i_{L_i} = \frac{nFD_iC_i}{\delta_e}. \tag{15}$$

To calculate the diffusion coefficient D_i, the model expects an external reference value $D_{i,\text{ref}}$ for the gas mixture. It therefore relies on the external calculation of gas concentration and diffusivity. The reference value will then be adjusted for electrode porosity ϵ and tortuosity τ as well as temperature T, pressure p and liquid water volume fraction s [9,22,23]:

$$D_i = \frac{\epsilon}{\tau^2}(1-s)^3 D_{i,\text{ref}}\left(\frac{T}{T_{\text{ref}}}\right)^{1.5}\frac{p_{\text{ref}}}{p}, \tag{16}$$

$$T_{\text{ref}} = 353.15 \text{ K}; \quad p_{\text{ref}} = 1.01325 \text{ bar}.$$

Porosity ϵ and tortuosity τ are used to approximately describe the geometry of the electrodes, while a value of 1 for the liquid water volume fraction s represents a fully flooded channel. As a simplification, it is assumed that the gas diffusivity in liquid water is 0. The approach of relying on $D_{i,\text{ref}}$ as a model input for the calculation of D_i ensures compatibility with arbitrary gas mixtures and composition of external models for fluid mechanics—for instance, when used in a vehicle model and being connected to components for the media supply.

A simple logarithmic average is used to account for nonhomogeneous distribution of the gas concentration C_{lm}. If desired, this can be disabled by supplying the same value for input and output [5]:

$$C_{\text{lm}} = \frac{C_{\text{in}} - C_{\text{out}}}{\ln \frac{C_{\text{in}}}{C_{\text{out}}}}. \tag{17}$$

2.6. Cell and Stack Performance

What is labeled as the electrical efficiency η_{electric} in this work relates to the lower heating value of hydrogen E_{LHV} [11]:

$$\eta_{\text{electric}} = \frac{E_{\text{cell}}}{E_{\text{LHV}}}, \tag{18}$$

$$E_{\text{LHV}} = 1.254 \text{ V},$$

and, furthermore, is used to calculate the heat flow $\dot{Q}$ of the cell/stack based on the power delivered P_{stack}. As a simplification, it is assumed that the product water fully evaporates before leaving the cell/stack [11]:

$$\dot{Q}_{\text{stack}} = \left(\frac{P_{\text{stack}}}{\mu_{\text{electric}}}\right) - P_{\text{stack}}. \tag{19}$$

2.7. Water Transport

In this model, the estimation of the flow of water j_w from anode to cathode is divided in three categories: osmotic j_{osmo} and diffusive j_{diff} flow as well as hydraulic permeation j_{hyd}. A positive value means an increase in water concentration:

$$j_{w_{\text{anode}}} = j_{\text{diff}} + j_{\text{hyd}} - j_{\text{osmo}}, \tag{20}$$

$$j_{w_{\text{cathode}}} = j_{\text{gen}} + j_{\text{osm}} - j_{\text{diff}} - j_{\text{hyd}}. \tag{21}$$

Three major simplifications are applied to reduce the complexity of water transport mechanisms. First, only the flow through the membrane is considered, whilst transport through the porous media of catalyst and gas diffusion layers is neglected. Second, it is assumed that liquid water is only present once the gas mixture is saturated. Finally, liquid and vapor phases are not directly considered. Instead, the chosen equations for diffusive and hydraulic flow are adjusted by several functions (Figure 2) created with the MATLAB curve fitting application and experimental data from Adachi et al. [24]. As a result, three cases are described in the following sections: vapor–vapor (VVP), liquid–vapor (LVP) and liquid–liquid permeation (LLP).

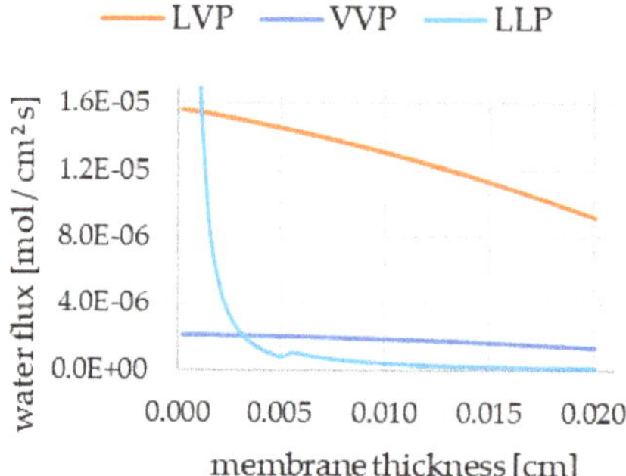

Figure 2. Estimated Water flux through the membrane at 70 °C (Equations (24)–(32), adjusted with data from Adachi et al. [24]); VVP: cathode 96%, anode 38% RH; LVP: cathode liquid volume fraction 100% (flooded), anode 38% RH; LLP: cathode and anode flooded, Δp 1 [bar] note: only a few data points were available to develop each function, which leads to some numerical inaccuracies.

2.7.1. Osmosis

Water transport due to osmosis j_{osmo} is expressed by a linear correlation composed of the Faraday constant F, the current I [11]

$$j_{osmo} = n_{osmo} \frac{I}{F} \tag{22}$$

and the osmotic coefficient n_{osmo}, which is dependent on the (average) water content of the membrane λ [25]:

$$n_{osmo} = 0.0029\, \lambda^2 + 0.05\, \lambda - 3.4 \cdot 10^{-19}. \tag{23}$$

2.7.2. Diffusion

In terms of diffusion j_{diff}, the abovementioned functions to differentiate between the liquid and gaseous phase are utilized. The basis for these calculations is given by [17]

$$j_{diff} = \frac{A D_\lambda \nabla C_w}{\delta_{mem}}, \tag{24}$$

which considers the cell area A, (average) diffusive coefficient of water D_λ, water concentration gradient ∇C_w—here: difference between cathode and anode—and membrane thickness δ_{mem}. The latter is treated as a constant (=201 µm) in the above equation and variations are instead considered by the adjustment-functions (Equations (26)–(32)).

To estimate the diffusion coefficient for water through the membrane D_λ, the following expression dependent on membrane temperature T_{mem} and (average) water content λ is applied [9,26]:

$$D_\lambda = \frac{3.842\, \lambda^3 - 32.03\, \lambda^2 - 67.74\, \lambda}{\lambda^3 - 2.115\, \lambda^2 - 33.013\, \lambda + 103.37} \cdot 10^{-6} \exp\!\left[20 \cdot \frac{20000}{R}\left(\frac{1}{T_{ref}} - \frac{1}{T_{mem}}\right)\right], \tag{25}$$

$$T_{ref} = 353.15\text{K}.$$

Subsequently, the diffusive flow j_{diff} is adjusted for the thickness of the membrane. VVP-correction is applied in the complete absence of liquid water, whilst LVP-correction is used if at least one side is flooded with liquid water. For mixtures of gaseous and liquid phases, a linear scaling is applied:

$$j_{\text{diff}_{\text{VVP}}} = 0.9178 \cdot j_{\text{diff}}(\delta_{\text{mem}}|_{0.0201})\left(-947.5\,\delta_{\text{mem}}^2 - 6.198\,\delta_{\text{mem}} + 1.508\right), \tag{26}$$

$$j_{\text{diff}_{\text{LVP}}} = 3.592 \cdot j_{\text{diff}}(\delta_{\text{mem}}|_{0.0201})\left(-687\,\delta_{\text{mem}}^2 - 21.73\,\delta_{\text{mem}} + 1.714\right). \tag{27}$$

2.7.3. Hydraulic Permeation

Beyond that, hydraulic permeation is only considered if liquid water is present on both sides of the membrane, since the pressure difference has a minor impact on water vapor transport [27]. The hydraulic flow j_{hyd} is estimated by a linear correlation with the pressure gradient ∇p—here: difference between cathode and anode—affected by the cell area A, dynamic viscosity of water $\mu_{\text{H}_2\text{O}}$ as well as water concentration inside the membrane C_w, its thickness δ_{mem} and hydraulic permeability K_λ [16]:

$$j_{\text{hyd}} = \frac{A C_w K_\lambda}{\mu_{\text{H}_2\text{O}} \delta_{\text{mem}}} \nabla p \cdot 10^5. \tag{28}$$

For the hydraulic permeability K_λ, a direct dependency on the membrane water content λ is assumed [16]:

$$K_\lambda = K_w \lambda. \tag{29}$$

Furthermore, the dynamic viscosity of water is approximated by a function of temperature T [28]

$$\mu_w = \mu_0 \exp\left[a_\mu p + \frac{d_\mu - b_\mu p}{R\left(T - \theta_\mu - c_\mu p\right)}\right], \tag{30}$$

where μ_0 denotes a reference value while a_μ, b_μ, c_μ and d_μ are constants. The pressure p has been neglected, since it has a minor impact on μ_w at typical PEMFC operating pressures.

Experimental data suggest a nonlinear relation between the hydraulic flow and membrane thickness [24], hence an adjustment-function is applied for the hydraulic permeability K_λ at a reference pressure difference of 0.025 [bar]:

$$K_{\lambda_{\text{LLP}}}\begin{cases} = 0.1158\,K_\lambda\,(5.749 \cdot 10^{-3}\,\delta_{\text{mem}}\,\exp[-1.326]) \\ \quad \text{for } \delta_{\text{mem}} \geq 0.0056\,\text{cm} \\ = 0.1158\,K_\lambda\,(2.518 \cdot 10^{-4}\,\delta_{\text{mem}}\,\exp[-1.872]) \\ \quad \text{for } \delta_{\text{mem}} < 0.0056\,\text{cm} \end{cases} . \tag{31}$$

Subsequently, the hydraulic flow is corrected for the actual pressure difference:

$$j_{\text{hyd}_{\text{LLP}}} = j_{\text{hyd}_{\text{ref}}}(32.41\,\Delta p + 0.06016). \tag{32}$$

3. Application

Two modes of operation are available for using this model, as displayed in Figure 3. First, the cell performance can be calculated solely based on physical parameters. Second, the cell performance can be estimated based on the voltage deviation from a supplied polarization table. In the latter case, a distinction can be made between using a single or multiple polarization curves as reference.

For this purpose, the inputs of the MATLAB-function are divided into two categories: one for the state variables mandatory to estimate the cell performance and another for the polarization table references. When running in mode *1*, the inputs from the second category are ignored. In mode *2a*, the model expects reference values for the experimental conditions of the polarization curve recording. Lastly, in the case of *2b*, internal calculations for voltage deviation can be disabled by supplying state variables as table references. e.g., when using polarization curves for different temperatures, supplying the current cell temperature will lead to no additional adjustments for this factor, since current and reference values are identical. The operation mode has no impact on the calculations for concentration overpotential and water transport, however, since these heavily depend on stack composition and media supply.

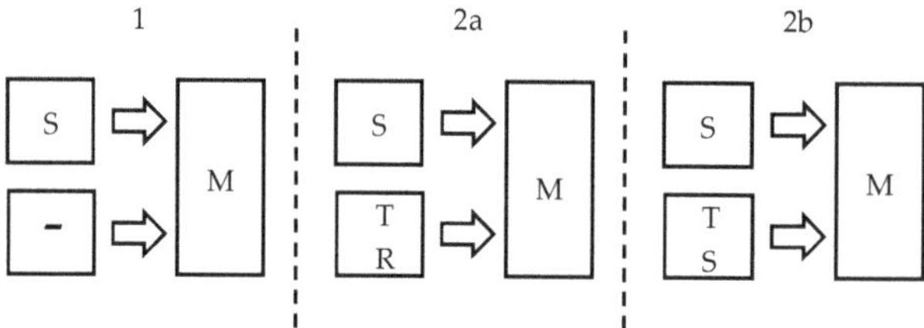

Figure 3. Modes of operation. 1: cell mode; 2a: table mode, single curve polarization table; 2b: table mode, multi curve polarization table; S: state variable; T: polarization table; R: polarization table reference values.

It is also possible to use this model outside of the MATLAB environment via Co-Simulation. As an example, the program code can be run inside of a MATLAB-Function-block within a Simulink model, which can be compiled as C/C++ code. Depending on the target software, the exact Simulink model composition—regarding inputs, outputs and parameters—and compilation procedure may be wary and has to be looked up in the corresponding documentation. Following this practice allows the fuel cell model to be used in any simulation platform with support for Simulink coupling.

4. Simulation Results

Figure 4 shows a set of polarization curves calculated by the presented model. A variation of the chosen parameters affects the overpotentials and therefore the overall cell voltage (Section 2). In terms of computational time, creating a polarization curve with hundreds of data points only takes a few s on a modern CPU. This shows the suitability of the model at hand for large system simulations or real-time applications. The inputs and most important parameters for the reference case are represented by Table 1, and a complete list of all model parameters is supplied as Supplementary Material.

Table 1. Inputs and parameters for the reference case.

Name/type	Value	Unit
Input		
Temperature	70	[°C]
Pressure (absolute)	1.013	[bar]
Relative humidity cat/an	96/38	[%]
O_2 concentration	1.2×10^{-6}	[mol/cm^3]
H_2 concentration	5×10^{-5}	[mol/cm^3]
Liquid water volume fraction	0	[%]
O_2 diffusive reference	0.36	[mol/cm^2 s]
H_2 diffusive reference	1.24	[mol/cm^2 s]
Parameter		
Membrane thickness	201	[μm]
Membrane density	1.97	[g/cm^3]
Membrane EW	1020	[g/mol]

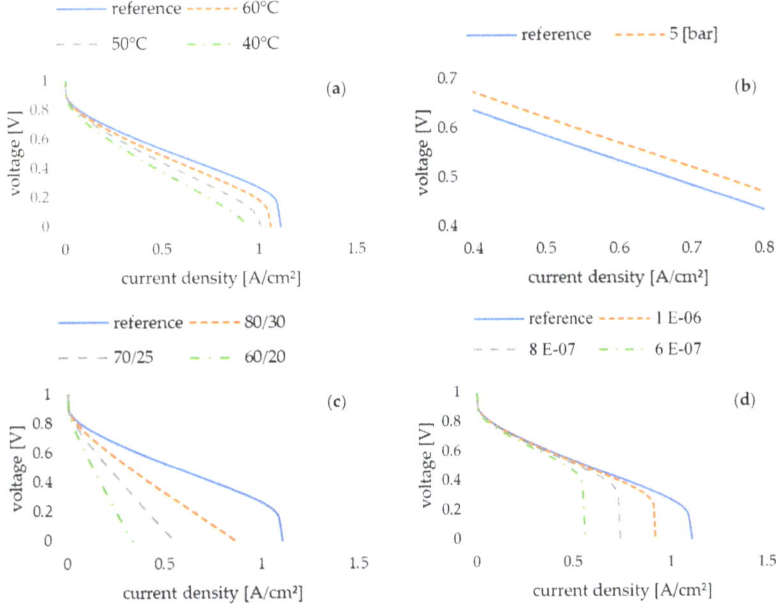

Figure 4. Simulation results for variations of (**a**) temperature; (**b**) pressure; (**c**) RH [%]; (**d**) O_2 concentration [mol/cm^3].

5. Conclusions

The model at hand represents a one-dimensional, dynamic proton exchange membrane fuel cell. To estimate the cell voltage, activation, ohmic and concentration overpotentials are calculated and subtracted from the ideal cell voltage in every time step. Furthermore, the water transport through the membrane by means of osmosis, diffusion and hydraulic permeation is considered. In order to reduce the complexity and computational load, simplified correlations are used to estimate the cell performance. This approach allows the model to be used in complex systems such as complete vehicle models or real-time applications. Two modes of operation allow for flexible use of the fuel cell model by supplying either polarization tables or physical cell parameters, which enables a quick model setup. The program code is written in MATLAB and designed to be used in a MATLAB-function-block inside of a Simulink model. By compiling the Simulink model as C/C++, this PEMFC model can also be used within any software tool that supports Simulink coupling. It is supplied under an open-access license to make it available to anyone for free.

However, the use of simplified equations for cell performance estimation also reduces the accuracy of the results. In particular, the consideration of the MEA can be further expanded because water and gas transport as well as concentration, can be significantly affected by the MEA composition. Additionally, cell geometry and local differences in the distribution of temperature, current density and reactants can also affect the overall performance. For future expansions, the computational load should be considered, in order to not increase the model's calculation time too much.

Supplementary Materials: The supplementary materials are available online at http://www.mdpi.com/1996-1073/12/18/3478/s1.

Author Contributions: Formal analysis, investigation, methodology, software, and visualization, writing—original draft, A.L.L.; data curation, S.C.K.; supervision, resources, H.R.; validation, A.L.L. and S.C.K.; writing—review and editing, S.C.K. and H.R.; conceptualization, project administration, A.L.L., S.C.K. and H.R.

Funding: This research received no external funding.

Conflicts of Interest: The authors declare no conflict of interest.

Nomenclature

a	water activity
a_i	catalyst-specific area, cm^2/mg
a_μ	constant in the calculation of μ_w, bar^{-1}
b_μ	constant in the calculation of μ_w, J/mol bar
c_μ	constant in the calculation of μ_w, K/mol bar
C_i	molar concentration, mol/cm^3
C_w	membrane water concentration, mol/cm^3
D_i	diffusion coefficient, cm^2/s
D_λ	diffusion coefficient of water through the membrane, cm^2/s
d_μ	constant in the calculation of μ_w, J/mol
E_i	voltage, V
EW	membrane equivalent weight, g/mol
f	membrane liquid water volume fraction
F	Faraday constant, A s/mol
ΔG	Gibbs free energy, J/mol
ΔG_i	activation energy, J/mol
H	enthalpy, J
i	current density, A/cm^2
i_0	exchange current density, A/cm^2
$i_{0,\text{ref}}$	reference exchange current density, A/cm^2
i_L	limiting current density, A/cm^2
j_i	water flow, mol/s
K_{hyd}	hydraulic permeability of the membrane (liquid water), cm^2
K_w	hydraulic permeability coefficient (liquid water)
L_i	catalyst loading, mg/cm^2
N_i	molar flux, mol/cm^2 s
n	number of electrodes involved
n_i	water transport coefficient
P	power, W
p	pressure, bar
p_i	partial pressure, bar
p_{ref}	reference pressure, bar
$\dot{Q}$	heat flow, W
R	universal gas constant, J/mol K
s	liquid water volume fraction
S	entropy, J/K
T_i	temperature, K
T_{ref}	reference temperature, K
V_m	acid equivalent volume of the membrane, cm^3/mol
V_W	molar volume of water, cm^3/mol
x_i	percentage

Greek Letters

α_i	charge transfer coefficient
γ	pressure dependency coefficient
δ_i	thickness, cm
ϵ	electrode porosity
η_i	efficiency
λ	membrane water content
$\mu_{w,0}$	reference water dynamic viscosity, Pa s
μ_w	water dynamic viscosity, Pa s
ρ_i	density, g/cm^3
σ_i	conductivity, A/V cm
τ	electrode tortuosity

References

1. Harvey, D.B.; Pharaoh, J.G.; Karan, K. FAST-FC: Open source fuel cell model: An Open Source Performance and Degradation Fuel Cell Model for PEMFCs. Available online: https://www.fastsimulations.com/source-code.html (accessed on 3 September 2019).
2. Falcão, D.S.; Pinho, C.; Pinto, A.M.F.R. Water management in PEMFC: 1-D model simulations. *Ciênc. Tecnol. dos Mater.* **2016**, *28*, 81–87. [CrossRef]
3. Secanell, M.; Putz, A.; Wardlaw, P.; Zingan, V.; Bhaiya, M.; Moore, M.; Zhou, J.; Balen, C.; Domican, K. OpenFCST: An Open-Source Mathematical Modelling Software for Polymer Electrolyte Fuel Cells. *ECS Trans.* **2014**, *64*, 655–680. [CrossRef]
4. Falcão, D.S.; Gomes, P.J.; Oliveira, V.B.; Pinho, C.; Pinto, A.M.F.R. 1D and 3D numerical simulations in PEM fuel cells. *Int. J. Hydrog. Energy* **2011**, *36*, 12486–12498. [CrossRef]
5. Spiegel, C. *PEM Fuel Cell Modeling and Simulation Using Matlab*; Elsevier: Amsterdam, The Netherlands, 2008; ISBN 978-0-12-374259-9.
6. Chakraborty, U.K.; Abbott, T.E.; Das, S.K. PEM fuel cell modeling using differential evolution. *Energy* **2012**, *40*, 387–399. [CrossRef]
7. Conti, B.; Bosio, B.; McPhail, S.J.; Santoni, F.; Pumiglia, D.; Arato, E. A 2-D model for Intermediate Temperature Solid Oxide Fuel Cells Preliminarily Validated on Local Values. *Catalysts* **2019**, *9*, 36. [CrossRef]
8. Chakraborty, U.K. A New Model for Constant Fuel Utilization and Constant Fuel Flow in Fuel Cells. *Appl. Sci.* **2019**, *9*, 1066. [CrossRef]
9. Vetter, R.; Schumacher, J.O. Free open reference implementation of a two-phase PEM fuel cell model. *Comput. Phys. Commun.* **2019**, *234*, 223–234. [CrossRef]
10. Chakraborty, U.K. Reversible and Irreversible Potentials and an Inaccuracy in Popular Models in the Fuel Cell Literature. *Energies* **2018**, *11*, 1851. [CrossRef]
11. Barbir, F. *PEM Fuel Cells. Theory and Practice*, 2nd ed.; Academic Press: Waltham, MA, USA, 2013; ISBN 9780123877109.
12. Yigit, T.; Selamet, O.F. Mathematical modeling and dynamic Simulink simulation of high-pressure PEM electrolyzer system. *Int. J. Hydrog. Energy* **2016**, *41*, 13901–13914. [CrossRef]
13. Carmo, M.; Fritz, D.L.; Mergel, J.; Stolten, D. A comprehensive review on PEM water electrolysis. *Int. J. Hydrog. Energy* **2013**, *38*, 4901–4934. [CrossRef]
14. Song, C.; Tang, Y.; Zhang, J.L.; Zhang, J.; Wang, H.; Shen, J.; McDermid, S.; Li, J.; Kozak, P. PEM fuel cell reaction kinetics in the temperature range of 23–120°C. *Electrochim. Acta* **2007**, *52*, 2552–2561. [CrossRef]
15. Springer, T.E. Polymer Electrolyte Fuel Cell Model. *J. Electrochem. Soc.* **1991**, *138*, 2334. [CrossRef]
16. Jiao, K.; Li, X. Water transport in polymer electrolyte membrane fuel cells. *Prog. Energy Combust. Sci.* **2011**, *37*, 221–291. [CrossRef]
17. Abdin, Z.; Webb, C.J.; Gray, E.M. PEM fuel cell model and simulation in Matlab–Simulink based on physical parameters. *Energy* **2016**, *116*, 1131–1144. [CrossRef]
18. Chemours. Chemours Product-Bulletin Nafion N115-N117-N1110. Available online: https://nafionstore-us.americommerce.com/Shared/Bulletins/N115-N117-N1110-Product-Bulletin-Chemours.pdf (accessed on 13 May 2019).
19. Weber, A.Z.; Newman, J. Transport in Polymer-Electrolyte Membranes. *J. Electrochem. Soc.* **2003**, *150*, A1008–A1015. [CrossRef]
20. Kersting, A.; Aeschbach-Hertig, W. Vorlesungen Physik aquatischer Systeme I. Vorlesung. Available online: http://www.iup.uni-heidelberg.de/institut/studium/lehre/AquaPhys/docMVEnv3_11/MVEnv3_2011_p1_2_Dichte.pdf (accessed on 13 May 2019).
21. Musio, F.; Tacchi, F.; Omati, L.; Gallo Stampino, P.; Dotelli, G.; Limonta, S.; Brivio, D.; Grassini, P. PEMFC system simulation in MATLAB-Simulink® environment. *Int. J. Hydrog. Energy* **2011**, *36*, 8045–8052. [CrossRef]
22. Rosén, T.; Eller, J.; Kang, J.; Prasianakis, N.I.; Mantzaras, J.; Büchi, F.N. Saturation Dependent Effective Transport Properties of PEFC Gas Diffusion Layers. *J. Electrochem. Soc.* **2012**, *159*, F536–F544. [CrossRef]
23. Bird, R.B. Transport phenomena. *Appl. Mech. Rev.* **2002**, *55*, R1–R4. [CrossRef]
24. Adachi, M.; Navessin, T.; Xie, Z.; Li, F.H.; Tanaka, S.; Holdcroft, S. Thickness dependence of water permeation through proton exchange membranes. *J. Membr. Sci.* **2010**, *364*, 183–193. [CrossRef]

25. Dutta, S.; Shimpalee, S.; van Zee, J.W. Numerical prediction of mass-exchange between cathode and anode channels in a PEM fuel cell. *Int. J. Heat Mass Transf.* **2001**, *44*, 2029–2042. [CrossRef]

26. Mittelsteadt, C.K.; Staser, J. Simultaneous Water Uptake, Diffusivity and Permeability Measurement of Perfluorinated Sulfonic Acid Polymer Electrolyte Membranes. *ECS Trans.* **2011**, *41*, 101–121. [CrossRef]

27. Karpenko-Jereb, L.; Innerwinkler, P.; Kelterer, A.-M.; Sternig, C.; Fink, C.; Prenninger, P.; Tatschl, R. A novel membrane transport model for polymer electrolyte fuel cell simulations. *Int. J. Hydrog. Energy* **2014**, *39*, 7077–7088. [CrossRef]

28. Likhachev, E.R. Dependence of water viscosity on temperature and pressure. *Tech. Phys.* **2003**, *48*, 514–515. [CrossRef]

Article

Analysis and Control of Fault Ride-Through Capability Improvement for Wind Turbine Based on a Permanent Magnet Synchronous Generator Using an Interval Type-2 Fuzzy Logic System

Altan Gencer

Department of Electrical and Electronics Engineering, Nevsehir H.B.V. University, Nevsehir 50300, Turkey; altangencer@nevsehir.edu.tr; Tel.: +90-5058-3962-85

Received: 24 April 2019; Accepted: 11 June 2019; Published: 15 June 2019

Abstract: Recently, wind energy conversion systems in renewable energy sources have attracted attention due to their effective application. Wind turbine systems have a complex structure; however, traditional control systems are inadequate in answering the demands of complex systems. Therefore, expert control systems are applied to wind turbines, such as type-1 and interval type-2 fuzzy logic control (IT-2 FLC) systems. An IT-2 FLC system is used to solve the complexity of the wind turbine system and increases the efficiency of the wind turbine. This paper proposes a new control approach using the IT-2 FLC method applied to a wind turbine based on a permanent magnet synchronous generator (PMSG) to improve the transient stability during grid faults. An IT-2 FLC was designed to enhance the fault ride-through performance of a wind turbine and was implemented to control the machine side converter and grid side converter of a wind turbine. The proposed algorithm performance of a wind turbine based on a PMSG was investigated for different types of grid fault. The analysis results verify that the interval type-2 fuzzy logic control system is robustly utilized under different operational conditions.

Keywords: permanent magnet synchronous generator; fault ride-through; interval type-2 fuzzy logic system; wind energy conversion

---

1. Introduction

Due to fluctuations in the price of fossil fuels, the interest in renewable energy sources is increasing day by day [1]. There are many methods of generating electricity from renewable sources, such as the wind turbine and solar panel. Wind turbines are among the most important of these methods. Wind turbines convert wind energy into electrical energy. Many types of generators are used in wind turbines [2]. A permanent magnet synchronous generator (PMSG) has recently begun to attract the attention of wind turbine manufacturers, due to its superior features. The PMSG is supplied to the electrical grid system by means of the grid side converter (GSC), machine side converter (MSC), and control systems [3,4]. The fault ride-through (FRT) capability is one of the important issues for the operation system of the wind turbine. The grid connection requirements (GCRs) involve the operational condition control of the distributed power system [5]. The GCRs have to provide efficiency and reliability to the electrical grid system. The wind turbine (WT) must remain connected to the electrical grid system during grid faults [6]. The fault ride-through is depicted in three stages [7]:

- In the first stage, a WT supplies an electrical power grid system during grid fault time;
- In the second stage, a wind turbine can inject reactive power to support the grid voltage recovery;
- In the third stage, a WT restarts the delivery of the active power after a grid fault.

All these requirements must be considered in the design of the controller and power converter of the WT. This design increases the stability of the WT during a grid fault [8]. Therefore, many methods are proposed in the literature for the FRT capability enhancement of the PMSG.

A braking chopper (BC) system has been implemented to enhance the FRT capability of a PMSG based on a wind energy conversion system (WECS) during a grid fault [9–12]. This method has some advantages, such as low cost and a simple control structure, but BC does not enhance the power quality in the WT's output. The static synchronous compensator (STATCOM) has been implemented to analyze a dynamic mechanism for a wind farm [13]. A coordinating control system for wind turbines was presented in Reference [13]. However, STATCOM has disadvantages, such as high cost and additional hardware needs. The peak current limitation for a high-power PMSG was realized in Reference [14]. Maximum power point tracking (MPPT) was implemented in the GSC and MSC. An active crowbar is kept the DC link voltage value by using this method. A superconducting fault current limiter (SFCL) was implemented for the FRT enhancement of a PMSG in Reference [15]. The presented method achieved reductions in the fault currents in DC systems.

Recently, soft computing methods have started to develop rapidly with the development of computer technology. Soft computing methods are applied in real-world applications, such as renewable energy and automotive and motor control. Soft computing methods are widely implemented in wind power applications, such as for MPPT control, pitch control, fault diagnosis, wind power generation, wind turbine power control, and prediction of wind speed and power. Soft computing methods consist of four computing algorithms such as predictive method, genetic algorithm, artificial neural networks, and fuzzy logic controllers (Type 1 and Type 2).

Interval type-2 fuzzy logic control (IT-2 FLC) is a type of soft computing method that overcomes the uncertainties of any system [16]. Uncertainty is a natural part of intelligent systems in many applications. Fuzzy Type 1 does not fully deal with the uncertainties of intelligent systems. An IT-2 FLC system is designed to minimize the uncertainties of any system. The IT-2 FLC system has been implemented in industrial applications. For example, interval type-2 hesitant fuzzy sets (IT2HFSs) have been implemented to cope with the underground hydrogen storage site selection problem in Romania [17]. The IT-2 FLC system has been applied in wind power system applications, such as diagnosis, pattern recognition decision, classification, control, and time series prediction.

Mokryani et al. [18] introduced a fault ride-through method using the FLC algorithm for several wind turbines. The presented method was applied to adjust the reactive and active power generated by the generator during grid faults. The proposed algorithm was utilized for all grid fault cases and different locations. Tahir et al. [19] developed a low voltage ride-through (LVRT) method using an adaptive FLC algorithm for wind turbines based on a wound field synchronous generator (WFSG). The proposed method was applied to both power converters of a WT. The proposed control system improved the reactive power of the grid system and regulated the current of the grid side converter. Morshed and Fekih [20] presented a design and analysis of the FRT method for a wind turbine. A new fuzzy second order integral terminal sliding mode control was developed for the power electronic converters. The proposed system was employed in a wind turbine. The overcurrents of the rotor and the stator with the presented control system did not exceed 10%. Therefore, the voltage and current values of the generator were within acceptable ranges. Rashid and Ali [21] proposed to achieve an improved FRT capability based on a fuzzy logic controlled parallel resonance fault current limiter (FLC-PRFCL) for doubly fed induction generator (DFIG)-based wind farms. The effectiveness of the presented protecting control system was compared with conventional proportional-integral (PI) control, the bridge-type fault current limiter, and the crowbar circuit system. Bechkaoui et al. [22] introduced online monitoring of grid faults based on the FLC method for wind plants. The proposed method provided a diagnosis of two grid faults. These faults were open phase and short-circuit. The proposed method was implemented to define the stator condition with high certainty. The data of the whole system were generated under both faulty and healthy conditions. The authors indicate that the proposed method is quite efficient. However, Fuzzy Type 1 does not fully cope with the uncertainties of complex systems, such as wind turbines. Several methods have been applied to deal with the uncertainties of complex systems in the literature [23–25]. In addition, in the literature, the interval type-2 FLC has started to be implemented to cope with the uncertainties of complex systems.

Yassin et al. [26] implemented a low voltage ride-through (LVRT) method using an interval type-2 FLC technique for a wind turbine. The input variables of the interval type-2 FLC technique were selected as the DC link voltage and rotor speed. There was irregularity between the delivered to the grid power and the generated active power. To protect from this harmful effect, the proposed method kept the DC link voltage constant. The authors indicate that the proposed algorithm is quite efficient. However, the interval type-2 FLC was implemented only to the MSC control during grid faults. In the normal operational condition, the system was controlled by a traditional control system (PI). The PI control system did not perform better than the IT-2 FLC in any aspect.

This paper proposes a new control approach using the IT-2 FLC method in the WT based on a PMSG to improve the transient stability during grid faults. An IT-2 FLC is designed to enhance the FRT performance of the PMSG. Unlike other studies in the literature, an IT-2 FLC was implemented to control the MSC and GSC of a PMSG during grid faults and normal operational conditions. The aim of the proposed control system is to maintain the generator connected with the grid system and to prevent the harmful effect of an overcurrent occurring during grid faults. The proposed IT-2 FLC is very simple, cost effective, and easy to implement in comparison to the conventional control system. All simulation results proved that the presented IT-2 FLC scheme has the capability to improve the FRT capability of a PMSG.

The rest of the paper is organized as follows: the wind energy conversion system is introduced in Section 2, the proposed protection control system of a PMSG is presented in Section 3, a review of the interval type-2 fuzzy logic system is given in Section 4, the simulation results validating the proposed methodology are presented in Section 5, and, finally, concluding remarks are provided in Section 6.

2. Wind Energy Conversion System

A WECS consists of a blade generator, control system, transformer, point of common coupling (PCC) system, and power electronics components, as shown in Figure 1. Wind turbines convert wind energy into electrical energy. Many types of generators are used in wind turbines. Recently, PMSGs started being used in WTs. Both the GSC and MSC are required to connect wind turbines to the PCC. These converters consist of a neutral point clamped (NPC) system in the study. The NPC three-level converter systems are more effective than conventional converter methods for high-power applications.

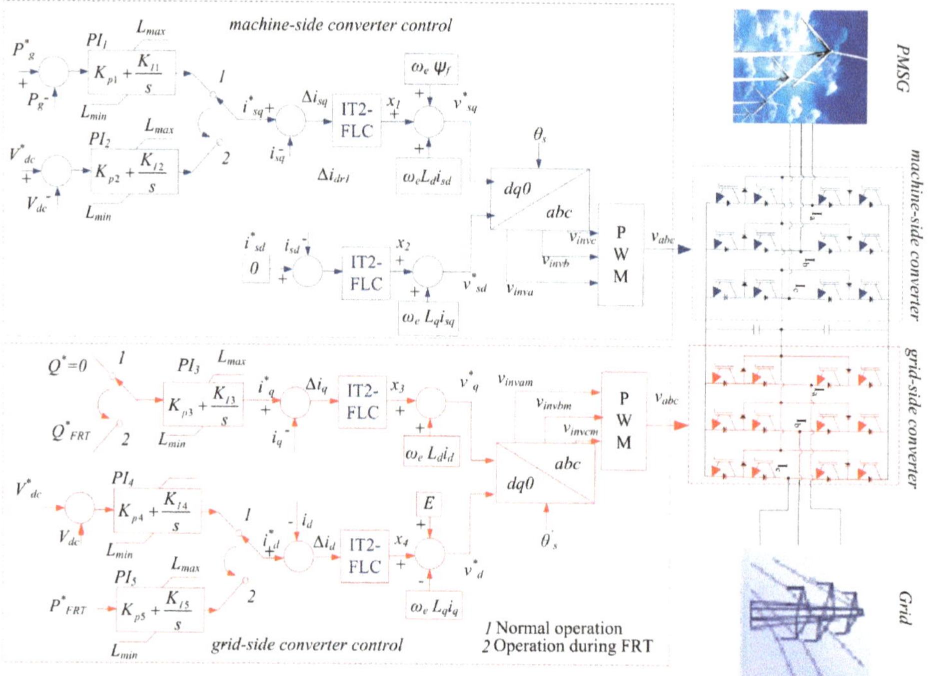

Figure 1. control structure of a permanent magnet synchronous generator (PMSG) based on a wind energy conversion system (WECS).

2.1. Wind Turbine Characteristics

The mechanical power of the WT, which converts wind energy into electric energy, is calculated by the following formula [27]:

$$P_m = \frac{1}{2}\rho A C_p(\lambda, \beta) V_\omega^3 \tag{1}$$

where C_p is the power coefficient and does not have a constant value. It varies with the tip speed ratio of the WT. λ, which is the tip speed ratio of the WT, varies the rotational speed of the WT. ρ, which is the air density, depends on both air pressure and temperature. β denotes the pitch angle. V_ω depicts the wind speed. A represents the area swept by a blade [27]. λ is calculated by the formula

$$\lambda = \frac{\omega_r R}{V_\omega} \tag{2}$$

where ω_r depicts the rotor speed and R depicts the blade radius.

The power coefficient (C_p), which is a function of the λ and β, is the most important parameter for the maximum power generated from a wind turbine. This parameter varies for each turbine type. The C_p value of each wind turbine is given as the table by the manufacturer. As shown in Equation (1), the maximum active power (P_m) changes linearly with the wind speed.

2.2. Mathematical Model of Permanent Magnet Synchronous Generator (PMSG)

The mathematical model of the PMSG is derived from the output voltage equations of the stator. The output voltage equations of the PMSG are illustrated in Equation (3) below. The windings are placed as balanced on the stator. The resistances of the windings are equal and depicted by $R_s = R_q = R_d$. R_s depicts the stator resistance, L_s depicts the stator inductance, and ω_e denotes the electrical angular frequency. v_s, which depicts an output voltage of the PMSG, is described in Equation (3) [27–30]. The mathematical model is very useful for the generator to operate at optimum values and most important for the safe operation of the generator.

$$\begin{pmatrix} v_{sd} \\ v_{sq} \end{pmatrix} = R_s \begin{pmatrix} i_{sd} \\ i_{sq} \end{pmatrix} + \frac{d}{dt}\begin{pmatrix} L_d i_{sd} \\ L_q i_{sq} \end{pmatrix} + \omega_e \begin{pmatrix} -\psi_{sq} \\ \psi_{sd} \end{pmatrix} \tag{3}$$

The stator flux linkages (*dq* frame) are defined in Equation (4):

$$\begin{pmatrix} \psi_{sd} \\ \psi_{sq} \end{pmatrix} = \begin{pmatrix} L_d & 0 \\ 0 & L_q \end{pmatrix}\cdot\begin{pmatrix} i_{sd} \\ i_{sq} \end{pmatrix} + \psi_f \begin{pmatrix} 1 \\ 0 \end{pmatrix} \tag{4}$$

where ψ_{sd} and ψ_{sq} are the flux linkages, L_d and L_q represent the stator inductances in the *dq* frame, i_{sd} and i_{sq} represent the generator's *dq* frame currents, and ψ_f represents the flux linkage in the permanent magnets. The *dq* frame stator voltages by means of the flux linkages are as follows:

$$\begin{pmatrix} v_{sd} \\ v_{sq} \end{pmatrix} = R_s \begin{pmatrix} i_{sd} \\ i_{sq} \end{pmatrix} + \frac{d}{dt}\begin{pmatrix} L_d i_{sd} \\ L_q i_{sq} \end{pmatrix} + \omega_e \begin{pmatrix} -L_q i_{sq} \\ L_d i_{sd} \end{pmatrix} + \omega_e \begin{pmatrix} 0 \\ \psi_f \end{pmatrix} \tag{5}$$

where v_{sq} and v_{sd} are the voltages of the q and d loops in the stator, respectively. The voltages of the v_{sd} and v_{sq} are utilized to generate the reference three-phase sinusoidal voltage, and the i_{sd} and i_{sq} depict the currents of the d and q loops in the stator, respectively. R_s is a stator resistor, L_q and L_d represent the inductances of the q and d loops in the stator, respectively, and ω_e is the electrical angular of the PMSG.

3. Proposed Protection Control System

The controller design for the PMSG is very important in high performance applications. The design procedure for synthesizing and applying the controllers is very similar to the controllers in other high performance generators. However, the IT-2 FLC has more features than conventional control systems, such as numerical uncertainties and modeling the uncertainties in linguistic variables. The parameters of the PI are difficult to adjust for a WECS based on high nonlinearity with uncertain operating conditions. The PI controller supplies the proper performance for a given operating point. However, new methods have been investigated to overcome numerical uncertainties and new linguistics. The type-2 fuzzy set is a new method with specific characteristics. Thus, the special characteristics of the IT-2 FLC are used to improve wind energy conversion systems in this study. The block diagram for the IT-2 FLC is given in Figure 1. An open source interval type-2 fuzzy logic system (IT2-FLS) Matlab/Simulink (R2106a, MathWorks, Natick, Massachusetts, USA) toolbox produced by Taskin and Kumbasar [31] was used in this study.

The line voltages are measured by the control system in the system runtime. The measurement voltages are compared with the reference voltage. When the measurement current value is higher than the reference current value, the switching mode is changed by the control system. A flowchart of the proposed control algorithm is given in Figure 2.

The q loop of the machine side converter has two switching modes. The first mode is active during normal operation. The first mode input is the active power during normal operation. An error signal is produced by comparing the measured active power (P^*_g) with the reference active power (P_g). To obtain the reference current (i^*_{sq}), this error signal is increased by the PI control.

The second mode is active during a grid fault. The input of the second mode is the DC link voltage. An error signal is produced by comparing the reference DC link voltage (V_{dc}) with the measured DC link voltage (V^*_{dc}). To obtain the reference current (i^*_{sq}), this error signal is increased by the PI control during a grid fault.

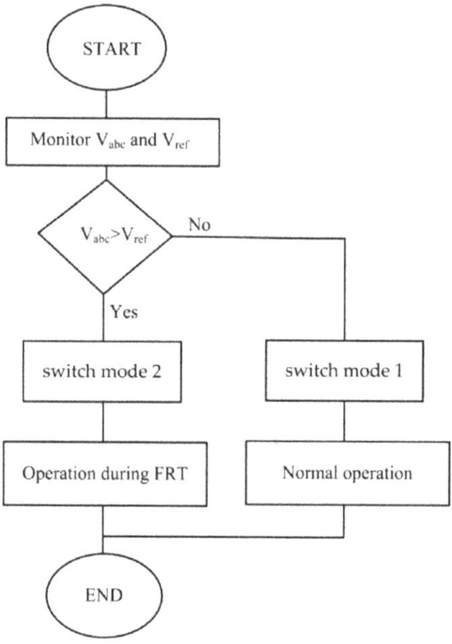

Figure 2. Flowchart of the proposed control algorithm.

Δi^*_{sq} is produced by comparing the reference current (i^*_{sq}) with the measured current (i_{sq}). Δi^*_{sq} is the input of the IT-2 FLC system for both the fault and normal operations. In this study, the IT-2 FLC was designed specifically to obtain efficient results in both the fault and normal operations. To obtain the required voltage (v^*_{sq}) for switching signals, the IT-2 FLC output signal (x_1) sum with both $\omega_e L_d i_{sd}$ and $\omega_e \psi_f$.

In addition, the d-axis current reference (i^*_{sd}) of the MSC is set to 0. Δi^*_{sd} is produced by comparing the reference current (i^*_{sd}) with the measured current (i_{sd}). Δi^*_{sd} is the input of the IT-2 FLC system for both the normal and fault operations. To obtain the required voltage (v^*_{sd}) for switching signals, the IT-2 FLC output signal (x_2) is added to $\omega_e L_q i_{sq}$.

In Figure 1, the reference voltages (v^*_{sd} and v^*_{sq}) are added to enhance the transient response, as expressed by

$$\begin{pmatrix} v^*_{sd} \\ v^*_{sq} \end{pmatrix} = \begin{pmatrix} V'_{sd} \\ V'_{sq} \end{pmatrix} + \omega_e \begin{pmatrix} -L_q i_{sq} \\ L_d i_{sd} \end{pmatrix} + \omega_e \begin{pmatrix} 0 \\ \psi'_f \end{pmatrix} \tag{6}$$

The q loop of the grid side converter has two switching modes. The first mode is active during normal operation. The first mode input is the reactive power during normal operation. The reactive power (Q^*) is set to 0. To obtain the reference current (i^*_q), this error signal is increased by the PI control. The input of the second mode is the fault reactive power (Q^*_{FRT}). To obtain the reference current (i^*_q), this error signal is increased by the PI control during a grid fault.

Δi^*_q is obtained by comparing the reference current (i^*_q) with the current measured (i_q). Δi^*_q is the input of the IT-2 FLC system for both the fault and normal operations. In this study, the IT-2 FLC was designed specifically to obtain efficient results in both fault operation and normal operation. To obtain the required voltage (v^*_q) for switching signals, the IT-2 FLC output signal (x_3) is added to $\omega_e L_d i_d$.

The second mode is active during the grid fault. The input of the second mode is the active power. To obtain the reference current (i^*_d), this error signal is increased by the PI control.

Δi^*_d is produced by comparing the reference current (i^*_d) with the measured current (i_d). Δi^*_d is the input of the IT-2 FLC system for both fault and normal operations. In this study, the IT-2 FLC was designed specifically to obtain efficient results in both the normal and fault operations. To obtain the required voltage (v^*_d) for the switching signals, the IT-2 FLC output signal (x_4) is combined with both the $\omega_e L_q i_q$ and E.

In Figure 1, the reference voltages (v^*_d and v^*_q) are added to enhance the transient response, expressed as

$$\begin{pmatrix} v^*_d \\ v^*_q \end{pmatrix} = \begin{pmatrix} V'_d \\ V'_q \end{pmatrix} + \omega_e \begin{pmatrix} -L_q i_q \\ L_d i_d \end{pmatrix} + \begin{pmatrix} E_s \\ 0 \end{pmatrix} \tag{7}$$

4. Overview of Interval Type-2 Fuzzy Logic Systems

The first interval type-2 method was proposed by Zadeh in 1975. Then, many other authors started to implement the IT-2 FLC system in many applications. The IT-2 FLC method consists of five components: fuzzifier, rules, inference engine, type reducer, and defuzzifier. The structure of the IT-2 FLC method is given in Figure 3. The crisp inputs of the IT-2 FLC are obtained from the input sensors. The fuzzifier converts the physical input values into a normalized fuzzy subset. The inference engine of the IT-2 FLC system uses the same rules as those used in a T-1 FLC system. Then, the type reducer converts the IT-2 FLC to a T-1 FLC. Finally, defuzzification is usually called output processing [32]. The fuzzifier is the first stage in applying fuzzy logic control. The fuzzifier converts the physical input values into a normalized fuzzy subset. The physical input values of the sensors are mapped to a set of input fuzzy values [0, 1] by the membership functions. Finally, the fuzzifier converts the physical input values to a fuzzy input of the inference engine [33]. The general membership functions of the IT-2 FLC are given in Figure 4.

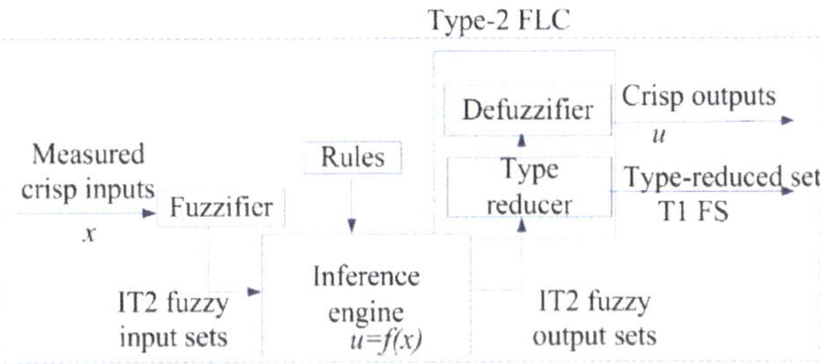

Figure 3. The general structure of the interval type-2 fuzzy logic control.

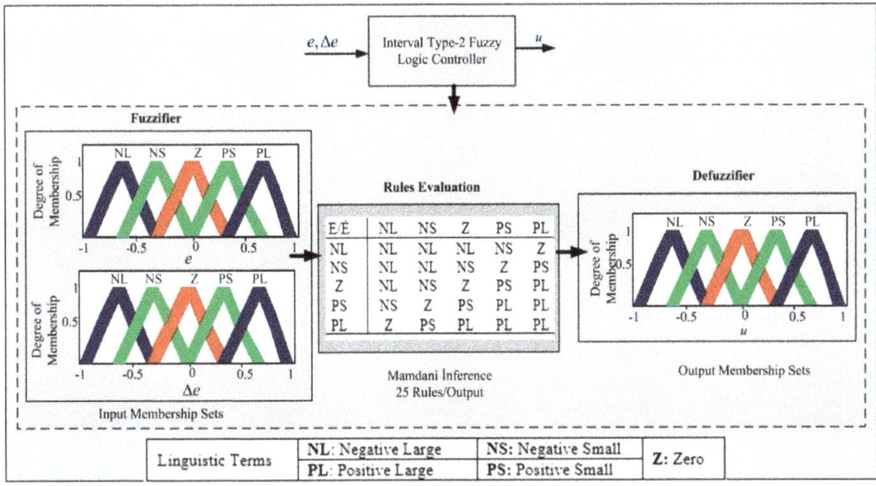

E/Ė	NL	NS	Z	PS	PL
NL	NL	NL	NL	NS	Z
NS	NL	NL	NS	Z	PS
Z	NL	NS	Z	PS	PL
PS	NS	Z	PS	PL	PL
PL	Z	PS	PL	PL	PL

Linguistic Terms	NL: Negative Large	NS: Negative Small	Z: Zero
	PL: Positive Large	PS: Positive Small	

Figure 4. Proposed structure of the interval type 2 fuzzy logic control (IT-2 FLC) system.

The inference engine is the second stage in applying fuzzy logic control. The inference engine is regarded as a transformer, which from a given input maps an output by using linguistic variables. The inference engine generates functional mapping between the output and the input using fuzzy mapping rules. The inputs of the inference engine are implemented by a set of fuzzy mapping rules (if/then). The fuzzy mapping rules (if/then) decide a given condition using linguistic variables, and, then, the fuzzy sets convert to a set of fuzzy outputs [34,35]. All 25 rules and membership functions for the IT-2 FLC are selected and given in Figure 4. Fuzzy rules are written as follows: if (input1 is membership function1) and/or then (output is output membership function).

Mamdani systems have widespread acceptance, are well suited to human input, and are intuitive. Therefore, the Mamdani system was used in this study.

The defuzzifier is the third stage in applying fuzzy logic control. The defuzzifier converts the fuzzy output available into the control objective. The output of the inference engine is still a linguistic variable. This linguistic variable is transformed into the crisp output by the defuzzifier stage. This stage is regarded as a conversion from the fuzzy output to the crisp output needed for real applications. The mean of the maximum technique, a center of gravity technique, and height techniques are commonly used in defuzzification. The Karnik–Mendel algorithm was implemented in this study as the defuzzification method. The Karnik–Mendel algorithm identifies the largest and smallest elements among the centroids [36]. This method converts fuzzy values into crisp system output values, expressed as

$$y(x) = \frac{y_1 + y_r}{2} \tag{8}$$

5. Simulation Results

The simulations in this study were realized in Matlab/Simulink to verify the effectiveness and analysis of the presented method. The sampling frequency of the presented Simulink system was modeled at 20 kHz. The parameters of the PMSG, turbine, and power converter systems are given in Tables A1–A3, respectively. An open source IT2-FLC toolbox in Matlab/Simulink was used in this study. The rules of the IT-2 FLC system were designed to maximize the power generation from the generator. The IT-2 FLC rules were adjusted to generate the optimized gains for the power performance of the wind system based on the PMSG. An IT-2 FLC control system was implemented for the analysis of three different cases of grid fault. The different types of symmetrical and asymmetrical faults were implemented separately at the proper time on the grid side of the PMSG. The simulated fault conditions were as follows:

(i) The 3/ϕ symmetrical fault was implemented at t = 4.0 s and was cleared at t = 4.5 s;

(ii) The 2/ϕ asymmetrical fault was implemented at t = 4.0 s and was cleared at t = 4.5 s;

(iii) The 1/ϕ asymmetrical fault was implemented at t = 4.0 s and was cleared at t = 4.5 s.

The generator with the proposed control method was connected to the grid during all grid fault types. The simulation results illustrate that the IT-2 FLC system gives an appropriate performance for the power generation of the wind system using a PMSG for different scenarios. The rated value of the DC link was 1150 V, the rated value of active power was 1 p.u., the rated value of the electromagnetic torque power was 1 p.u., and the rated value of the reactive power was 0 p.u. in the study.

Scenario 1

The 3/ϕ symmetrical fault was implemented at t = 4.0 s, and then it was cleared at t = 4.5 s, as shown in Figure 5a. The 3/ϕ symmetrical fault was the severest fault type used. Therefore, the control of this fault type is vital. The IT-2 FLC and T-1 FLC were separately implemented in the WT. The parameters of the WT that were measured were the rotor speed, DC link (Vdc), electromagnetic torque (Te), and reactive and active power. The maximum value of the rotor speed with the T-1 FLC system was 1.4 p.u., and its drop value was 70% p.u., as shown in Figure 5b. However, the maximum value of the rotor speed with the IT-2 FLC system was near the nominal value, and the drop value of the rotor speed was 1 p.u., as shown in Figure 5b. The DC link voltage of the system is given in

ripple of the electromagnetic torque value with the proposed control system was smaller than
_C system. The electromagnetic torque value with the IT-2 FLC system was the nominal value
ng and after the grid fault.

_ario 2

2/ϕ asymmetrical fault was implemented at t = 4.0 s, and then it was cleared at t = 4.5 s, as
. Figure 6a. The 2/ϕ asymmetrical fault was lighter compared to the 3/ϕ symmetrical fault.
, the 2/ϕ asymmetrical fault was more severe than the 1/ϕ asymmetrical fault. The IT-2 FLC
_LC were separately implemented in the system during the 2/ϕ asymmetrical fault, as shown in
The rotor speed, DC link (V_{dc}), electromagnetic torque (T_e), and the reactive and active power
_tem with the IT-2 FLC had near nominal values and are given in Figure 6b–f, respectively.
_ speed value with the T-1 FLC system increased to 1.4 p.u. in Figure 6b. The overshoot
_he DC link with the T-1 FLC system was 1400 V. The ripples in the DC link voltage value
_-1 FLC system were higher than the proposed control system even after the grid fault time.
 value of the active power with the T-1 FLC was 63% p.u., and the overshoot value of the
_wer with the T-1 FLC was 1.21 p.u. The ripple in the active power was greatly reduced by
_LC. The reactive power with the T-1 FLC did not track the ideal zero value during the grid
_ reactive power value with the T-1 FLC was 40% p.u. during the grid fault. The reactive
_ue with the proposed control system matched the ideal value perfectly both during and after
_ault. The ripple of the electromagnetic torque value with the proposed control system was
_an with the T-1 FLC system.

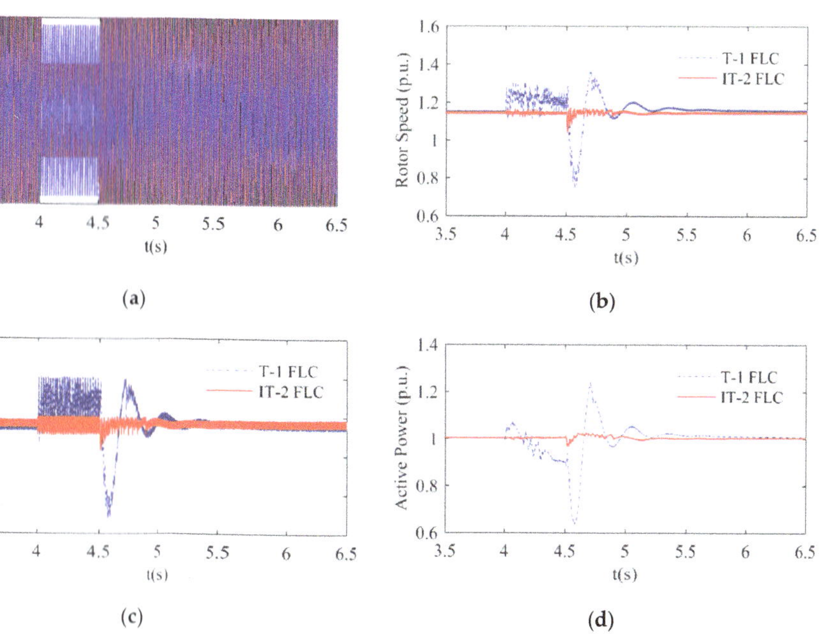

Figure 6. *Cont.*

Figure 5c. The DC link voltage value using a T-1 FLC system was ˙
the DC link voltage value with the IT-2 FLC system was near the no
DC link voltage value with the T-1 FLC system was higher than ˙
after the grid fault time. The active power of the system is given i˙
active power with the T-1 FLC was 60% p.u. and the overshoot val
FLC was 1.25 p.u. However, the active power value with the IT-˙
The reactive power of the system is given in Figure 5e. The oversh˙
the T-1 FLC system was 40% p.u. The ripple of the reactive pow˙
system was smaller than the T-1 FLC system. The reactive power v
nominal value both during and after the grid fault. The electrom˙
given in Figure 5f.

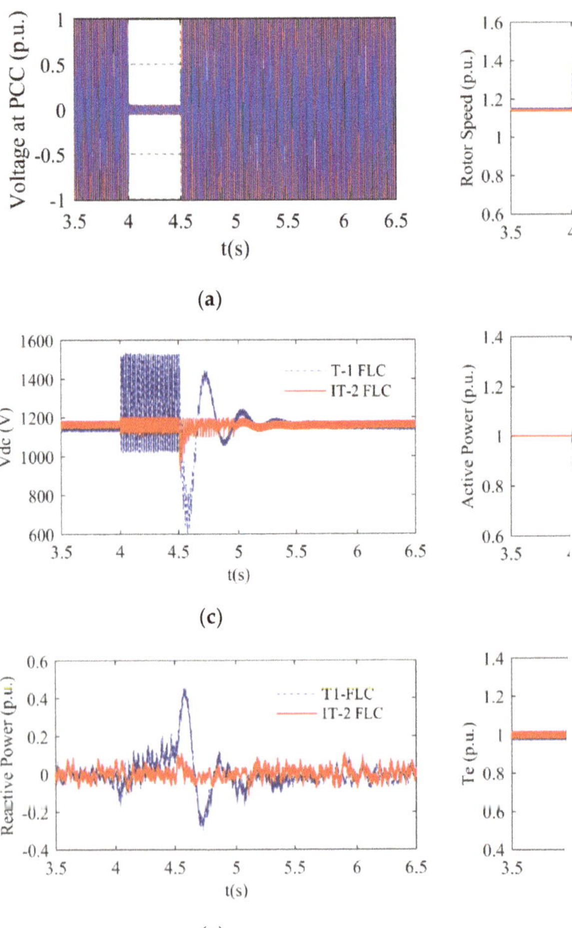

Figure 5. (a–f). Dynamic response of a 1.5 MVA permanent mag
with the type-1 fuzzy logic control (T-1 FLC) and interval type-2 f
during a 3/ϕ symmetrical fault.

Figure 5c. The DC link voltage value using a T-1 FLC system was 1400 V during the grid fault, while the DC link voltage value with the IT-2 FLC system was near the nominal value. The oscillation of the DC link voltage value with the T-1 FLC system was higher than the proposed control system even after the grid fault time. The active power of the system is given in Figure 5d. The drop value of the active power with the T-1 FLC was 60% p.u. and the overshoot value of the active power with the T-1 FLC was 1.25 p.u. However, the active power value with the IT-2 FLC was near the nominal value. The reactive power of the system is given in Figure 5e. The overshoot value of the reactive power with the T-1 FLC system was 40% p.u. The ripple of the reactive power value with the proposed control system was smaller than the T-1 FLC system. The reactive power value of the IT-2 FLC system was the nominal value both during and after the grid fault. The electromagnetic torque (Te) of the system is given in Figure 5f.

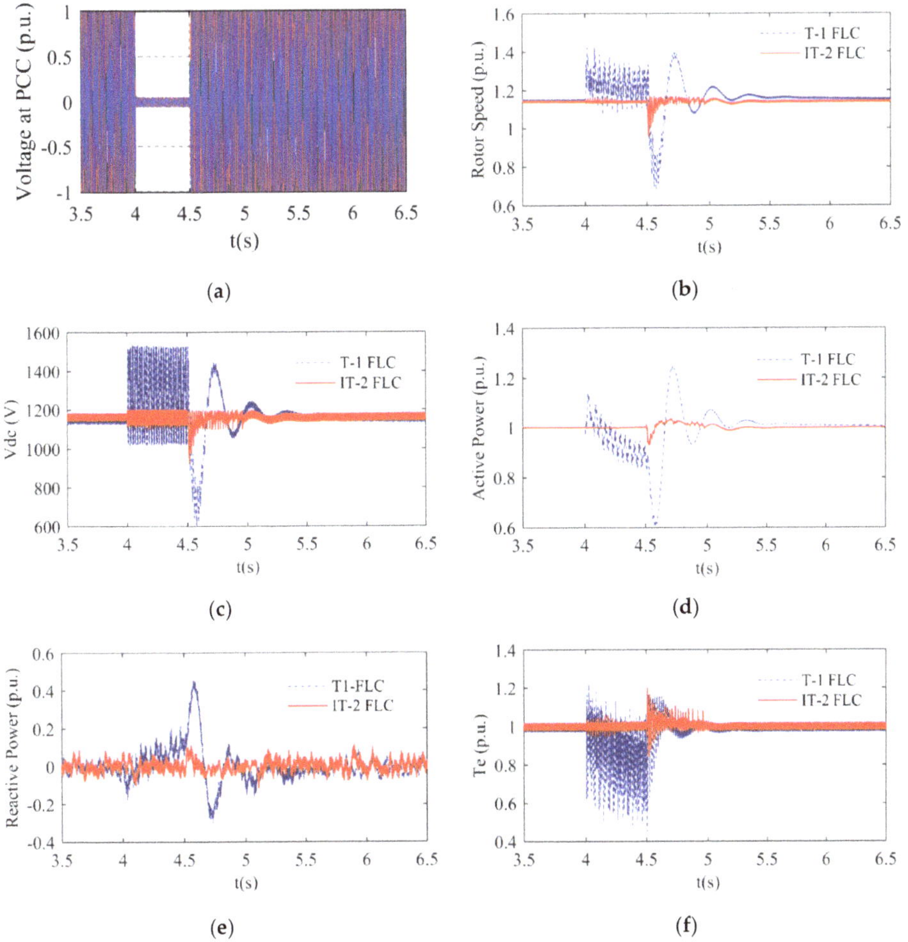

Figure 5. (a–f). Dynamic response of a 1.5 MVA permanent magnet synchronous generator (PMSG) with the type-1 fuzzy logic control (T-1 FLC) and interval type-2 fuzzy logic control (IT-2 FLC) systems during a 3/ϕ symmetrical fault.

The ripple of the electromagnetic torque value with the proposed control system was smaller than the T-1 FLC system. The electromagnetic torque value with the IT-2 FLC system was the nominal value both during and after the grid fault.

Scenario 2

The 2/ϕ asymmetrical fault was implemented at t = 4.0 s, and then it was cleared at t = 4.5 s, as shown in Figure 6a. The 2/ϕ asymmetrical fault was lighter compared to the 3/ϕ symmetrical fault. However, the 2/ϕ asymmetrical fault was more severe than the 1/ϕ asymmetrical fault. The IT-2 FLC and T-1 FLC were separately implemented in the system during the 2/ϕ asymmetrical fault, as shown in Figure 6a. The rotor speed, DC link (V_{dc}), electromagnetic torque (T_e), and the reactive and active power of the system with the IT-2 FLC had near nominal values and are given in Figure 6b–f, respectively. The rotor speed value with the T-1 FLC system increased to 1.4 p.u. in Figure 6b. The overshoot value of the DC link with the T-1 FLC system was 1400 V. The ripples in the DC link voltage value with the T-1 FLC system were higher than the proposed control system even after the grid fault time. The drop value of the active power with the T-1 FLC was 63% p.u., and the overshoot value of the active power with the T-1 FLC was 1.21 p.u. The ripple in the active power was greatly reduced by the IT-2 FLC. The reactive power with the T-1 FLC did not track the ideal zero value during the grid fault. The reactive power value with the T-1 FLC was 40% p.u. during the grid fault. The reactive power value with the proposed control system matched the ideal value perfectly both during and after the grid fault. The ripple of the electromagnetic torque value with the proposed control system was smaller than with the T-1 FLC system.

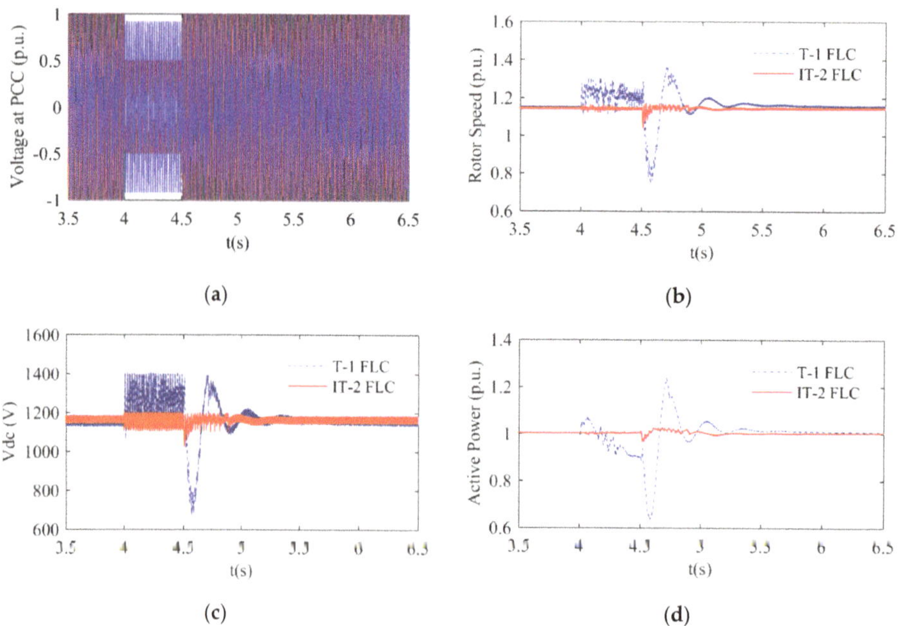

Figure 6. *Cont.*

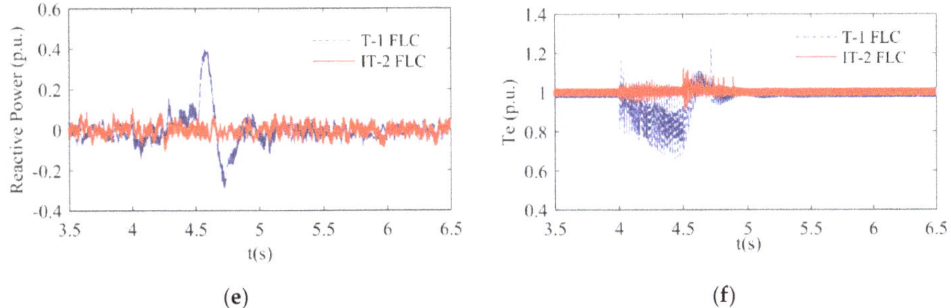

(e)   (f)

Figure 6. (a–f). Dynamic response of a 1.5 MVA permanent magnet synchronous generator (PMSG) with the type-1 fuzzy logic control (T-1 FLC) and interval type-2 fuzzy logic control (IT-2 FLC) systems during a 2/ϕ unsymmetrical fault.

The electromagnetic torque value with the IT-2 FLC system was the nominal value both during and after the grid fault.

Scenario 3

The 1/ϕ asymmetrical fault was implemented at t = 4.0 s, and then it was cleared at t = 4.5 s, as shown in Figure 7a. The 1/ϕ asymmetrical fault was lighter than the other fault types. However, the 1/ϕ asymmetrical fault is the most common type of grid fault. Therefore, the control of this fault type is vital. The IT-2 FLC and T-1 FLC were separately implemented in the system. The parameters of the system that were measured were the rotor speed, DC link (Vdc), electromagnetic torque (Te), and reactive and active power. The maximum value of the rotor speed with the T-1 FLC system was 1.2 p.u., and its drop value was 1 p.u. value, as shown in Figure 7b. The DC link voltage with the T-1 FLC system was 1300 V during the grid fault. The drop value of the active power with the T-1 FLC was 82% p.u., and the overshoot value of the active power with the T-1 FLC was 1.05 p.u. The parameters of the system that were observed were the rotor speed, DC link (V_{dc}), electromagnetic torque (T_e), and reactive and active power. The rotor speed, DC link (V_{dc}), active power, reactive power, and electromagnetic torque (T_e) of the system with the IT-2 FLC had near nominal values and are given in Figure 7b–f, respectively. All the parameters of the system with the IT-2 FLC closely tracked the rated values both during and after the grid fault.

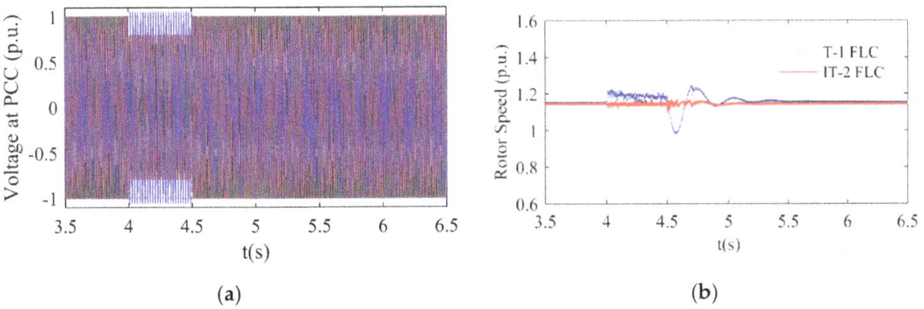

(a)   (b)

Figure 7. *Cont.*

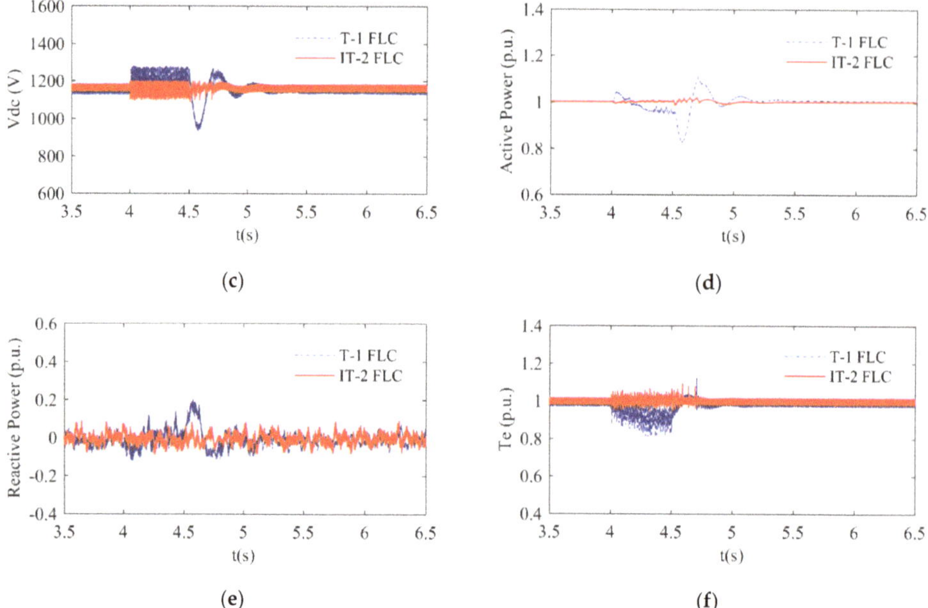

Figure 7. (**a**–**f**). Dynamic response of a 1.5 MVA permanent magnet synchronous generator (PMSG) with the type-1 fuzzy logic control (T-1 FLC) and interval type-2 fuzzy logic control (IT-2 FLC) systems during a $1/\phi$ unsymmetrical fault.

All the parameters with the proposed control system settled to the rated value within comparatively less time than with the T-1 FLC system. The proposed control system reduced the ripples of all the parameters in the system due to the appropriate selection of IT-2 FLC parameters.

6. Conclusions

The FRT capability of wind turbines is a vital issue for the electrical energy generated from wind energy and contributes to reliable grid integration. This paper proposes a new control approach using the IT-2 FLC method that is implemented in the WT based on a PMSG to improve the transient stability during grid faults. The main contributions of this work are:

(i)   An IT-2 FLC was designed to enhance the fault ride-through performance of the PMSG in order to obtain effective results during grid faults;

(ii) The proposed control system was applied to control the MSC and GSC of the PMSG;

(iii) It was observed by measurements that the proposed control system protects the power electronic devices from the harmful effect of overvoltage during grid faults;

(iv) The simulation results have confirmed that the proposed control system can effectively reduce the ripples of all the parameters in the system. The proposed system also improved the conversion efficiency of the wind turbine system due to the appropriate selection of IT-2 FLC parameters.

The parameters of the system that were observed were the rotor speed, DC link (V_{dc}), electromagnetic torque (T_e), and reactive and active power. The rotor speed, DC link (V_{dc}), active power, reactive power, and electromagnetic torque (T_e) of the system with the IT-2 FLC had near nominal values. All the parameters of the system with the IT-2 FLC system closely tracked the rated values during and after the grid faults. All the parameters with the proposed control system settled to the rated value within comparatively less time than the T-1 FLC system. All the simulation results proved that the presented IT-2 FLC scheme has the capability to improve the FRT capability of the PMSG and performs better than the T-1 FLC in all aspects.

In the future, the other rule evaluation method (Sugeno) should be applied to the system. In addition, different interval type-2 fuzzy logic control systems, such as the IT-2 FLC PI, can be adapted to improve the FRT performance of wind turbines, and the results can be compared with the results in this paper.

Funding: This research received no external funding.

Conflicts of Interest: The authors declare no conflicts of interest.

Nomenclature

C_p	power coefficient
λ	tip speed ratio of WT
ρ	air density
β	pitch angle
V_ω	wind speed
A	area swept by a blade
ω_r	rotor speed
R	blade radius
R_s	stator resistance
L_s	stator inductance
ω_e	electric angular of the PMSG
v_s	output voltage of the PMSG
ψ_{sd}, ψ_{sq}	flux linkages of the stator windings
ψ_f	flux linkage in the permanent magnets
L_d, L_q	stator inductances in the dq frame
i_{sd}, i_{sq}	generator dq frame currents
v_{sq}, v_{sd}	voltages of the q and d loops in the stator

Abbreviations

Acronym	Definition
PMSG	Permanent magnet synchronous generator
IT-2 FLC	Interval type-2 fuzzy logic control system
FRT	Fault ride-through
GSC	Grid side converter
MSC	Machine side converter
GCRs	Grid connection requirements
STATCOM	Static synchronous compensator
SFCL	Superconducting fault current limiter
MPPT	Maximum power point tracking
DFIG	Doubly fed induction generator
WT	Wind turbine

Appendix A

Table A1. Parameters of the PMSG.

Parameter	Symbol	Value	Unit
Base power	P_b	1.5	MVA
Base voltage	V_b	690	V
Base frequency	f_b	60	Hz
Pole pairs of PMSG	n_p	48	
Rated generator power	P_g	1	pu
Rated generator speed	ω_r	1	pu
Rated generator line voltage	V_{abc}	1	pu
Stator resistance	R_s	0.006	Ω
Stator inductance	L_s	0.000835	H
Permanent magnet flux	ψ_f	1.48	wb

Table A2. Parameters of the turbine.

Parameter	Symbol	Value	Unit
Air density	ρ	1.225	Kg/m^3
Area swept by blades	A	4775.94	m^2
Base wind speed	v	12	m/s
Cut-in wind speed	v_c	3	m/s
Optimal power coefficient	C_{pmax}	0.4412	
Optimal tip speed ratio	λ_{opt}	6.9	

Table A3. Parameters of the power converter systems.

Parameter	Symbol	Value	Unit
DC capacitance	C	10,000	μF
DC link voltage	V_{dc}	1150	V
Grid side filter resistance	R_f	0.027	Ω
Grid side filter inductance	L_f	1.65	mH

References

1. Li, Y.; Xu, Z.; Wong, K.P. Advanced Control Strategies of PMSG-Based Wind Turbines for System Inertia Support. *IEEE Trans. Power Syst.* **2017**, *32*, 3027–3037. [CrossRef]
2. Zhang, Z.; Wang, F.; Wang, J.; Rodríguez, J.; Kennel, R. Nonlinear Direct Control for Three-Level NPC Back-to-Back Converter PMSG Wind Turbine Systems: Experimental Assessment with FPGA. *IEEE Trans. Ind. Inform.* **2017**, *13*, 1172–1183. [CrossRef]
3. Abdelrahem, M.; Hackl, M.C.; Kennel, R. Simplified model predictive current control without mechanical sensors for variable-speed wind energy conversion systems. *Electr. Eng.* **2017**, *99*, 367–377. [CrossRef]
4. Gencer, A. Analysis and control of low-voltage ride-through capability improvement for PMSG based on an NPC converter using an interval type-2 fuzzy logic system. *Elektronika ir Elektrotechnika* **2019**, *25*, 3.
5. Alepuz, S.; Calle, A.; Busquets, M.S.; Kouro, S.; Wu, B. Use of Stored Energy in PMSG Rotor Inertia for Low-Voltage Ride-Through in Back-to-Back NPC Converter-Based Wind Power Systems. *IEEE Trans. Ind. Electron.* **2013**, *60*, 1787–1796. [CrossRef]
6. Hossain, M.E. A non-linear controller based new bridge type fault current limiter for transient stability enhancement of DFIG based Wind Farm. *Electr. Power Syst. Res.* **2017**, *152*, 466–484. [CrossRef]
7. Pulido, A.B.; Romero, J.; Enriquez, H.C. Robust Active Disturbance Rejection Control for LVRT capability enhancement of DFIG-based wind turbines. *Control Eng. Pract.* **2018**, *77*, 174–189. [CrossRef]
8. Jerin, A.R.A.; Kaliannan, P.; Subramaniam, U. Improved fault ride through capability of DFIG based wind turbines using synchronous reference frame control based dynamic voltage restorer. *ISA Trans.* **2017**, *70*, 465–474. [CrossRef]

9. Gencer, A. Analysis and Control of Fault Ride Through Capability Improvement PMSG Based on WECS Using Active Crowbar System During Different Fault Conditions. *Elektronika ir Elektrotechnika* **2018**, *24*, 64–69. [CrossRef]

10. Conroy, J.F. Watson Low-voltage ride-through of a full converter wind turbine with permanent magnet generator. *IET Renew. Power Gener.* **2007**, *1*, 182–189. [CrossRef]

11. Yang, S.; Zhou, T.; Sun, D.; Xie, Z.; Zhang, X. A SCR crowbar commutated with power converter for DFIG-based wind turbines. *Int. J. Electr. Power Energy Syst.* **2016**, *1*, 87–103. [CrossRef]

12. Nasiri, M.; Milimonfared, J.; Fathi, S.H. A review of low-voltage ride-through enhancement methods for permanent magnet synchronous generator based wind turbine. *Renew. Sustain. Energy Rev.* **2015**, *4*, 399–415. [CrossRef]

13. Geng, H.; Liu, L.; Li, R. Synchronization and Reactive Current Support of PMSG-Based Wind Farm during Severe Grid Fault. *IEEE Trans. Sustain. Energy* **2018**, *9*, 1596–1604. [CrossRef]

14. Nasiri, M.; Mohammadi, R. Peak Current Limitation for Grid Side Inverter by Limited Active Power in PMSG-Based Wind Turbines during Different Grid Faults. *IEEE Trans. Sustain. Energy* **2017**, *8*, 3–12. [CrossRef]

15. Yehia, D.M.; Mansour, D.A.; Yuan, W. Fault Ride-Through Enhancement of PMSG Wind Turbines with DC Microgrids Using Resistive-Type SFCL. *IEEE Trans. Appl. Supercond.* **2018**, *28*, 1–5. [CrossRef]

16. Zadeh, L.A. The concept of a linguistic variable and its application to approximate reasoning. *Inf. Sci.* **1975**, *8*, 43–80. [CrossRef]

17. Iordache, M.; Schitea, D.; Deveci, M.; Akyurt, İ.Z.; Iordache, I. An integrated ARAS and interval type-2 hesitant fuzzy sets method for underground site selection: Seasonal hydrogen storage in salt caverns. *J. Pet. Sci. Eng.* **2019**, *175*, 1088–1098. [CrossRef]

18. Mokryani, G.; Siano, P.; Piccolo, A.; Chen, Z. Improving Fault Ride-Through Capability of Variable Speed Wind Turbines in Distribution Networks. *IEEE Syst. J.* **2013**, *7*, 713–722. [CrossRef]

19. Tahir, K.; Belfedal, C.; Allaoui, T.; Champenois, G. A new control strategy of WFSG-based wind turbine to enhance the LVRT capability. *Int. J. Electr. Power Energy Syst.* **2016**, *79*, 172–187. [CrossRef]

20. Mohammad, J.M.; Afef, F. A new fault ride-through control for DFIG-based wind energy systems. *Electr. Power Syst. Res.* **2017**, *146*, 258–269.

21. Gilmanur, R.; Mohd, H.A. Fault ride through capability improvement of DFIG based wind farm by fuzzy logic controlled parallel resonance fault current limiter. *Electr. Power Syst. Res.* **2017**, *146*, 1–8.

22. Bechkaoui, A.; Ameur, A.; Bouras, S.; Hadjadj, A. Open-circuit and Inter-turn Short-circuit Detection in PMSG for Wind Turbine Applications Using Fuzzy Logic. *Energy Procedia* **2015**, *74*, 1323–1336. [CrossRef]

23. Chybowski, L. Qualitative and Quantitative Multi-Criteria Models of the Importance of the Components in Reliability structure of a Complex Technical system. *J. Konbin* **2013**, *24*, 33–44. [CrossRef]

24. Chybowski, L.; Gawdzińska, K. On the Present State-of-the-Art of a Component Importance Analysis for Complex Technical Systems. *New Adv. Inf. Syst. Technol.* **2016**, *445*, 691–700.

25. Marczyk, J. *Practical Complexity Management*; C. Ontonix pub.: Torino, Italy, 2009.

26. Yassin, H.M.; Hanafy, H.H.; Hallouda, M.M. Enhancement low-voltage ride through capability of permanent magnet synchronous generator-based wind turbines using interval type-2 fuzzy control. *IET Renew. Power Gener.* **2016**, *10*, 339–348. [CrossRef]

27. Hong, C.-M.; Chen, C.-H.; Tu, C.S. Maximum power point tracking-based control algorithm for PMSG wind generation system without mechanical sensors. *Energy Convers. Manag.* **2013**, *69*, 58–67. [CrossRef]

28. Yao, J.; Yu, M.; Gao, W.; Zeng, X. Frequency regulation control strategy for PMSG wind-power generation system with flywheel energy storage unit. *IET Renew. Power Gener.* **2017**, *11*, 1082–1093. [CrossRef]

29. Eltamaly, A.M.; Farh, H.M. Maximum power extraction from wind energy system based on fuzzy logic control. *Electr. Power Syst. Res.* **2013**, *97*, 144–150. [CrossRef]

30. Calle, P.A.; Alepuz, S.; Bordonau, J.; Cortes, P.; Rodriguez, J. Predictive Control of a Back-to-Back NPC Converter-Based Wind Power System. *IEEE Trans. Ind. Electron.* **2016**, *63*, 4615–4627. [CrossRef]

31. Taskin, A.; Kumbasar, T. An Open Source Matlab/Simulink Toolbox for Interval Type-2 Fuzzy Logic Systems. In Proceedings of the IEEE Symposium Series on Computational Intelligence, Cape Town, South Africa, 7–10 December 2015; pp. 1561–1568.

32. Zadeh, L.A. *The Concept of a Linguistic Variable and Its Application to Approximate Reasoning*; Springer: Berlin, Germany, 1974.

33. Mendel, J.M. Advances in type-2 fuzzy sets and systems. *Inf. Sci.* **2007**, *177*, 84–110. [CrossRef]
34. Mendel, J.M.; Hagras, H.; Tan, W.W.; Melek, W.W.; Ying, H. *Introduction to Type-2 Fuzzy Logic Control: Theory and Applications*; Wiley: New York, NY, USA, 2014.
35. Mendel, J.M. *Uncertain Rule-Based Fuzzy Logic System: Introduction and New Directions*; Prentice Hall PTR: Upper Saddle River, NJ, USA, 2001; pp. 131–184.
36. Dongrui, W. Approaches for reducing the computational cost of interval type-2 fuzzy logic systems: Overview and comparisons. *IEEE Trans. Fuzzy Syst.* **2013**, *21*, 80–99.

Article

PLC Automation and Control Strategy in a Stirling Solar Power System

Dan-Adrian Mocanu [1,2,*], **Viorel Bădescu** [2], **Ciprian Bucur** [1], **Iuliana Ștefan** [1], **Elena Carcadea** [1], **Maria Simona Răboacă** [1,3,4] and **Ioana Manta** [1,5,*]

[1]   National R&D Institute for Cryogenic and Isotopic Technologies, 240050 Rm. Valcea, Romania; ciprian.bucur@icsi.ro (C.B.); iulia.stefan@icsi.ro (I.Ș.); elena.carcadea@icsi.ro (E.C.); simona.raboaca@icsi.ro (M.S.R.)

[2]   Department of Engineering Thermodynamics, University POLITEHNICA of Bucharest, 060042 Bucharest, Romania; badescu@theta.termo.pub.ro

[3]   Faculty of Electrical Engineering and Computer Science, University "Stefan cel Mare" of Suceava, 720229 Suceava, Romania

[4]   Faculty of Building Services, Technical University of Cluj-Napoca, 400114 Cluj-Napoca, Romania

[5]   Energetics Faculty, University POLITEHNICA of Bucharest, 060042 Bucharest, Romania

*   Correspondence: dan.mocanu@icsi.ro (D.-A.M.); ioana.manta@icsi.ro (I.M.); Tel.: +40-726-709-585 (D.-A.M.)

Received: 23 March 2020; Accepted: 12 April 2020; Published: 14 April 2020

Abstract: The Stirling engine together with a solar concentrator represents a solution for increasing energy efficiency. Thus, within the National Research and Development Institute for Cryogenic and Isotopic Technologies, an automation system was designed and implemented in order to control the processes inside the solar conversion unit using a programmable logic controller from Schneider Electric. The acquired parameters from the installed sensors were monitored using Unity Pro L software. The main objective of this paper is to solve the starting, operating, and shut-down sequences in safe conditions, as well as monitor the working parameters.

Keywords: Stirling engine; solar concentrator; automation system; human–machine interface; sensors

---

1. Introduction

The impact of integrating renewable energy sources from the environment solves one of the major problems regarding global warming and CO_2 emissions. Fossil fuel resources, hydrocarbons, coal, and natural gas are all limited, cause pollution, and have a strong influence on carbon emissions, affecting the environment irreversibly [1].

Green, renewable energy sources, as an alternative to fossil fuels, are used by hybrid energy systems [2] based on high-efficiency technologies. For example, the solar concentrator that provides the energy needed to drive a Stirling engine [3], with a working cycle having the same name, created by the Scotsman Robert Stirling and patented in 1816 [4]. Thus, the Stirling engine appears as a promising option in the supply of energy compared to other traditional solutions based on renewable energy sources, with a reduced polluting effect similar to that of geothermal, solar, or biomass energy [5,6].

In the specialist literature, Zabalaga et al. presented in [7] an analysis in terms of energy efficiency and economic profitability of a hybrid energy system consisting of photovoltaic panels, a Stirling engine, and a battery system [8]. The dimensioning, using Homer software and, later, the simulation [9–11] of the starting and cooling sequences in MATLAB/Simulink T.M. 8.9 demonstrated a 69% reduction in greenhouse gas emissions, an 11% improvement in annual costs, and a 5% increase of energy efficiency compared to conventional diesel engine systems [12–14].

In [15], Islas et al. conducted a research study on the influence of several design variables and operating parameters over the performance of a 2 kW alpha-type Stirling engine, as well as methods for optimizing the major factors affecting engine efficiency. An important aspect is temperature and pressure control in the primary and secondary circuits of the Stirling engine.

Malali et al. discuss in [16] the effects of circumsolar radiation on the thermal efficiency of a Stirling engine–solar concentrator system, the mathematical model being realized in MATLAB. In the optimal case, the parameters thus obtained are used in the design of the hybrid system.

In [17], Zare and Saleh used the mathematical model of the Stirling engine and the Lyapunov method to predict the optimal operating conditions of a Stirling engine [18,19]. The starting sequences and the oscillations that may occur during operation were especially studied.

Buscemi et al. presented in [20] a hybrid energy system consisting of a solar concentrator and a 32 kW Stirling engine installed in Southern Italy. Their model is capable of predicting annual energy production, especially taking into account the level of dirt on the solar concentrator mirrors [21,22].

In [23], Arora et al. developed a genetic evolutionary algorithm in MATLAB for choosing the optimum values of the decision variables in order to obtain maximum power in terms of energy and heat, as well as an increased economic efficiency [24,25].

Based on the scientific literature, the novelty of this paper consists of the design and implementation of an automation and control system for the operation of a Stirling engine—this system being used so far only in other types of industrial installations.

Even though the use of programmable logic controllers (PLCs) and industrial automations is widely presented in literature, there are few to no cases where these are used in Stirling engine systems for the generation of both electrical and thermal energy, which increased the research team's interest in approaching this topic.

One particular element of this automation, in comparison to others from the industrial sector, is related to the specificity of the Stirling engine system, namely the adaptation to operating requirements, starting sequences, normal operation, control and valve actuation for loading and unloading the working fluid, operation of the fan and cooling pump, controlled shut-down, overload protection, actuation of the safety curtain of the primary exchanger, monitoring of working parameters, temperature, and speed.

Another aspect was to ensure the recording of the parameters, the global automation transmission through bidirectional ethernet connection, including the positioning of the solar concentrator on the maximum flux of solar thermal irradiation, according to the data collected by the Solys2 weather station, which determines the solar coordinates of the Stirling engine solar concentrator location.

In this context, the installation developed in the National Research and Development Institute for Cryogenic and Isotopic Technologies in Romania has integrated a Stirling engine working together with a solar concentrator with the purpose of cogeneration of electricity (10 kW electric) and thermal energy (25 kW thermal) at a low cost [26].

In this paper, we have designed, tested, and optimized the starting, operating, and stopping sequences of a Tedom-modified Stirling alpha V191 engine, representing the conversion unit of the Sunflower 35 solar concentrator [27], through an automation system. In the laboratory, we designed the electrical installation and the automation system, we implemented practically the automation system [28,29], and we programmed the PLCs in order to realize the starting sequences and the opening of the valves for both the cooling system of the compressor and the charging system with helium. The automation also provides the protection and alarm system, control of the protection curtain, and command and control of the electric generator, the heat exchanger, and fan control for the Stirling engine.

Programmable logic controllers (PLCs) emulate the electrical scheme and the computer equipment built around a processor and are capable of controlling, through digital and analog input/output modules, one or more pieces of industrial equipment based on dedicated software [30].

The specific objectives of the present study, based on the automation design, are the following:

- real-time monitoring of the parameters given by temperature, pressure, and rotation sensors;

- recording of the operation results—electrical and thermal energy produced by the Stirling engine with solar concentrator, operating times, and various reports;
- automation of start and stop sequences of the Stirling engine with an integrated protection system;
- optimization of energy production according to the solar irradiation;
- alarm management and security procedures (depending on various weather factors, on a working gas pressure drop, or on a rise in temperature outside the working domain);
- remote transmission of the generated and acquired signals;
- integration of the installation in complex cogeneration systems, as well as in energy optimization systems.

The main objective and focus of this paper is to solve the starting, operating, and shut-down sequences in safe conditions, as well as monitor the working parameters. The performance indicators are given by the stabilization of the temperature of the heat exchanger towards the working temperature, of the engine speed, and of the working pressure according to the values given by the producer [27]. The main findings are as follows:

- implementation of the start and stop sequences of the test stand in safe conditions;
- development of a human–machine interface (HMI) and an application for monitoring and data acquisition for the purpose of reporting safe operation;
- automatization of the maximum solar radiation tracking system.

The solar concentrator using a Stirling engine offers a promising solution compared to other thermal installations that use solar energy, which is an unlimited renewable energy. The most-developed solar solutions currently used are the following:

- photovoltaic panels;
- tower solar concentrators;
- thermal power stations, having solar collectors with parabolic trough or Fresnel systems (parabolic trough collectors (PTC) or linear Fresnel reflectors (LFR));
- solar concentrators with a Stirling engine.

Even though photovoltaic panels are a great development, the solar concentrators using a Stirling engine accomplish a higher conversion efficiency of solar energy into electricity—29.4% [31]. By comparison, solar collectors that have a parabolic trough or Fresnel-type system can achieve a solar-to-electricity conversion efficiency of only 18–22% [32]. We observe a significant advantage for the Stirling solar concentrators. The tower solar concentrators achieve a solar-to-electricity conversion efficiency of about 20–27% [32], closer to that of the Stirling solar concentrator, but with considerably higher execution costs.

Another revealing analysis is related to the cost of obtaining a kWh, as presented by Zayeda [33]. The Stirling solar concentrator has a promising cost of USD 0.2565/kWh, compared to USD 0.4/kWh for the solar tower concentrators, and USD 0.14–0.16/kWh for solar collectors with a parabolic trough or Fresnel system. The best cost is for the photovoltaic panels, at USD 0.06/kWh.

Awana has shown in [34] the advantages of solar concentrators over photovoltaic systems. That is, a 35.7% higher electricity production, but also a better capacity-use factor and a better solar-to-electric conversion efficiency.

It was also shown that Stirling solar concentrators have led to a significant reduction in the cost of electricity production and, in addition, that the use of Stirling solar concentrator systems working on cogeneration of electric and thermal energy achieved an efficiency of over 60%.

The article was structured as follows: in the first part, in the introduction, the main objectives of the work were stated, up -to-date information was presented relating to PLC automation and control strategies used in Stirling solar power systems and a comparative study was conducted regarding the conversion efficiency and the cost of the proposed system with respect to other solar solutions.

The second chapter, "Structure of the Automation and Control system", presented two automation systems and identified the most suitable one, according to the equipment to be monitored/controlled. The third chapter, "Operating Principle", presented the electrical schemes of the main and secondary circuits of the Stirling engine, as well as the automation schemes of the proposed system. In chapter four, "Case Study", the experimental test stand was presented together with the data acquisition, control, and monitoring of the process parameters in tandem with the HMI interface. The results and discussions were presented in the fifth chapter and, finally, in the sixth chapter, the work was summarized and future perspectives were presented.

2. Structure of the Automation and Control System

2.1. Proposed Automation Systems: National Instruments and Schneider Electric

In order to realize the proposed installation, two solutions have been taken into account for the automation of the system, highlighting the advantages and disadvantages of each solution.

The first solution considered is to use the LabVIEW 19.0 HMI (human–machine interface) graphics interface provided by National Instruments, which offers various hardware platforms on which LabVIEW program developers implement operator interfaces, using programmable controllers that allow reconfigurable inputs/outputs (RIO), combining an improved, robust architecture with small industrial input/output modules, and time-sensitive networking (TSN). For monitoring the processes, either a PC (Personal Computer) can be connected remotely through a data transmission interface or a touchscreen can be used to enter the sequences directly. Figure 1 shows the hardware and software platform offered by National Instruments.

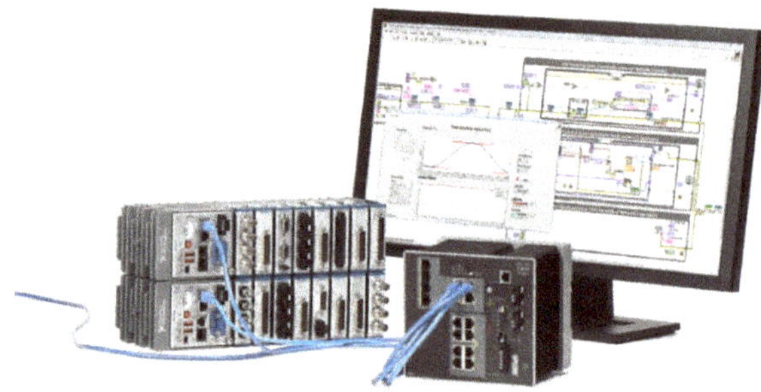

Figure 1. National Instruments hardware and software platform [35].

The second solution proposed is to use a system which has a Schneider Electric programmable automaton (Figure 2), a PLC control application made in Unity Pro L software (Figure 3), and the SCADA (HMI) graphical control and monitoring interface. In order to program these PLCs, the ladder diagram (LD) and function block diagram (FBD) programming languages are used, according to IEC (International Electrotechnical Commission) standard 61131-3.

Figure 2. Schneider Electric programmable automaton [36].

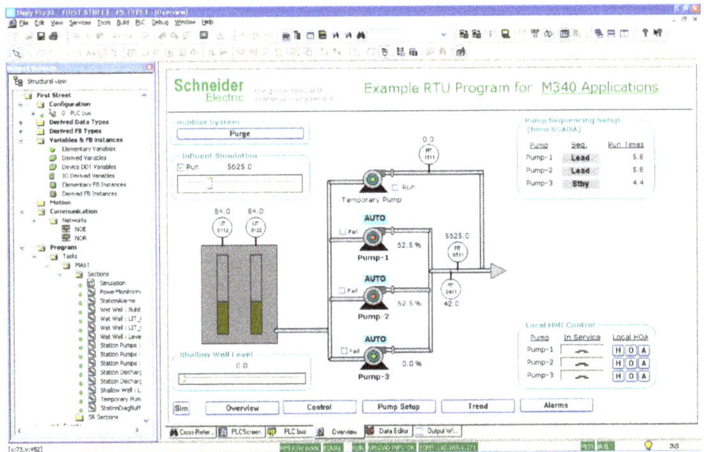

Figure 3. Unity Pro L working interface.

Following a multicriterial analysis, taking into account the availability, the quality-price ratio, the software interface, and the applicability for the developed installation, the second option (Schneider Electric) was chosen.

2.2. Equipment Monitored/Controlled by the Automation System

The signals entering the automation system are given by temperature, pressure, and frequency sensors and the controlled equipment is represented by solenoid valves, fan, pump motors, electromagnets, an actuator motor, and a start motor impulse. Hardware components include connectors, input/output modules, PLCs with a processing unit, a multi-rack motherboard, and analog processing modules. Table 1 presents the main equipment to be monitored/controlled:

Table 1. Monitored/controlled equipment.

Equipment	Model
Pressure sensors	Drucksensor BA 520, Huba Control AG $I = 4....20$ mA; $U_1 \leq 30$ V; $I_1 \leq 100$ mA; $P_1 \leq 750$ mW $T = -30 \div 120\,°C$
Fan	Ebm-papst Mulfingen GmbH&Co, W3G630GQ;3721-BA-ENU Single phase; 230V; Speed: 1000 rot/min; $P = 720$ W; 140 Pa; $T = -250 \div 60\,°C$
Rotation sensor	BESM08EH-PSC15B-S04G, BALLUF $U_e = 24$ Vcc; $I_e = 200$ mA; $H = 15\%$; $I_r = 20\ \mu A$; $f = 3000$ Hz; IP68; $T = -25 \div 70\,°C$
Frequency signal adapter	ZKFD2-UFC-11.D produced by PEPPER+FUCHS which converts the frequency into proportional current $4 \div 20$ mA 24 V DC supply (Power Rail); 1 mHz … 10 kHz; I_e 0/4 mA … 20 mA
Slot extension rack	Modicon Premium TSXRKY12EX
Communication module processor	TSXP573634M
Power supply module	TSXPSY2600M
ETHERNET	TSX ETY 110 module
One analog input module	TSX AEY 1600—16 voltage/current inputs
One analog input module	TSX AEY 1614–16 thermocouple inputs
One analog output module	TSX ASY 800—8 voltage/current outputs
One discrete output module	TSX DSY 16R5—16 discrete relay outputs
One discrete input module	TSX DEY 64D2K—64 discrete inputs

3. Operating Principle

In order to achieve and optimize the proposed installation, the Stirling solar conversion unit was studied and two main circuits were identified, depending on the working fluid, functionality, and destination:

- main circuit of the Stirling engine (working fluid—Helium);
- secondary circuit of the Stirling engine—cooling circuit (working fluid—water).

Figure 4 shows the main components of the Sunflower 35 solar concentrator, Figure 5 presents the Stirling solar conversion unit, and Figure 6 describes the test stand.

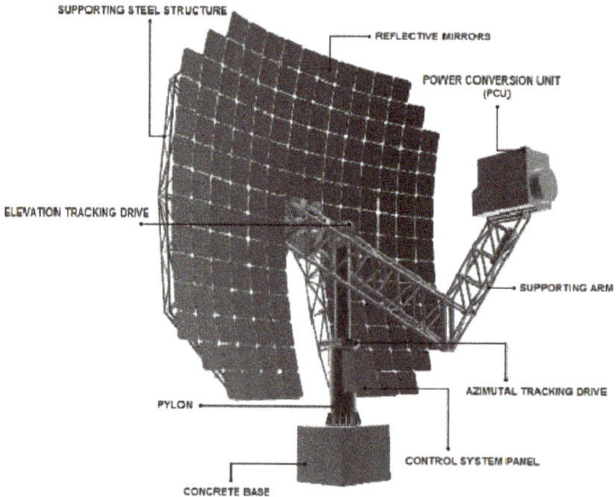

Figure 4. Sunflower 35 solar concentrator components [27].

The test stand presented in Figure 6 contains the following components:

- burner Gulliver BS2D, made by Riello, Italy, having the following characteristics: thermal power, $35/40 \div 91$ kW; propane tank, $1.3/1.6 \div 3.5$ Nm^3/h; volume 26 L; burner control MG569; electrical power: 0.180 kW;
- fan CA/line—10, made by SODECA, Spain, having the following characteristics: power 0.095 kW; speed $2460 \div 2720$ s^{-1};
- thermocouple type K, made by CAOM, Pascani, Romania;
- mixing tube, Ø300 mm, length 800 mm, material 304 L;
- gas exhaust pipe.

The use of the stand was justified by the numerous interactions, adjustments, modifications, and tests that would have been much more difficult and would not have allowed a fast reaction if the solar concentrator had been used as the heat source.

Figure 5. Solar conversion unit.

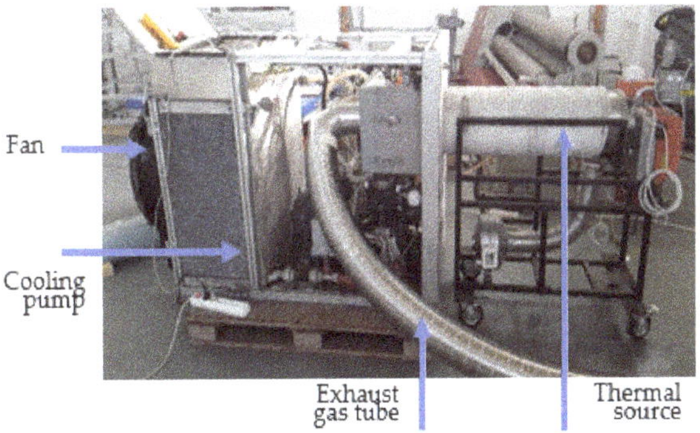

Figure 6. Test Stand.

3.1. Main Circuit of the Stirling Engine (Working Fluid—Helium)

The main circuit of the Stirling engine is the most complex, having the following main blocks of equipment:

- temperature sensors positioned on the main exchanger (hot bulb). Physically, there are 20 sensors, of which only 10, BT1–BT10, are used for temperature monitoring on the main exchanger and are placed in the connector in the local connection box, from where they are interconnected in the input/output connectors from the automation cabinet. The other 10 sensors are kept as back-up in case the used ones fail;
- two pressure switches, one on the oil circuit, BP1, and the other on the water circuit, BP2;
- generator speed sensor BR1;
- two pressure transmitters, one on the water heat-exchanger and the other on the oil circuit BP3, BP4;
- the rotation sensor, BR1, which reads the rotations on the Stirling engine flywheel;
- solenoid actuators, motors, etc.

Figure 7 shows the main circuit of the Stirling engine (working fluid—Helium)

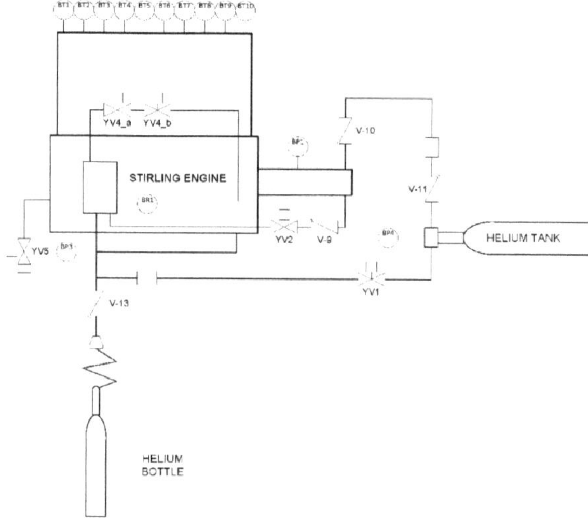

Figure 7. Main circuit of the Stirling engine. Working fluid—Helium. BT1–BT10: temperature sensors; BP1 and BP2: pressure switches; BR1: rotation sensor and generator speed sensor; BP3 and BP4: pressure transmitters.

3.2. Secondary Circuit of the Stirling Engine: Cooling Circuit (Working Fluid—Water)

The secondary circuit of the Stirling engine not only performs the cooling of the Stirling engine, but also supplies the thermal agent, being able to generate a power of up to 25 kW, identifying the following main blocks of equipment:

- temperature sensors mounted on the cooling water circuit, BT11 and BT12;
- pressure switch BP2 on the water circuit;
- actuators, valves, motors, pulse compensator, etc.

Figure 8 presents the block diagram of the secondary circuit of the Stirling engine: cooling circuit (working fluid—water):

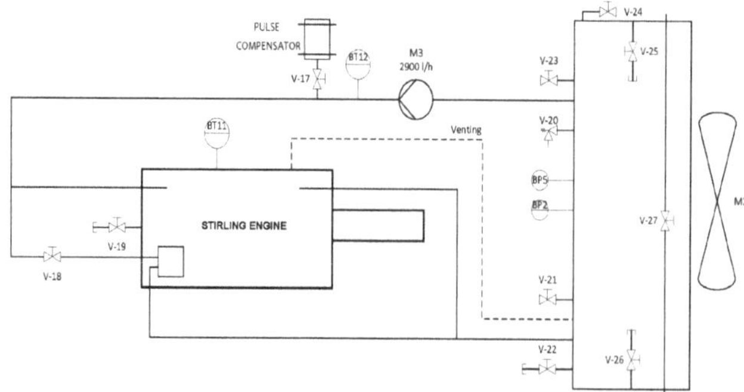

Figure 8. Secondary circuit of the Stirling engine, water-cooling. BT11 and BT12: temperature sensors.

3.3. Electrical Automation Schemes

Starting from the two circuits, in which the components to be controlled by the Stirling engine automation were presented, the following schemes were developed:

- connection diagrams of the temperature sensors from the primary exchanger BT1–BT10, from the cooling circuit BT12, rotation sensors BR1, pressure sensors BP3 and BP4, and the temperature sensor BT11, made in the rack modules two and three;
- connection diagram of the actuation of the motors M1 – M4, realized in rack modules 4, 5, and 6;
- power supply diagram for 24 VDC and 230 VAC;
- automation loops of temperature sensors BT1–BT10, and BT12, rack module two (Figure 9);
- automation loops of the pressure sensors BP3, BP4, the rotation sensor BR1, and the temperature sensor on the housing of the motor crankcase BT11, rack module three (Figure 10);
- automation loops of pressure sensors BP1, BP2, and actuator M2, rack modules four and five;
- automation loops for actuating the solenoid valves YV1, YV2, YV4_a, YV4_b, YV5, and YV6, rack module six;
- automation loops for actuating the motors M1, M2, M3, and M4, rack module six.

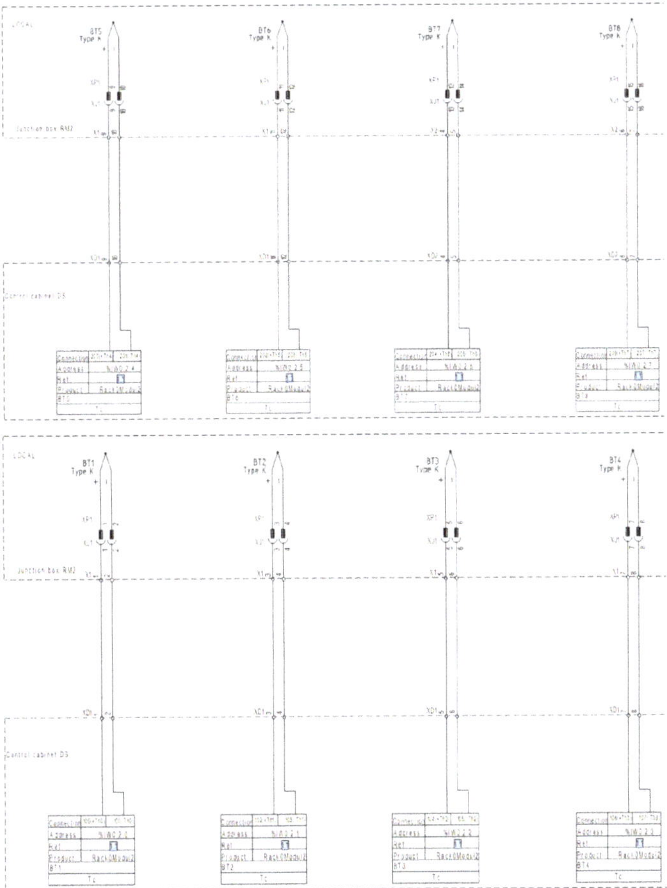

Figure 9. Automation loops of temperature sensors BT1 - BT8.

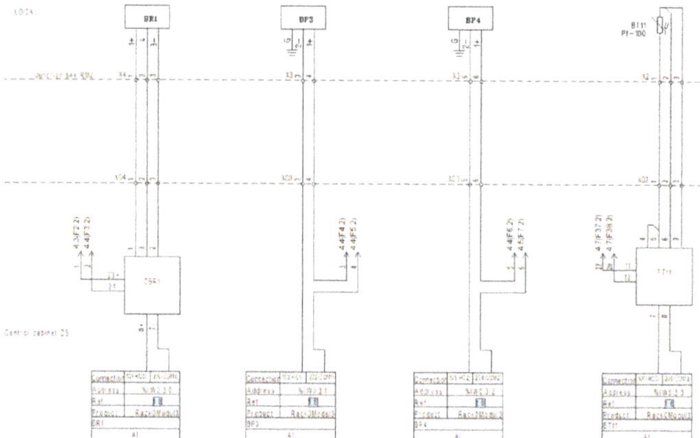

Figure 10. Automation loops of pressure sensors BP3 and BP4, rotation sensor BR1, and temperature sensor on the motor crankcase BT11.

4. Case Study

The monitored parameters of the experimental set-up which are presented in Table 2 are the following: temperature, pressure, and rotation.

Table 2. Parameters of the experimental set-up.

No.	TAG	Type	Domain	Measured Parameter
1	BT1	Type K temperature sensor	0–1000 °C	Helium
2	BT2	Type K temperature sensor	0–1000 °C	Helium
3	BT3	Type K temperature sensor	0–1000 °C	Helium
4	BT4	Type K temperature sensor	0–1000 °C	Helium
5	BT5	Type K temperature sensor	0–1000 °C	Helium
6	BT6	Type K temperature sensor	0–1000 °C	Helium
7	BT7	Type K temperature sensor	0–1000 °C	Helium
8	BT8	Type K temperature sensor	0–1000 °C	Helium
9	BT9	Type K temperature sensor	0–1000 °C	Helium
10	BT10	Type K temperature sensor	0–1000 °C	Helium
11	BT11	Type PT100 temperature sensor	0–250 °C	Cooling water
12	BT12	Type K temperature sensor	0–1000 °C	Helium
13	BP3	Pressure transmitter	0–250 bar (g)	Helium
14	BP4	Pressure transmitter	0–250 bar (g)	Helium
15	BR1	Rotation sensor	0–3000 Hz	Rotation

The experimental stand is composed of: the CPU (Central Processing Unit) with automation panel assembly (Figure 11), the local CPU connections box (Figure 12), and the automation panel box (Figure 13). Figure 14 shows the positioning of the temperature sensors on the primary exchanger.

Figure 11. CPU and automation panel.

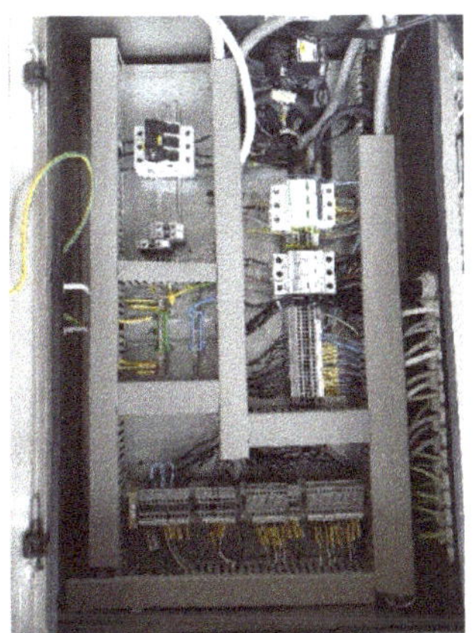

Figure 12. Local CPU connections.

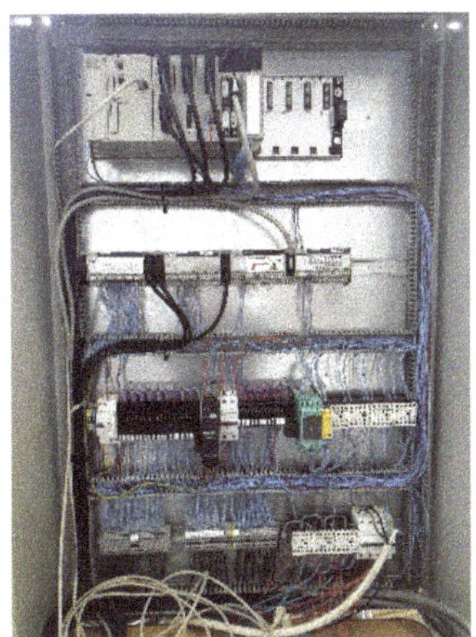

Figure 13. Automation panel.

Figure 14. Positioning of temperature sensors on the primary exchanger.

The main objective of this study was to ensure the operation and functioning of the Stirling engine in safe conditions with the achievement of the designed performances.

The sensors mounted on the engine generate the following types of signal:

- for continuous measurements—analogic signal 4–20 mA;
- for discrete signals—zero potential contact.

The control signals generated from the control interface are of the following type:

- for continuous signal—analogic signal 0–10 V;
- for discrete signal—digital signal 24 VDC.

The signals from the motor are acquired in the connection box mounted on the Stirling engine and are transmitted to the control panel via multilayer cables. The control of the installation is implemented on a programable automaton—Modicon premium type.

The Stirling engine instrumentation system provides the following functions:

- temperature measurement.

Twelve temperatures are measured through 11 thermocouple type-K sensors and one PT100-type sensor [37–39]. Minimum- or maximum-type alarms are generated in the PLC in order to keep the process within normal limits.

- Pressure measurement.

Two relative pressures are measured in the mixing tank and the compression chamber through pressure transducers with an output signal of 4–20 mA; minimum- or maximum-type alarms are generated in the PLC in order to keep the process within normal limits.

- Engine rotation measurement.

Engine rotation is measured by a frequency sensor connected to a signal interface that generates an output signal of 4–20 mA corresponding to the engine rotation.

- Pump/fan control.

The control of the pump and of the fan is done as needed, either manually or automatically. Thus, in the distributed control system (DCS), command keys are virtually implemented for choosing manual or automatic man/auto mode, as well as for the manual on/off control. In manual mode, the pump and the fan are switched on/off directly by the operator by choosing the on or off position. In automatic mode, the pump and the fan start and stop according to the logic implemented in the DCS.

- Control of automatic valves.

Valve control is done as needed, either manually or automatically. Thus, in the DCS, command keys are implemented virtually for choosing the manual/automatic (man/auto) mode, as well as for the manual control open/close. In manual mode, the valves are opened/closed directly by the operator by choosing the open or close position. In automatic mode, the valves open and close according to the logic implemented in the DCS.

4.1. Acquisition, Control, and Monitoring of the Process Parameters

The control system is a distributed type and consists of a Schneider Electric (PLC) programable automaton and a PLC control application done in Unity Pro software using the SCADA (HMI) human–machine control and monitoring interface.

The signals from the transducers and from the system equipment are taken over by the PLC where they are processed according to the program loaded in the internal memory of the programable automaton. The PLC transmits the data to the SCADA application through an OPC (Open Platform Communication) server (OLE (Object Linking and Embedding) for process control) via ethernet communication.

The architecture and structure of the program loaded in the PLC is realized in the programming language ladder diagram (LD) and function block diagram (FBD), according to the IEC standard 61131-3 (Figure 15) using the Unity Pro application.

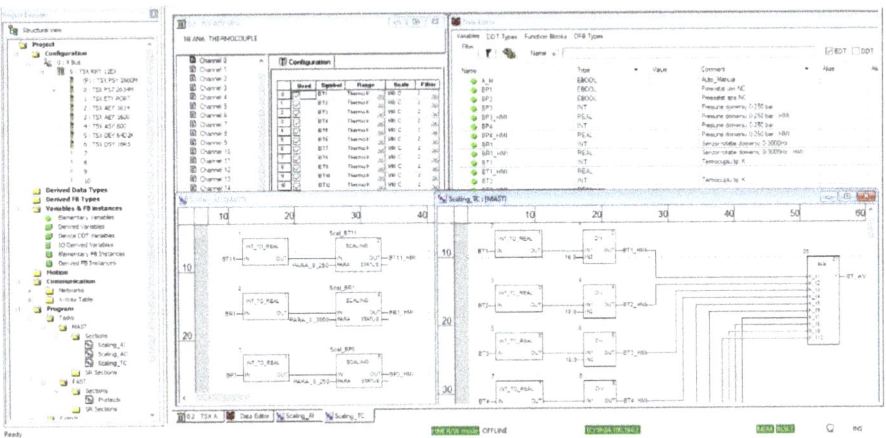

Figure 15. Unity PRO Software—Stirling engine.

4.2. Human-Machine Interface HMI

The human-machine interface (HMI) for controlling and monitoring the parameters of the Stirling engine system is realized on the Citect Scada platform from Schneider Electric (Figure 16).

The HMI provides an interface between the system and its human user. This graphical interface converts the data received from the control system into human-intuitive representations of the process systems. The operator visualizes the scheme of the system, and can control the on/off of the dynamic equipment (cooling-water recirculation pump, fan, and solar radiation shutter motor), as well as adjusting the fan speed depending on the cooling-water temperature of the engine.

Two operating modes were realized: manual (MAN) and automatic (AUTO). In MAN mode, each equipment can be operated individually for single tests. When the AUTO mode is active (during normal engine operation), the protection program starts (protection related to minimum and maximum pressures, minimum and maximum temperatures, cooling pump on, cooling fan on, and state of pressure switches for cooling water pressure and helium pressure). The AUTO mode is made in Unity Pro and loaded into the PLC memory, each process parameter being automatically controlled by the PLC.

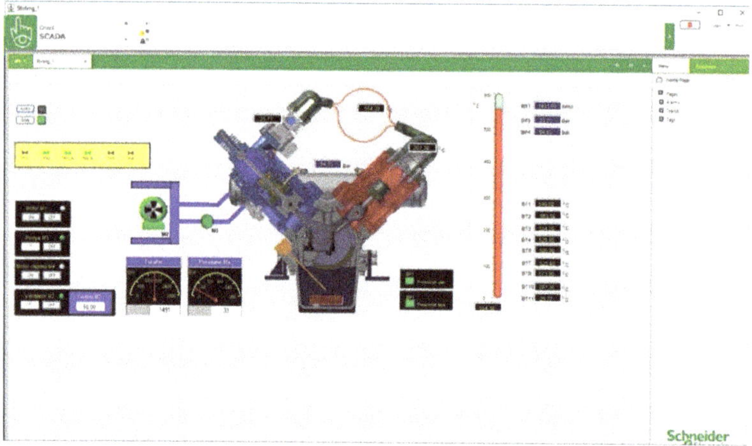

Figure 16. HMI control and monitoring interface. Process control screen.

The graphic symbols for the equipment are created so that the condition of the equipment is indicated at any time by colors representing normal operation, start/stop, as well as the discrepancy between the command and the state of the equipment.

The collected data are saved in Excel files using the trend and process analyst functions in the Citect Scada application.

The significance of the colors on the graphical interface is presented in Table 3.

Table 3. Graphic interface color-assignment interpretation.

Valves		Dynamic Equipment (Pumps, Compressors)		Electric Heaters	
Color	**State**	**Color**	**State**	**Color**	**State**
Gray	Closed	Gray	On	Gray	Unpowered
Green	Open	Green	Off	Green	Powered
Red	Discrepancy/missing	Red	No voltage		
Orange	In motion/partially open				

5. Results and Discussions

The testing procedure is the following:

➤ powering the automation installation;

➤ operation of the YV1 valve for charging the He circuit of the Stirling engine at 80 bar;

➤ starting the supply of thermal energy with a propane thermal burner, with the following steps:

    ○ setting the temperature to 450–500 °C;

    ○ positioning of the type-K thermocouple for the primary exchanger of the Stirling engine;

    ○ propane supply by opening the 26 L cylinder tap;

    ○ ignition of the burner;

    ○ adjustment of air flow through the mixing chamber;

○   reading of the temperature transmitted by the type-K thermocouple.

➙   When the minimum operating temperature exceeds 450 °C, start the Stirling engine using the Siemens electric generator connected to the engine, in order to start the rotation.

➙   At this moment, the Stirling engine quickly enters the operating mode. The data monitored by the automation which are presented in this section confirm the stability of its operation. The small variations are due to the variation of the heat source over time—repeated tests were performed, the values collected were maintained in the same ranges.

➙   For the shut-down sequence, the Riello burner heat source is reduced and, when the temperature drops to 350 °C, the engine diminishes its speed and stops.

Following the functioning tests, the automation and control installation has achieved stable operation under secure conditions.

The experimental data derived from the acquired and monitored parameters are further presented in the form of graphs using the trend and process analyst functions of the Citect Scada application.

In the graph presenting the variation of temperatures monitored by the automation (Figure 17), it can be clearly observed that the temperatures collected by the sensors located on the primary exchanger of the Stirling engine have a fast stabilization towards the working temperature, according to thermal power provided by the heat source. Each iteration is done at 330 ms.

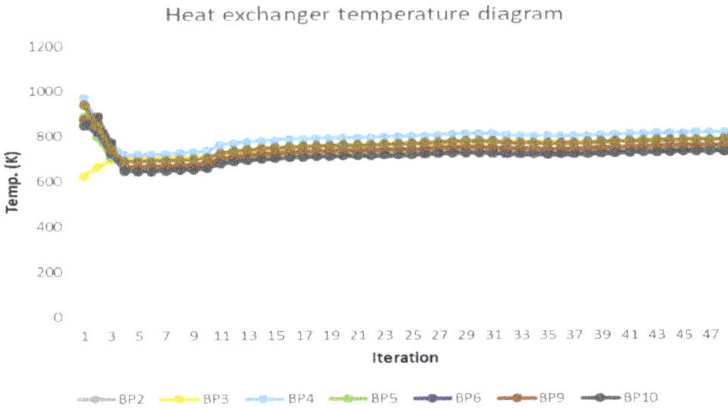

Figure 17. Heat exchanger temperature diagram.

Similarly, in the engine rotations/minute graph (Figure 18), the speed of the engine quickly reaches the synchronization speed of 1475 rot/min, keeping it constant.

These observations lead to the conclusion that the automation system performs the control of the operating processes of the Stirling engine.

The presentation of pressure monitoring on the Stirling BP3 engine and of the Helium cylinder pressure BP4 is shown in Figure 19.

By realizing the automation of the power generation unit using a Stirling engine, we performed the first stage for starting up the main module.

Another main stage is the implementation of the automation and control of the solar concentrator positioning. The supply of the sun position was made by a Solys2 installation (Figure 20) manufactured by Kipp & Zonen in order to determine the sun's position. It uses two pyranometers, one for global radiation measurement and the other for diffused radiation, as well as a pyrheliometer with temperature sensor for direct solar radiation measurement.

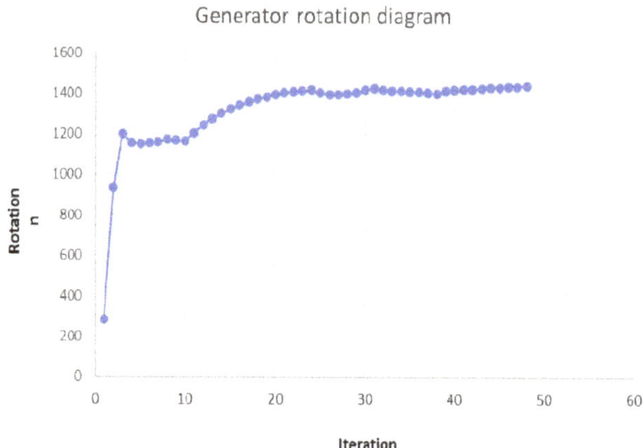

Figure 18. Generator rotation diagram.

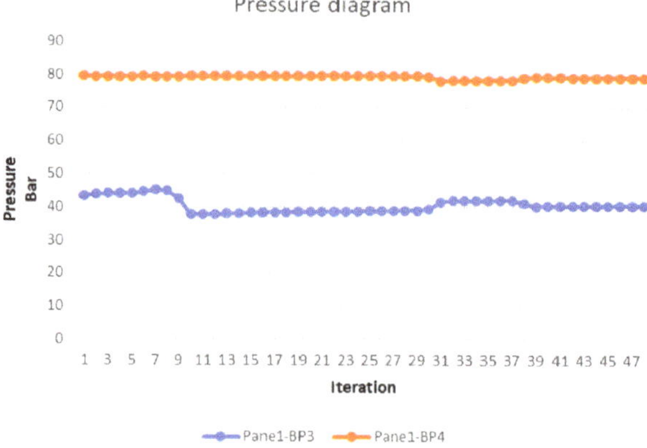

Figure 19. Helium Stirling engine and bottle pressure diagram.

Figure 20. Solys2 solar detection station.

The acquisition system transmits data to the control unit of the solar concentrator director that will position the axis of the concentrator in the direction of the sun through the two motors that control the rotations for elevation (vertical plane) and azimuth (horizontal plane), in order to obtain the maximum radiation.

According to [40], a computational analysis was completed using Ansys Fluent software, creating a model on which some simulations were performed for the exposure to thermal radiation around 1000 W/m^2, a model validated by the experimental results.

Regarding the use of the engine with the realized automation and other energy sources, the tests performed confirmed the operation in the following cases:

➢ the previously presented test which used a thermal energy source represented by a Riello gas burner;

➢ another test which used a battery of electrical resistors with a nominal power of 12 kW, which ensured the operation at a lower, but stable level.

The Stirling engine has worked with different heat sources, with stable and repeatable operation under safe conditions.

6. Conclusions

The purpose of this study was to realize the automation of the solar energy conversion unit with respect to the system having a Stirling engine, electric generator, and secondary heat exchanger, which provides the electrical and thermal energy produced by the Stirling engine solar concentrator.

This automation aims to solve the starting sequences, the functioning in safe conditions, monitoring of the parameters during operation, protecting it, and safely stopping it.

These objectives have been met, however, depending on the behavior of the installation over time, they will be improved and completed as a result of the process of optimizing the functioning of the automation. By using an auxiliary source that supplies 16 kW, the functioning of the engine was obtained and control and security was ensured throughout the operation of the Stirling engine.

Since the integration and promotion of renewable energy sources is in continual growth, this attracted our attention to the use of a solar concentrator with a Stirling engine.

The advantages of automating the Stirling solar concentrator consist of:

- creation of start and stop sequences of the test stand in safe conditions;
- improved security protocols;
- development of a monitoring application and data acquisition for reporting purposes;
- positioning of the solar concentrator in the direction of the sun in order to obtain the maximum radiation.

The next steps will be to improve the security protocols in operation, as well as their succession.

Author Contributions: Conceptualization, D.-A.M., V.B., and I.M.; methodology, D.-A.M., C.B., and I.Ş.; software, D.-A.M., C.B., and I.M.; validation, V.B., M.S.R., and E.C.; formal analysis, D.-A.M. and C.B.; investigation, D.-A.M. and V.B.; resources, D.-A.M., C.B., and I.Ş.; data curation, V.B. and I.M.; writing—original draft preparation, D.-A.M.; writing—review and editing, M.S.R. and I.M.; visualization, D.-A.M., C.B., and I.Ş.; supervision, V.B., M.S.R., E.C., and I.M.; project administration, V.B.; funding acquisition, D.-A.M. All authors have read and agreed to the published version of the manuscript.

Funding: This research received no external funding.

Acknowledgments: This work was supported by a grant from the Romanian Ministry of Research and Innovation, CCCDI-UEFISCDI, project numbers PN-III-P1-1.2-PCCDI-2017-0776/No. 36 and PCCDI/15.03.2018, within PNCDI III. This work was carried out through the Nucleus Program, financed by the Ministry of Education and Research, project number PN 19 11 02 02.

Conflicts of Interest: The authors declare no conflict of interest.

References

1. Kuban, L.; Stempka, J.; Tyliszczak, A. A 3D-CFD study of a g-type Stirling engine. *Energy* **2019**, *169*, 142–159. [CrossRef]
2. Burke, M.J.; Stephens, J.C. Political power and renewable energy futures: A critical review. *Energy Res. Soc. Sci.* **2018**, *35*, 78–93. [CrossRef]
3. Mancini, T.; Heller, P.; Butler, B.; Osborn, B.; Schiel, W.; Goldberg, V.; Buck, R.; Diver, R.; Andraka, C.; Moreno, J. Dish-Stirling systems: An overview of development and status. *J. Sol. Energy Eng.* **2003**, *125*, 135–151. [CrossRef]
4. Stirling engine. Available online: https://en.wikipedia.org/wiki/Stirling_engine (accessed on 17 January 2020).
5. Kongtragool, B.; Wongwises, S. A review of solar-powered stirling engines and low temperature differential stirling engine. *Renew. Sustain. Energy Rev.* **2003**, *7*, 131–154. [CrossRef]
6. CSP World Website. Available online: http://www.cspworld.org/cspworldmap (accessed on 17 January 2020).
7. Zabalaga, P.J.; Cardozo, E.; Campero, L.C.; Ramos, J.A. Performance Analysis of a Stirling Engine Hybrid Power System. *Energies* **2020**, *13*, 980. [CrossRef]
8. Toro, C.; Rocco, M.V.; Colombo, E. Exergy and Thermoeconomic Analyses of Central Receiver Concentrated Solar Plants Using Air as Heat Transfer Fluid. *Energies* **2016**, *9*, 885. [CrossRef]
9. Badea, G.; Felseghi, R.-A.; Varlam, M.; Filote, C.; Culcer, M.; Iliescu, M.; Răboacă, M.S. Design and Simulation of Romanian Solar Energy Charging Station for Electric Vehicles. *Energies* **2019**, *12*, 74. [CrossRef]
10. Raboaca, M.S. Sustaining the passive house with hybrid energy photovoltaic panels—Fuel cell. *Prog. Cryog. Isot. Sep.* **2015**, *18*, 65–72.
11. Cannistraro, M.; Mainardi, E.; Bottarelli, M. Testing a Dual-Source Heat Pump. *Math. Modeling Eng. Probl.* **2018**, *5*, 205–221. [CrossRef]
12. González-Roubaud, E.; Pérez-Osorio, D.; Prieto, C. Review of commercial thermal energy storage in concentrated solar power plants: Steam vs. molten salts. *Renew. Sustain. Energy Rev.* **2017**, *80*, 133–148. [CrossRef]
13. Dowling, A.W.; Zheng, T.; Zavala, V.M. Economic assessment of concentrated solar power technologies: A review. *Renew. Sustain. Energy Rev.* **2017**, *72*, 1019–1032. [CrossRef]
14. Felseghi, R.A.; Carcadea, E.; Raboaca, M.S.; Trufin, C.N.; Filote, C. Hydrogen Fuel Cell Technology for the Sustainable Future of Stationary Applications. *Energies* **2019**, *12*, 4593. [CrossRef]
15. Islas, S.; Beltran-Chacon, R.; Velazquez, N.; Leal-Chavez, D.; Lopez-Zavala, R.; Aguilar-Jimenez, J.A. A numerical study of the influence of design variable interactions on the performance of a Stirling engine System. *Appl. Therm. Eng.* **2020**, *170*, 115039. [CrossRef]
16. Malali, P.; Chaturvedi, S.; Agarwala, R. Effects of circumsolar radiation on the optimal performance of a Stirling heat engine coupled with a parabolic dish solar collector. *Appl. Therm. Eng.* **2019**, *159*, 113961. [CrossRef]
17. Zare, S.; Tavakolpur-Saleh, A.R. Predicting onset conditions of a free piston Stirling engine. *Appl. Energy* **2020**, *262*, 114488. [CrossRef]
18. Vieira de Souza, L.E.; Gilmanova Cavalcante, A.M. Concentrated Solar Power deployment in emerging economies: The cases of China and Brazil. *Renew. Sustain. Energy Rev.* **2017**, *72*, 1094–1103. [CrossRef]
19. Mohammadnia, A.; Ziapour, B.; Sedaghati, F.; Rosendhal, L.; Rezania, A. Utilizing thermoelectric generator as cavity temperature controller for temperature management in dish-Stirling engine. *Applied Thermal Engineering* **2020**, *165*, 114568. [CrossRef]
20. Buscemi, A.; Lo Brano, V.; Chiaruzzi, C.; Ciulla, G.; Kalogeri, C. A validated energy model of a solar dish-Stirling system considering the cleanliness of mirrors. *Appl. Energy* **2020**, *260*, 114378. [CrossRef]
21. Pelay, U.; Luo, L.; Fan, Y.; Stitou, D.; Rood, M. Thermal energy storage systems for concentrated solar power plants. *Renew. Sustain. Energy Rev.* **2017**, *79*, 82–100. [CrossRef]
22. Liu, M.; Tay, N.I.I.S.; Bell, S.; Belusko, M.; Jacob, R.; Will, G.; Saman, W.; Bruno, F. Review on concentrating solar power plants and new developments in high temperature thermal energy storage technologies. *Renew. Sustain. Energy Rev.* **2016**, *53*, 1411–1432. [CrossRef]

23. Arora, R.; Kaushik, S.C.; Kumar, R.; Arora, R. Multi-objective thermo-economic optimization of solar parabolic dish Stirling heat engine with regenerative losses using NSGA-II and decision making. *Electr. Power Energy Syst.* **2016**, *74*, 25–35. [CrossRef]
24. Kannan, N.; Vakeesan, D. Solar energy for future world: A review. *Renew. Sustain. Energy Rev.* **2016**, *62*, 1092–1105. [CrossRef]
25. Lovegrove, K.; Wyder, J.; Agrawal, A.; Borhuah, D.; McDonald, J.; Urkalan, K. *Concentrating Solar Power in India*; Report Commissioned by the Australian Government and Prepared by ITP Power; ITP Power: Bangkok, Thailand, 2011; ISBN 978-1-921299-52-0.
26. Răboacă, M.S.; Badea, G.; Enache, A.; Filote, C.; Răsoi, G.; Rata, M.; Lavric, A.; Felseghi, R.A. Concentrating Solar Power Technologies. *Energies* **2019**, *12*, 1048. [CrossRef]
27. Manual Instruction—V1.0—Sunflower 35 CONCENTRATION SOLAR PLANT—STROJÍRNY BOHDALICE. Available online: http://www.concentrating.eu/index.php/power-conversion-unit (accessed on 30 November 2014).
28. Raboaca, M.S.; Dumitrescu, C.; Manta, I. Aircraft Trajectory Tracking Using Radar Equipment with Fuzzy Logic Algorithm. *Mathematics* **2020**, *8*, 207. [CrossRef]
29. Aschilean, I.; Rasoi, G.; Raboaca, M.S.; Filote, C.; Culcer, M. Design and Concept of an Energy System Based on Renewable Sources for Greenhouse Sustainable Agriculture. *Energies* **2018**, *11*, 1201. [CrossRef]
30. Alphonsus, E.R.; Abdullah, M.O. Mohammad Omar Abdullah. A review on the applications of programmable logic controllers (PLCs). *Renew. Sustain. Energy Rev.* **2016**, *60*, 1185–1205. [CrossRef]
31. Punnathanam, V.; Koetcha, P. Effective multi-objective optimization of Stirling engine systems. *Appl. Therm. Eng.* **2016**, *108*, 261–276. [CrossRef]
32. Pitz-Paal, P. Parabolic Trough—Linear Fresnel—Power Tower: A Technology Comparison. Available online: http://www.iass-potsdam.de/sites/default/files/files/12.5-iass_pitz-paal.pdf (accessed on 6 April 2020).
33. Zayeda, M.; Zhaoa, J.; Lia, W.; Elsheik, A.; Zhaoa, Z.; Lia, K. Performance prediction and techno-economic analysis of solar dish/stirling system for electricity generation. *Appl. Therm. Eng.* **2020**, *164*, 114427. [CrossRef]
34. Awana, A.; Zubaira, M.; Praveena, R.P.; Bhattib, A. Design and comparative analysis of photovoltaic and parabolic trough based CSP plants. *Sol. Energy* **2019**, *183*, 551–565. [CrossRef]
35. CompactRIO. Available online: http://www.ni.com/ro-ro/shop/compactrio.html (accessed on 17 January 2020).
36. Where is an Example MODICON RTU Unity Pro Project using DNP3 Protocol. Available online: https://www.schneider-electric.com/en/faqs/FA339010/ (accessed on 17 January 2020).
37. Kabir, E.; Kumar, P.; Kumar, S.; Adelodun, A.A.; Kim, K.H. Solar energy: Potential and future prospects. *Renew. Sustain. Energy Rev.* **2018**, *82*, 894–900. [CrossRef]
38. Law, E.W.; Kay, M.; Taylor, R.A. Calculating the financial value of a concentrated solar thermal plant operated using direct normal irradiance forecasts. *Sol. Energy* **2016**, *125*, 267–281. [CrossRef]
39. Ehtiwesh, I.A.S.; Coelho, M.C.; Sousa, A.C.M. Exergetic and environmental life cycle assessment analysis of concentrated solar power plants. *Renew. Sustain. Energy Rev.* **2016**, *56*, 145–155. [CrossRef]
40. Mocanu, D.A.; Badescu, V.; Carcadea, E.; Armeanu, A. Computational analysis of the tubular heat exchanger of an integrated energy generation system using a solar Stirling engine. *U.P.B. Sci. Bull.* **2020**, *82*, 139–150.

MDPI

St. Alban-Anlage 66

4052 Basel

Switzerland

Tel. +41 61 683 77 34

Fax +41 61 302 89 18

www.mdpi.com

Energies Editorial Office

E-mail: energies@mdpi.com

www.mdpi.com/journal/energies